Ecophysiology of Coniferous Forests

Physiological Ecology
A Series of Monographs, Texts, and Treatises

Series Editor
Harold A. Mooney
Stanford University, Stanford, California

Editorial Board
Fakhri Bazzaz   F. Stuart Chapin   James R. Ehleringer
Robert W. Pearcy   Martyn M. Caldwell   E.-D. Schulze

T. T. KOZLOWSKI (Ed.). Growth and Development of Trees, Volumes I and II, 1971
D. HILLEL (Ed.). Soil and Water: Physical Principles and Processes, 1971
V. B. YOUNGER and C. M. McKELL (Eds.). The Biology and Utilization of Grasses, 1972
J. B. MUDD and T. T. KOZLOWSKI (Eds.). Responses of Plants to Air Pollution, 1975
R. DAUBENMIRE (Ed.). Plant Geography, 1978
J. LEVITT (Ed.). Responses of Plants to Environmental Stresses, 2nd Edition.
Volume I: Chilling, Freezing, and High Temperature Stresses, 1980
Volume II: Water, Radiation, Salt, and Other Stresses, 1980
J. A. LARSEN (Ed.). The Boreal Ecosystem, 1980
S. A. GAUTHREAUX, JR. (Ed.). Animal Migration, Orientation, and Navigation, 1981
F. J. VERNBERG and W. B. VERNBERG (Eds.). Functional Adaptations of Marine Organisms, 1981
R. D. DURBIN (Ed.). Toxins in Plant Disease, 1981
C. P. LYMAN, J. S. WILLIS, A. MALAN, and L. C. H. WANG (Eds.). Hibernation and Torpor in Mammals and Birds, 1982
T. T. KOZLOWSKI (Ed.). Flooding and Plant Growth, 1984
E. L. RICE (Ed.). Allelopathy, Second Edition, 1984
M. L. CODY (Ed.). Habitat Selection in Birds, 1985
R. J. HAYNES, K. C. CAMERON, K. M. GOH, and R. R. SHERLOCK (Eds.). Mineral Nitrogen in the Plant–Soil System, 1986
T. T. KOZLOWSKI, P. J. KRAMER, and S. G. PALLARDY (Eds.). The Physiological Ecology of Woody Plants, 1991
H. A. MOONEY, W. E. WINNER, and E. J. PELL (Eds.). Response of Plants to Multiple Stresses, 1991

The list of titles in this series continues at the end of this volume.

Ecophysiology of Coniferous Forests

Edited by

William K. Smith
Department of Botany
University of Wyoming
Laramie, Wyoming

Thomas M. Hinckley
College of Forest Resources
University of Washington
Seattle, Washington

Academic Press

San Diego   New York   Boston   London   Sydney   Tokyo   Toronto

QK
494
.E36
1995

Cover photograph: Conifer tree species exhibit a unique shoot and leaf architecture. The needle-like geometry of individual leaves enables a high number of leaves per unit stem length with little self-shading, reduced irradiance per leaf, and the mechanical strength to withstand a severe climate.

This book is printed on acid-free paper. ∞

Copyright © 1995 by ACADEMIC PRESS, INC.

All Rights Reserved.
No part of this publication may be reproduced or transmitted in any form or by any means, electronic or mechanical, including photocopy, recording, or any information storage and retrieval system, without permission in writing from the publisher.

Academic Press, Inc.
A Division of Harcourt Brace & Company
525 B Street, Suite 1900, San Diego, California 92101-4495

United Kingdom Edition published by
Academic Press Limited
24-28 Oval Road, London NW1 7DX

Library of Congress Cataloging-in-Publication Data

Ecophysiology of coniferous forests / edited by William K. Smith,
   Thomas M. Hinckley.
         p.   cm. -- (Physiological ecology series)
     Includes bibliographical references and index.
     ISBN 0-12-652875-6 (case)
       1. Conifers--Ecophysiology. I. Smith, William K. (William
Kirby), date. II. Hinckley, Thomas M. III. Series:
   Physiological ecology.
   QK494.E36   1994
   585' .2041--dc20
                                                                94-27804
                                                         CIP

PRINTED IN THE UNITED STATES OF AMERICA
94 95 96 97 98 99 QW 9 8 7 6 5 4 3 2 1

Contents

Contributors   ix
Preface   xi

1. Genetics and the Physiological Ecology of Conifers
Jeffry B. Mitton

    I. Conifers Have High Levels of Genetic Variation   2
   II. Patterns of Geographic Variation   5
  III. Ecological and Evolutionary Significance of the Genetic Variation   12
  IV. Natural Selection   13
    V. Summary   26
       References   27

2. Long-Term Records of Growth and Distribution of Conifers: Integration of Paleoecology and Physiological Ecology
Lisa J. Graumlich and Linda B. Brubaker

    I. Introduction   37
   II. Climatic Variation   38
  III. Vegetation Responses   41
  IV. Implications for Ecophysiological Studies   51
    V. Conclusions   57
       References   58

3. Plant Hormones and Ecophysiology of Conifers
W. J. Davies

    I. Introduction   63
   II. Quantification of Plant Growth Regulators   64
  III. A Role for Plant Growth Regulator Biologists in Physiological Plant Ecology?   66
  IV. Chemical Signaling in Woody Plants   66
    V. A General Model for Chemical Regulation of Stomatal Behavior, Water Relations, and Development of Plants in the Field   68
  VI. Importance of Sensitivity Variation and Involvement of ABA in Stomatal Response to Climatic Variables   71

VII. Regulation of Growth and Development: A Conclusion 73
References 76

4. Ecophysiological Controls of Conifer Distributions
F. I. Woodward
I. Introduction 79
II. Climatic Limits 80
III. The Xylem 83
IV. Growth 84
V. Beyond the Boreal Zone 87
VI. Conclusions 90
References 91

5. Physiological Processes during Winter Dormancy and Their Ecological Significance
Wilhelm M. Havranek and Walter Tranquillini
I. Introduction 95
II. The Coniferous Forest Zone 96
III. Frost Resistance 100
IV. Winter Water Relations 105
V. Carbon Metabolism in Winter 109
VI. Conclusions 116
References 117

6. Ecophysiology and Insect Herbivory
Karen M. Clancy, Michael R. Wagner, and Peter B. Reich
I. Acquisition and Allocation of Nutrients (Sugars, Nitrogen, Minerals) 125
II. Water Relations 151
III. Carbon Acquisition and Allocation 159
IV. Conclusions: Effects of Herbivory on Conifer Forest Ecology 169
References 170

7. Leaf Area Dynamics of Conifer Forests
Hank Margolis, Ram Oren, David Whitehead, and Merrill R. Kaufmann
I. Introduction 181
II. Leaf Area: Structural and Functional Relationships 182
III. Leaf Area Dynamics 195
IV. Conclusions 215
References 216

8. Causes and Consequences of Variation in Conifer Leaf Life-Span
Peter B. Reich, Takayoshi Koike, Stith T. Gower, and Anna W. Schoettle
I. Introduction 225
II. Variation in Leaf Life-Span 227

III. Relationship of Leaf Life-Span to Leaf, Plant,
and Ecosystem Traits 238
IV. Summary 249
References 250

9. Response Mechanisms of Conifers to Air Pollutants
Rainer Matyssek, Peter Reich, Ram Oren, and William E. Winner

I. Introduction 255
II. O_3 Uptake and Impacts at Leaf to Cellular Tissue Scales 261
III. Organ Differentiation and Senescence in the Presence of O_3 270
IV. Scaling from the Leaf to the Whole-Plant Level 272
V. Responses to O_3 under Multiple Stress Interaction Scenarios 278
VI. Scaling to Understand Mature Tree or Stand Level Responses 288
VII. Models That Incorporate O_3 Effects 295
VIII. Conclusions 299
References 300

10. Potential Effects of Global Climate Change
Hermann Gucinski, Eric Vance, and William A. Reiners

I. Introduction 309
II. Effects on and Responses of Conifers and Coniferous Forests 311
III. Direct Effects of Elevated CO_2 312
IV. Large-Scale Responses of Coniferous Forests
to Climate Change 314
V. Pools and Flux of Carbon in Coniferous Biomes 324
VI. Final Comments 326
References 327

Index 333

Contributors

Numbers in parentheses indicate the pages on which the authors' contributions begin.

Linda B. Brubaker (37), College of Forest Resources, University of Washington, Seattle, Washington 98195

Karen M. Clancy (125), Rocky Mountain Forest and Range Experiment Station, United States Department of Agriculture Forest Service, Flagstaff, Arizona 86001

W. J. Davies (63), Division of Biological Sciences, Institute of Environmental and Biological Sciences, Lancaster University, Lancaster LA1 4YQ, United Kingdom

Stith T. Gower (225), Department of Forestry, University of Wisconsin–Madison, Madison, Wisconsin 53706

Lisa J. Graumlich (37), Laboratory of Tree Ring Research, University of Arizona, Tucson, Arizona 85721

Hermann Gucinski (309), United States Forest Service Ecosystems Processes Program, Oregon State University, Corvallis, Oregon 97331

Wilhelm M. Havranek (95), Forstliche Bundesversuchsanstalt, A-6020 Innsbruck, Austria

Merrill R. Kaufmann (181), Rocky Mountain Forest and Range Experiment Station, United States Department of Agriculture Forest Service, Fort Collins, Colorado 80526

Takayoshi Koike (225), Hokkaido Research Center, Forest and Forestry Products Research Institute, Sapporo 06Z, Japan

Hank Margolis (181), Department of Forest Science, University of Laval, Sainte-Foy, Quebec, Canada G1K 7P4

Rainer Matyssek[1] (255), Swiss Federal Institute for Forest, Snow, and Landscape Research, CH-8903 Birmensdorf, Switzerland

Jeffry B. Mitton (1), Department of Environmental, Population, and Organismic Biology, University of Colorado, Boulder, Colorado 80309

Ram Oren (181, 255), School of the Environment, Duke University, Durham, North Carolina 27706

---

[1]Present address: Department of Forest Botany, University of Munich, D-85354 Freising, Germany.

Peter B. Reich (125, 225, 255), Department of Forest Resources, University of Minnesota, St. Paul, Minnesota 55108

William A. Reiners (309), United States Forest Service Ecosystems Processes Program, Oregon State University, Corvallis, Oregon 97331

Anna W. Schoettle (225), Rocky Mountain Forest and Range Experiment Station, United States Department of Agriculture Forest Service, Fort Collins, Colorado 80526

Walter Tranquillini (95), Institut für Botanik, Universität Innsbruck, A-6020 Innsbruck, Austria

Eric Vance (309), National Council of the Paper Industry for Air/Stream Improvement, Gainesville, Florida 32607

Michael R. Wagner (125), School of Forestry, Northern Arizona University, Flagstaff, Arizona 86011

David Whitehead (181), New Zealand Forest Research Institute, Christchurch, New Zealand

William E. Winner (255), Department of Botany and Plant Pathology, Oregon State University, Corvallis, Oregon 97331

F. I. Woodward (79), Department of Animal and Plant Sciences, University of Sheffield, Sheffield S10 2UQ, United Kingdom

Preface

The field of physiological ecology emphasizes the functioning of the whole organism in its environment, but it also provides an evolutionary perspective from which more mechanistic studies can be addressed at the cellular and molecular levels. In addition, information at the organism level gives mechanistic explanations of stand and ecosystem level processes. Thus, the study of physiological ecology is a natural bridge between the ultimate mechanisms of molecular genetics and evolution and the mechanisms inherent in ecosystem processes. This coupling between molecular and ecosystem processes may be nowhere more important than within our coniferous forests, one of our most economically and socially important natural resources. Moreover, it is now widely recognized that the conifer forests of the world play an essential role in the health of the global ecosystem. Expressions such as "new forestry," "sustainable ecosystems," "biodiversity," and "ecosystem management" are becoming an integral part of environmental, natural resource, and public discussions throughout the world. Conifer forests are no longer managed only as commodities of harvestable wood, but as natural resources that must be sustained for their multiple uses including aesthetic, scientific, and global values.

A comprehensive knowledge of the environmental response capabilities of conifer tree species is fundamental to the sustainability of these forests. Integration of the genetic mechanisms involved in the response of the organism to its environment with the impact of species changes on the community and ecosystem levels spans the full breadth of the biological spectrum. It is this biological scaling that is currently recognized as crucial to our ultimate understanding of such complex issues as the impact of natural and anthropogenic changes in the environment on a global level. The chapters in this book include topics that range from genetics and plant hormones to paleoecology and global distribution. One chapter addresses the importance of conifer ecophysiology for evaluating future scenarios of global climate change.

The ideas and organization of this book were derived from a 3-day workshop at the University of Wyoming's and the U.S. National Park Service's Research Station near Jackson Hole, Wyoming (September,

1991). Researchers from around the world attended. In addition to the current book, a companion volume entitled *Resource Physiology of Conifers: Acquisition, Allocation, and Utilization* was generated from the workshop. To add breadth to both books, authors with similar interests, but at different biological scales, were assigned to the same chapter. This was not universally possible. In a few cases where areas of expertise were recognized as missing, appropriate authors not in attendance were invited to contribute chapters. Our intent was to provide a current, comprehensive, state-of-the-art treatment of the field of conifer tree ecophysiology.

We thank the staff at the UW–NPS Jackson Hole Research Station for their help and hospitality, and for letting us share such a marvelous natural setting. We also thank the National Science Foundation, the U.S. Department of Energy, the U.S. National Park Service, the U.S. Forest Service, the Wood and Paper Products Industry, and the National Council on Air and Stream Improvement for their financial support of the workshop and production of these two books.

WILLIAM K. SMITH
THOMAS M. HINCKLEY

1

Genetics and the Physiological Ecology of Conifers

Jeffry B. Mitton

Natural selection acts on the diversity of genotypes, adapting populations to their specific environments and driving evolution in response to changes in climate. Genetically based differences in physiology and demography adapt species to alternate environments and produce, along with historical accidents, the present distribution of species. The sorting of conifer species by elevation is so marked that conifers help to define plant communities arranged in elevational bands in the Rocky Mountains (Marr, 1961). For these reasons, a genetic perspective is necessary to appreciate the evolution of ecophysiological patterns in the coniferous forests of the Rocky Mountains.

The fascinating natural history and the economic importance of western conifers have stimulated numerous studies of their ecology, ecological genetics, and geographic variation. These studies yield some generalizations, and present some puzzling contradictions. This chapter focuses on the genetic variability associated with the physiological differences among genotypes in Rocky Mountain conifers. Variation among genotypes in survival, growth, and resistance to herbivores is used to illustrate genetically based differences in physiology, and to suggest the mechanistic studies needed to understand the relationships between genetic and physiological variation.

I. Conifers Have High Levels of Genetic Variation

Evolutionary theory predicts that species that have large population sizes and broad geographic ranges will have high levels of genetic variation. This prediction is based on several relationships. First of all, the number of new mutations arising each generation increases with population size. In addition, the rate of genetic drift is inversely proportional to population size, so that more mutations are expected to accumulate in large populations. Jointly, these relationships predict that genetic variation will increase with population size. For example, the expected number of alleles at a locus k is $k = 4Nu + 1$, where N is the population size and u is the neutral mutation rate (Hartl and Clark, 1989). Similarly, the predicted heterozygosity, H, also increases with population size:

$$H = 4Nu/(4Nu + 1).$$

In addition to population size, many conifers have other life-history traits and ecological circumstances that promote high levels of genetic variation. For example, many species of conifers, such as ponderosa pine (*Pinus ponderosa*), lodgepole pine (*Pinus contorta*), and Douglas fir (*Pseudotsuga menziesii*), have immense geographical ranges that place them in a wide variety of natural environments. The degree of environmental variation experienced by a species is expected to increase with the geographic range of the species, and the selection among heterogeneous environments is expected to enhance genetic variation (Hedrick *et al.*, 1976; Hedrick, 1986; Gillespie and Turelli, 1989). We can easily imagine that the Douglas fir growing in sympatry with pinyon pine (*Pinus edulis*) in the Chiricahua Mountains of southern Arizona will be selected to have traits very different from those favored in the Douglas fir growing in sympatry with Sitka spruce (*Picea sitchensis*) in the temperate rain forest of the Olympic Peninsula.

Forest biologists detect genetic variation with diverse methods. Trees planted as seeds into common gardens in provenance studies yield estimates of heritability for virtually any character that can be measured. Heritability is the proportion of total phenotypic variation attributable to genetic variation. By planting seeds from diverse sources into a common garden, the environmental differences among collection sites are eliminated, and therefore apparent differences can be attributed to genetic differences among collection localities. The use of several common gardens in contrasting environments allows inferences of how genotypes respond to a diversity of environmental conditions. Although common garden studies reveal differences among genotypes and patterns of performance along environmental gradients, individual genes are not identified. Specific genes or sequences of DNA from seeds and needles collected in the field can be examined with electrophoretic surveys of proteins and with a variety of molecular techniques. Surveys of electrophoretically detectable genetic variation of proteins, or allozyme variation, have been used to measure the genetic variation and describe the geographic variation of many species of conifers (Hamrick *et al.*, 1979; Hamrick and Godt, 1990; Loveless and Hamrick, 1984; Mitton, 1983). The DNA from the nucleus, from mitochondria, and from chloroplasts can be sequenced, or examined for restriction fragment length polymorphisms (RFLPs) (Wagner, 1992a,b), including examination of the variable number of tandem repeats (VNTRs), also popularly referred to as DNA fingerprints.

A. Provenance Studies

Provenance studies of many species of conifers have identified substantial proportions of genetic variation for characters such as germinability, growth, disease resistance, times of bud break and bud set, susceptibility to frost damage, and growth form (Wright, 1976). One indication of the magnitude of variation among individuals is the range of values for maximum photosynthetic capacity, gleaned from the review of Ceulemans and Saugier (1991) (Table I). Clear differences distinguish species, but these data illustrate that there is substantial variation within species as well. In both sweet chestnut and black cottonwood, the maximum photosynthetic capacity varies by a factor of two among individuals.

Table I Range of Average Photosynthetic Capacities for a Variety of Trees[a]

Species	Common name	Range (μmol/sec/m^2)
Castanea sativa	Sweet chestnut	8–16
Fraxinus pennsylvanica	Green ash	20–25
Populus tremuloides	Quaking aspen	20–22
Populus trichocarpa	Black cottonwood	14–30
Populus euramericana	Euramerican poplar	20–25
Acer saccarinum	Silver maple	6–7
Betula pendula	White birch	9–15
Quercus ilex	Evergreen oak	3–7
Quercus suber	Cork oak	5–6
Triplochiton scleroxylon	Obeche	4–7
Larix decidua	European larch	6–7
Lagarostrobos franklini	Huon pine	5–7
Pinus halepensis	Alep's pine	5–7

[a] Photosynthetic capacity values refer to maximum CO_2 exchange rates at saturating light, 20–25°C, and 330 ppm CO_2. From Ceulemans and Saugier (1991).

In addition to demonstrating substantial proportions of genetic variation underlying phenotypic variation, provenance studies have revealed geographic variation of the genetically determined variation. Thus, populations have accumulated genes that promote survival, growth, and reproduction in their local environments. These sorts of observations are so broadly based that foresters have defined seed collection zones for species such as Douglas fir and ponderosa pine (Namkoong et al., 1988). The seed zones specifically recognize adaptation to local environments, and they provide guidelines for use of seed sources to be used for restocking stands.

B. Allozyme Studies

Among plants, conifers stand out as a group that is wind-pollinated, long-lived, and highly fecund. All of these character traits were found to be associated with high levels of genetic variation in a multivariate study of 113 species of plants (Hamrick et al., 1979). Wind pollination increases gene flow and neighborhood size, effectively increasing population size. The opportunity for balancing selection is expected to increase with fecundity, and the environmental heterogeneity experienced by an individual is expected to increase with generation time. Furthermore, the widespread conifers are known to be hosts of large numbers of parasites and herbivores (Sturgeon and Mitton, 1982). For example, the number of insects utilizing ponderosa pine exceeds 200 (Furniss and Carolin, 1977). It is likely that the large numbers of parasites and predators impose diversifying selection on their hosts (Sturgeon and Mitton, 1986; Linhart et al., 1989, 1994; Linhart, 1991; Snyder 1992, 1993; Snyder and Linhart, 1994), further enhancing genetic variation.

Surveys of genetic variation of proteins have identified conifers as the most genetically variable group of species (Hamrick et al., 1979, 1992a,b; Hamrick and Godt, 1990). For example, the proportion of loci polymorphic in gymnosperms, dicots, and monocots was 0.71, 0.59, and 0.45, respectively ($P < 0.001$) (Hamrick and Godt, 1990). Similarly, the proportion of loci heterozygous within populations in gymnosperms, dicots, and monocots was 0.16, 0.14, and 0.10, respectively ($P < 0.001$). A survey of genetic variation within populations of annuals, short-lived herbaceous species, short-lived woody species, long-lived herbaceous species, and long-lived woody plants reported expected heterozygosities of 0.101, 0.098, 0.096, 0.082, and 0.148, respectively ($P < 0.001$) (Hamrick and Godt, 1990).

C. DNA Studies

In addition to the genes residing in the nucleus, conifers have genes in the chloroplasts and the mitochondria. The mitochondrial and cytoplasmic genomes are much smaller than the nuclear genomes, 1–2 ×

10^5 base pairs (bp), as opposed to approximately $30-45 \times 10^9$ bp in the nuclear genome. Both of the organellar genomes are circular and they replicate independently, but the mitochondrial DNA (mtDNA) is usually maternally inherited, whereas the chloroplast DNA (cpDNA) is inherited paternally in conifers (Neale *et al.*, 1986; Strauss *et al.*, 1989). Interesting exceptions to this pattern are found in coast redwood (*Sequoia sempervirens*) (Neale *et al.*, 1989) and incense cedar (*Calocedrus decurrens*) (Neale *et al.*, 1991), in which both organelles are paternally inherited.

cpDNA and mtDNA have been used in relatively few studies of conifers, so any generalizations are necessarily tentative. An earlier review suggested that cpDNA had sufficient variation for phylogenetic studies, but insufficient variation within species for studies of geographic variation (Clegg, 1990). Range-wide studies of cpDNA revealed little variation in jack pine (*Pinus banksiana*) but high levels of variation within and among populations in lodgepole pine (Dong and Wagner, 1993). Studies of mtDNA in conifers are not common, but there is evidence that indicates that cpDNA variation in conifers is common (Wagner *et al.*, 1987; Palmer, 1988; Wagner, 1992a,b; Matos, 1992; Strauss *et al.*, 1993).

II. Patterns of Geographic Variation

Patterns of geographic variation have been described with provenance studies and surveys of allozyme, mtDNA, and cpDNA variation. These different genetic perspectives reveal different magnitudes and patterns of geographic variation.

A. Provenance Studies

Campbell (1979) used a common garden study to describe genetic variation of Douglas fir [*P. menziesii* (Mirb.) Franco var. *menziesii*] in the H. J. Andrews Experimental Forest on the western slopes of the Cascade Range in Oregon. This site is a 6100-ha watershed with approximately 1100 m of vertical relief. To estimate the genotypic value of 193 trees, 13 variables were measured from 30 to 40 seedlings per family in a common garden. The variables estimated seed weight, germination, growth, and survival. Genetic differentiation within the watershed was surprisingly large, and much of it was associated with elevation and slope aspect. For example, variance among families due to geographic source was 67, 46, and 54% for height, diameter, and survival, respectively. Isograms on trend maps describing patterns of height, diameter, dry weight, and bud set closely resembled the elevational contours. For several characters, differences among elevations and slope aspects were great. For example, Campbell estimated that about half of the seeds moved ±300 m in elevation and none of the seedlings moved ±670 m in elevation would survive. Within this watershed, these vertical distances were covered by moving 4–5 km horizontally.

Population differentiation of white fir (*Abies concolor*) was examined in a provenance study of four populations sampled along an elevational transect extending 40 km from 1280 to 2073 m in the central Sierra of California (Hamrick, 1976). Thirteen variables were measured from families of seedlings, and analyses of variance were used to apportion variation into components within families, among families within populations, and among populations. Of the 13 growth, size, and needle morphology variables, 10 had significant differences among populations. Variables differed from 0 to 50% in the total variance that was distributed among populations. Seeds from high elevations produced smaller seedlings that had smaller needles, fewer adaxial stomatal rows, blunter needles, and a shorter growing season. Variables reflecting size and growth differed most among the populations, with 50% of the growth of the epicotyl varying among populations. The trends described in this elevational gradient were consistent with trends seen along a latitudinal gradient (Hamrick and Libby, 1972), and thus as responses to components of the environment that change consistently with increasing elevation and latitude.

Genetic differentiation of lodgepole pine was examined in a provenance study of seedlings for 173 populations in central and eastern Idaho, western Wyoming, and Utah (Rehfeldt,1988). Seedlings were tested in common gardens at 900 and 1500 m in the Priest River Experimental Forest in northern Idaho. Five variables estimating components of growth, timing of growth, and freeze tolerance were measured on each seedling. Multiple regression utilized 26 variables, such as elevation, latitude, and longitude, to maximize the proportion of variance explained in the variables measured on the seedlings. The regressions explained from 43 to 77% of the variation at the five dependent variables. Variation at each of the five variables was linear with elevation within regions; regions with different mean elevations had similar slopes but different intercepts. Differentiation of lodgepole pine occurs along gradients in the frost-free period, and, consequently, clines are geographically gentle and elevationally steep. Elevational clines are so steep that populations in close proximity but differing in elevation by as little as 300 m (equivalent to a difference of 24 frost-free days) are significantly different. Adaptation to heterogeneous environments is a compromise between selection for high growth potential in mild environments and high tolerance to fall frosts in cold environments.

Population differentiation of western larch (*Larix occidentalis*) was measured from differences among 82 populations planted out into common gardens (Rehfeldt, 1982). Each population was represented by seeds from 10 trees selected for the Inland Empire Cooperative Tree Improvement Program. The common gardens in Moscow and the Priest River Experimental Forest were at elevations of 700, 800, and 1500 m. Variables measured from seedlings estimated growth rate, phenology, and cold hardiness. Regression demonstrated that all variables except cold hardiness were significantly associated with elevation. Multiple regression using independent variables such as elevation, latitude, and longitude accounted for 39, 47, and 53% of the variation among populations in bud burst, bud set, and height, respectively. With increasing elevation, bud burst was delayed, bud set was earlier, and growth potential declined. These analyses led Rehfeldt (1982) to recommend that seed transfer could be fairly broad geographically, but not more than ±225 m in elevation.

A common garden study of ponderosa pine in the southwestern Rocky Mountains revealed substantial genetic variation (Rehfeldt, 1993), consistent with preceding provenance studies (Squillace and Silen, 1962; Hanover, 1963; Wells, 1964; Read, 1980; Rehfeldt, 1990, 1991). Variation among populations was studied by measuring 28 variables from 10 seedling families from 97 populations in Colorado, Utah, Arizona, and New Mexico. Seedlings were planted out into common gardens in the Priest River Experimental Forest in Idaho, Window Rock in Arizona, and Silver City in New Mexico. Variation among populations was highly significant for variables such as foliage damage due to winter desiccation, shoot diameter, duration of shoot elongation, and height after 3 and 4 years of growth (Rehfeldt, 1993). When the values for these characters were plotted against their geographical collection sites, striking patterns emerged that were consistent with adaptation to broad environmental gradients involving temperature and precipitation. Examples of two of these, duration of elongation and death of foliage due to winter desiccation, are presented in Fig. 1. This variation is important to the biologists who manage ponderosa pine, to the conservation biologists who are interested in maintaining biodiversity and genetic resources, and to the ecologists who are trying to understand the patterns of variation.

Although the pattern of geographic variation of ponderosa pine in the southwest is complex, much of the variation among populations follows variation in frost-free days. In the southwest, an elevational difference of 1000 m translates to a difference of 90 frost-free days (Baker, 1944). Trees from lower elevations are programmed to grow for a long period, whereas trees from higher elevations cease growth early and are consequently resistant to frost damage, but they have a low potential for growth.

The differentiation of populations of Rocky Mountain conifers can be measured in the common currency of elevation. Populations of Douglas

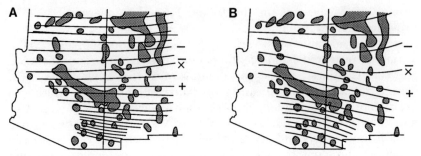

Figure 1 Patterns of geographic variation for duration of elongation (A) and foliage death due to winter desiccation (B). Duration of elongation is the number of days between initiation and cessation of growth. Isopleths are taken from a multiple regression describing variation at the common elevation of 2400 m. The interval between isopleths is equal to 0.5 times the least significant difference at the 0.20 probability level. (From Rehfeldt, 1993.)

fir separated by 200 m of elevation differ genetically (Rehfeldt, 1989), and the study of southwestern ponderosa pine (Rehfeldt, 1993) revealed significant differences at 220 m. The estimated critical elevations for lodgepole pine and western larch are 250 m (Rehfeldt, 1988) and 450 m (Rehfeldt, 1982), respectively. In contrast to these results, western white pine exhibits very little differentiation among populations (Rehfeldt, 1979; Rehfeldt et al., 1984).

A common garden measured photosynthetic rate, stomatal conductance, and photosynthetic water use efficiency in ponderosa pine seedlings of diverse parentage (Monson and Grant, 1989). Some of the seeds were produced by crosses between individuals of *P. ponderosa* var. *ponderosa*, and others utilized *P. ponderosa* var. *ponderosa* as the seed parent and *P. ponderosa* var. *scopulorum* as the pollen parent. *Pinus ponderosa* var. *ponderosa*, the coastal ponderosa pine, grows in environments moister than the semiarid environments of *P. ponderosa* var. *scopulorum*, which grows in the southern Rocky Mountains. Seedlings with genes from the more arid environments had lower stomatal conductances, lower stomatal densities, lower maximum photosynthesis rates, and lower needle nitrogen concentrations. These genetic differences have clear adaptive value (Monson and Grant, 1989). Genes from trees from the relatively arid interior convey higher water use efficiencies and lower transpiration rates, at the expense of reduced maximum photosynthesis rates.

B. Allozyme Studies

Studies of geographic variation in conifers have revealed that most have high levels of genetic variation within populations, and little differentiation among populations (Yeh and El-Kassaby, 1980; Yeh and O'Malley, 1980; Wheeler and Guries, 1982; Hiebert and Hamrick, 1983; Ledig *et al.*, 1983; Steinhoff *et al.*, 1983; Loveless and Hamrick, 1984; Hamrick *et al.*, 1989). For example, Conkle and Critchfield (1988) sampled ponderosa pine in northern California, eastern Washington, and southern Wyoming. They reported a Nei's genetic distance of $D = 0.01$ between California and Washington, $D = 0.06$ between Wyoming and California, and $D = 0.08$ between Wyoming and Washington. These values reveal only slight variation. O'Malley *et al.* (1979) surveyed allozyme variation in Idaho and Montana, and found only 12% of the variation to be among populations. They concluded that their stands were "not strongly differentiated." In contrast, Madesen and Blake (1977), working in the same geographic region, estimated that 28.5% of the variation in 2-year height of ponderosa pine was among populations. Thus, growth rate data revealed more than twice the differentiation among populations compared to the allozyme data.

Loveless and Hamrick (1984) examined the apportionment of genetic diversity within and among plant populations as a function of the mating system. They employed the G_{st} of Nei (1973), which measures the proportion of the total genetic diversity that is distributed among populations. When species were grouped by mating system, values of G_{st} were 0.523, 0.243, and 0.118 for groups that had autogamous, mixed, and outcrossed mating systems, respectively ($P < 0.01$). They also grouped species by the length of their life cycles, and found values of G_{st} of 0.430, 0.262, and 0.077 for annuals, short-lived species, and long-lived species. The values of G_{st} for annuals, short-lived herbaceous, short-lived woody, long-lived herbaceous, and long-lived woody plants were 0.355, 0.253, 0.155, 0.278, and 0.084 ($P < 0.001$), respectively (Hamrick *et al.*, 1992a,b). Thus, long-lived woody perennials, including conifers, stand apart from annuals, short-lived perennials, and long-lived herbaceous species for their higher levels of variation within populations and their lower degrees of differentiation among populations.

A comprehensive study of the differentiation among populations of species of conifers (El-Kassaby, 1991) revealed that the average of the values of G_{st} was 0.058. Values of G_{st} ranged from 0.010 to 0.061 for lodgepole pine, from 0.015 to 0.120 for ponderosa pine, and from 0.001 to 0.068 for Douglas fir.

C. Discordant Patterns of Variation

Provenance studies and allozyme surveys appear to reveal discordant patterns of variation. Striking patterns of phenology, often associated with environmental variation, necessitate seed zones for the major timber species (Namkoong *et al.*, 1988). The provenance studies summarized in this review generally revealed strong differentiation associated with elevational gradients. In contrast, differentiation of enzymes, at least measured with single loci, is typically slight (El Kassaby, 1991; Guries and Ledig, 1982; Yeh and Layton, 1979; Yeh and O'Malley, 1980; O'Malley *et al.*, 1979; Hiebert and Hamrick, 1983; Loveless and Hamrick, 1984; Yeh, 1988). We know that important patterns of phenological variation are genetically determined, but the intensity and patterning of phenological variation do not appear to match the pattern of enzyme variation, although multivariate analysis of enzyme variation improves the correspondence of the data sets (Westfall and Conkle, 1992a,b; Millar and Westfall, 1992a,b).

Karl and Avise (1992) examined patterns of geographic variation in enzymes, nuclear RFLPs, and mtDNA of the American oyster, *Crassostrea virginica*. They found that mtDNA and nuclear RFLPs provided very similar patterns of geographic variation, and that these patterns were strikingly different from the pattern of variation of enzymes. Enzyme frequencies in oysters exhibit very little variation from Maine to Texas, and this geographic pattern was initially interpreted as evidence for extensive gene flow among populations of oysters. However, DNA markers from both the mitochondrial and the nuclear genomes revealed an abrupt discontinuity in northeastern Florida, but little genetic variation in other areas. After considering alternate explanations, they tentatively concluded that balancing selection acting on enzyme loci caused this set of loci to diverge from the pattern of geographic variation exhibited by the other sets of loci. This empirical study demonstrates that different compartments of the genome can exhibit strikingly different patterns of geographic variation, and that evolutionary forces may impact alternate compartments of the genome in different ways.

Following the example in oysters, it may be fruitful to compare and contrast the geographic variation in the various compartments—nuclear, mitochondrial, chloroplast—of the genome of conifers. The nuclear genes coding for proteins are biparentally inherited. In most conifers, the mitochondrial genome is inherited maternally, and the chloroplast genome is inherited paternally. Because these sets of loci have different patterns of inheritance, they have different potentials for gene flow, for the paternally inherited genes are moved much further by pollen than maternally inherited genes are moved by seeds. That is, if all of these sets of genes were adaptively neutral, we would expect cpDNA to show the least geographic variation, mtDNA to show the most, and for allozymes and nuclear RFLPs to be intermediate in their degrees of geographic variation.

Allozyme surveys of genetic variation in knobcone pine (*Pinus attenuata*), Monterey pine (*Pinus radiata*), and Bishop pine (*Pinus muricata*) revealed the typical apportionment of genetic variation for pines—high amounts of variation within populations and relatively little differentiation among populations within a species. Millar *et al.* (1988) reported that 12–22% of the variation was among populations within a species. However, analyses of cpDNA (Hong *et al.*, 1993) and mtDNA (Strauss *et al.*, 1993) yielded very different patterns of variation. cpDNA variation was revealed as RFLPs detected with probes from Douglas fir. Neither knobcone pine nor Monterey pine has much variation, either within or between populations. Bishop pine has little variation within populations, but marked differentiation (G_{st} = 87 ± 8%) among populations. Similarly, although allozyme data reveal only 24% of the variation to be among species, the estimate from cpDNA is twice as great, 49%. mtDNA variation was revealed as RFLPs detected with a probe for the cytochrome oxidase I gene amplified from knobcone pine with the polymerase chain reaction. Variability for mtDNA is low within populations, but G_{st} varied from 75% in Monterey pine to 96% in Bishop pine.

Allozymes and cpDNA present contrasting patterns of variation in jack pine (*P. banksiana*) and lodgepole pine (Dong and Wagner, 1993). Two mtDNA polymorphisms were used to describe patterns of variation in 741 individuals throughout the ranges of jack pine and lodgepole pine. Probes for cytochrome oxidases I and II (COXI and COXII) identified cpDNA RFLPs from total DNA extracted from needle samples. A probe for the COXI gene revealed a diagnostic difference between species but almost no variation within populations or species. In contrast, a probe for the COXII gene revealed a small amount of variation within jack pine, but a high degree of variation within lodgepole pine. Large proportions of the variation within lodgepole pine were among populations within subspecies (F_{st} = 66%) and among subspecies (F_{st} = 31%). These patterns contrast sharply with the apportionment of genetic variation detected with allozymes. Specifically, values of F_{st} (similar to G_{st}) estimated from allozymes rarely exceeded 6% in lodgepole pine (Wheeler and Guries, 1982) , and are generally less than 10% in conifers (Hamrick and Godt, 1989).

Allozymes and DNA markers reveal contrasting patterns of genetic variation in limber pine (*Pinus flexilis*). Allozyme variation was used to describe variation within and among populations from lower tree line (1650 m) to upper tree line (3350 m) in Colorado (Schuster *et al.*, 1989). The average F_{st} for these loci was 0.02, or 2%, indicating little differentiation among sites and suggesting gene flow on the order of 10 migrants between populations per generation. This estimate of gene flow was taken from a relationship presented by Wright (1931), $F_{st} = 1/(1 + 4Nm)$, where N is the population size and m is the rate of migration. However, this estimate of gene flow did not seem to be possible, for the pollination phenology of limber pine varies dramatically with elevation; most sites along the elevational transect differing by 400 m or more do not have overlapping pollination periods, and therefore cannot exchange genes in a single generation. Preliminary surveys of random amplified polymorphic DNA (RAPD) revealed a fixed difference between the elevational extremes of this transect (R. Latta, personal communication). In this particular case, allozyme loci, pollination phenology, and RAPD markers suggest very high, intermediate, and very low levels of gene flow, respectively.

The pattern seen in these first studies of conifers is similar to that seen in oysters (Karl and Avise, 1992); allozyme loci reveal high levels of variation within populations and little variation among populations, whereas DNA markers reveal relatively little variation within populations, but a high proportion of variation among populations.

The disparity between the magnitudes of geographic variation revealed by provenance studies and allozyme surveys is not wholly unexpected. Lewontin (1984) argued that it is generally easier to detect differences among population means than among allelic frequencies. Although this matter of statistical sensitivity may contribute to the disparity between the two sets of observations, I do not believe that statistical sensitivity is the most important explanation. From the available data, it appears that allozyme loci reveal only slight differences among samples, but DNA markers reveal dramatic differentiation of populations. The same procedures are used to test the homogeneity of allozyme and DNA markers, so statistical sensitivity is not an issue here.

III. Ecological and Evolutionary Significance of the Genetic Variation

In one sense, the biochemical and molecular studies of genetic variation in conifers have been superficial. Most studies have used genetic variation to address questions in population biology or ecology, but relatively few studies have examined the nature and significance of the genetic variation. That is, population biologists have used genetic variation to study gene flow in natural populations (Adams, 1992a,b; Schuster and Mitton, 1994) or in seed orchards (Adams and Birkes, 1991; Wheeler *et al.*, 1992; Adams *et al.*, 1992), or to describe patterns of geographic variation across the range of a species; evolutionary biologists use genetic variation to infer phylogenetic relationships among species (Prager *et al.*, 1976; Strauss and Doerksen, 1990; Smith and Klein, 1994). But relatively few studies have examined, in any great detail, the physiological and demographical consequences of genetic variation for enzymes or mtDNA or cpDNA in conifers. Nevertheless, many studies of genetic variation in conifers have hinted of an association between genetic variation and either physiological variation or some component of life-history variation.

To understand more fully the ecology and evolution of coniferous forests, we need to better understand the physiological and demographic consequences of the alternate genotypes at a locus. Furthermore, we need a fuller understanding of geographic variation of all components of the genome, and the relative impacts of the evolutionary forces influencing genetic variation. The following sections review studies that report geographic variation or differences in growth or viability or susceptibility to herbivores, that suggest some physiological differences among genotypes. In addition, a few observations are brought in from other fields to suggest important phenomena for which forest biologists should be alert.

IV. Natural Selection

Natural selection is defined as the differential reproduction of genotypes. Natural selection can either maintain a population at an equilibrium or it can cause evolution, the change in the genotypic constitution of a population. Although natural selection is defined as the differential reproduction of genotypes, significant differences among genotypes are not limited to fecundity and fertility. Differences among genotypes in any aspect of life-history variation can produce natural selection.

Fitness is defined as either absolute or relative reproductive success. But due to the difficulties of measuring lifetime reproductive success, especially in long-lived species, biologists usually measure a component of fitness, such as germinability, growth, viability, or fecundity. Physiological differences among genotypes are probably significant in all components of fitness. Physiological variation may reflect genetic variation, environmental variation, or, most commonly, a mixture of the two.

If physiological variation reflects only environmental variation, then differential reproduction has no genetic consequences, and does not produce natural selection. Thus, it is the physiological variation that is caused at least partially by genetic variation that is of major importance to our understanding of the ecology and evolution of coniferous forests.

A. Opportunity for Selection

The very high fecundity of forest conifers and their numerous sources of mortality combine to produce the opportunity for natural selection to be intense. Campbell (1979) described the potential for natural selection in Douglas fir forests by estimating mortality at various stages in the life cycle. Approximately 31% of seeds landing on appropriate substrates start to germinate, and only 76% of these reach the seedling stage. Although the life-span of Douglas fir is estimated to be 500 years, the average life expectancy of a seedling is 400–500 days. Thus, germination failure and early mortality reduce approximately 10,000 seeds to 2000 seedlings, and mortality reduces these to a single mature breeding individual. Campbell (1979) concluded that this amount of mortality could sustain an intensity of natural selection sufficient to produce the degree of genetic differentiation associated with elevation and slope aspect in Douglas fir. Although the life tables of conifers may differ, the magnitude of the opportunity for natural selection cited for Douglas fir is probably representative of species such as ponderosa pine, Engelmann spruce, subalpine fir, lodgepole pine, and pinyon pine. For example, after fires, lodgepole pine reproduces profusely from serotinous cones, so that the number of seedlings the summer following a fire may be 500,000/ha, but this number is reduced by natural thinning to approximately 1000 mature trees per hectare (Rehfeldt, 1988). Similarly, during years of successful recruitment and moderately favorable conditions, approximately 100 ponderosa pine seeds produce a single seedling (Schubert, 1974). Within natural stands of ponderosa pine, a majority of seeds (>80%) are produced by a minority (<20%) of trees (Linhart and Mitton, 1994), and recruitment in stands of ponderosa pine is episodic (Linhart, 1988), typically with decades of zero recruitment punctuated by an occasional year class that enjoys high success.

B. Microgeographic Variation

The potential for gene flow in conifers is great, for they are wind pollinated, and the majority of species have winged seeds. The potential for high levels of gene flow leads to the expectation that for neutral genes, or genes not visible to natural selection, there should be little geographic variation. However, if genes are subject to natural selection, high gene flow can be overcome by selection among heterogeneous environments, producing clines associated with environmental gradients. Although geographic variation for enzymes is generally slight, there are some examples of microgeographic genetic differences associated with environmental gradients. These observations suggest but do not prove that differences in enzymatic genotypes are associated with adaptation.

Enzyme kinetic variation and geographic variation of the *Per-2* alleles in ponderosa pine provide insight into one of the causes of microgeographic variation. Inheritance studies demonstrated that there are two common alleles segregating at the *Per-2* locus, that they are inherited in a Mendelian fashion, and that the enzyme genotypes can be detected from homogenates of mature needle tissue (Mitton *et al.*, 1977). Peroxidase is involved in one of the photosystems, and it commonly has antifungal activity. Enzyme kinetic studies revealed that the three most common genotypes, the 23 heterozygote and the 22 and 33 homozygotes, differed in both maximum velocities and Michaelis constants (Beckman, 1977). The 22 homozygote was identified as being most efficient at cold temperatures, whereas the 33 homozygote was most efficient at high temperatures. The 23 heterozygote was a blend of the two homozygotes, and was efficient over a broad range of temperatures. The enzyme kinetics explain the microgeographic variation identified at this locus (Mitton *et al.*, 1977; Beckman and Mitton, 1984). Consistent variation in both allelic and genotypic frequencies was found between north- and south-facing slopes. Allele 2, which performs optimally in cool temperatures, was at higher frequency at high elevations than at low elevations, and it was also higher in frequency on north-facing than on south-facing slopes. Furthermore, there were excesses of heterozygotes on south-facing but not on north-facing slopes. The differences in allelic frequencies with aspect were consistent with selection on the warm south-facing slopes favoring heterozygous genotypes, which are efficient over a wide range of temperatures. The 22 homozygote is favored at high elevations and on north-facing slopes, where high temperatures are rare. The 33 homozygote is relatively more common at low elevations and on south-facing slopes, but this genotype probably suffers during low winter temperatures.

Very similar evolutionary dynamics occur during the colonization of grassland sites by ponderosa pine. Fire suppression that began with the settlement of the area around Boulder, Colorado, has changed the environment sufficiently to favor trees over grasses. Consequently, ponderosa pine has been invading the grasslands for the last century. The establishment of new stands has been studied by aging the trees from increment cores and identifying their peroxidase genotypes. Peroxidase genotypes do not colonize randomly, but arrive in an order that is consistent with their enzyme kinetics. The grassland on Shanahan Mesa was colonized predominantly by the 23 heterozygote, the genotype with the wide temperature range of enzymatic efficiency. The advantage switches to the 22 homozygote about the time that the canopy closes over the stand. At this time, the site becomes cooler, due to increased shade, needle litter, and the accumulations of snow on a shady site.

Enzyme kinetic variation is also consistent with geographical variation of isocitrate dehydrogenase alleles in silver fir (*Abies alba*) (Bergmann and Gregorius, 1992). Seeds were sampled from 45 populations spread across seven countries and 12 degrees of latitude, covering most of the natural range of silver fir. Two common alleles segregate at this locus, and the homozygotes bearing the alleles differ in thermostability. The allele with higher thermostability is more common in warmer environments. Allele 2, which has the higher thermostability, varies in frequency from approximately 0.90 in the south to 0.10 in the north. Although thermostability is not likely to be the kinetic parameter directly relevant to the geographic variation, thermostability probably reflects the flexibility of the enzyme, which may influence other kinetic variables.

Although enzyme kinetic data are available only for the studies summarized above, other reports of differentiation of enzyme loci associated with environmental variation suggest that there are additional cases in which differences among enzyme genotypes produce physiological differences that are acted on by natural selection. For example, peroxidase allelic frequencies differ dramatically across tree lines in both Engelmann spruce (*Picea engelmannii*) and subalpine fir (*Abies lasiocarpa*) (Grant and Mitton, 1977). Allelic frequencies at phosphoglucomutase vary clinally along a steep elevational gradient in ponderosa pine (Mitton *et al.*, 1980). Genotypic frequencies of Engelmann spruce differ between adjacent wet and dry sites; the differentiation appears to be driven by selection against heterozygotes in very wet sites, after establishment of young trees (Stutz and Mitton, 1988; Mitton *et al.*, 1989; Shea, 1990). Pinyon pines are differentiated at the glycerate dehydrogenase locus between cinder soils and adjacent sandy-loam soils near Sunset Crater, Arizona (Mopper *et al.*, 1991b), and the allele that is more common on the cinder soils is associated with higher survival and growth rates (Cobb *et al.*, 1994). This enzyme may be important for the synthesis of glycine betaine, an osmolyte that helps many plants endure drought stress (Somero, 1992).

C. Growth Rate

Because their growth is recorded in annual rings, temperate trees are convenient subjects for studies of the association between genotype and growth rate. The first relationship between protein heterozygosity and growth in trees was reported for quaking aspen (*Populus tremuloides*) (Mitton and Grant, 1980); growth rate increased with allozyme heterozygosity in a sample of 104 clones in the Front Range of Colorado. Similarly, growth rate increases with enzyme heterozygosity in aspen in Waterton Lakes National Park, Alberta (Jelinski, 1993). Relationships between heterozygosity and growth are complex in ponderosa pine and lodgepole pine (Knowles and Mitton, 1980; Knowles and Grant, 1981; Linhart and Mitton, 1985), for heterozygosity is associated not with mean growth rate, but with the variance of growth. Growth increases with heterozygosity in pitch pine (*Pinus rigida*), but only in older stands, probably as a consequence of competition for light and water (Ledig *et al.*, 1983; Bush *et al.*, 1987). The relationship between seedling radicle length and allozyme heterozygosity was tested in jack pine under nine experimental conditions (Govindaraju and Dancik, 1986, 1987). Radicle length increased significantly with heterozygosity in three of the nine experimental conditions, and most clearly under conditions that produced water stress. Growth rate increases with heterozygosity in knobcone pine (*Pinus attenuata*) (Strauss, 1986); the relationship is most pronounced in selfed progeny, but it is apparent in outcrossed progeny as well. Bongarten *et al.* (1985) found no relationship between genotype and growth in a plantation of 15-year-old Douglas fir. Although some loci appear to be related to growth rate in Norway spruce (*Picea abies*), growth rate is not correlated with enzyme heterozygosity (von Wuehlische and Krusche, 1991).

The growth rate of pinyon pine is influenced by the genotype at the glycerate dehydrogenase locus (Cobb *et al.*, 1994). The growth rates of 103 trees were estimated by the weight of new growth added to a branch in the last four growing seasons. The average weights were 10, 14, and 20 g for the 11, 12, and 22 genotypes, respectively ($P < 0.05$). The 22 homozygotes were accumulating biomass at twice the rate of the 11 homozygotes on the cinder soils around Sunset Crater, Arizona. The soil at Sunset Crater imposes a moisture stress on plants, and the glycerate dehydrogenase locus may be related to moisture stress (Jelinski, 1993; Cobb *et al.*, 1994). Estimates of relative viability taken from a separate set of trees were 0.43, 0.62, and 1.00 for the 11, 12, and 22 genotypes, respectively ($P < 0.001$). Thus, both growth and survival favor the 22 homozygotes on cinder soils; no differences among genotypes were detected on the nearby sandy-loam soil, which is a more typical and less stressful environment for pinyon pine.

Clearly, studies of the relationship between allozyme heterozygosity and growth rate in conifers have yielded heterogeneous results. One reason for this heterogeneity is that some studies have been conducted in plantations, where trees are evenly spaced, watered, and fertilized; these conditions are so salubrious that they may hide potential genotypic differences that would be expressed during the stresses typical in natural populations. Another reason for the heterogeneous results is inherent in growth rate, which may not accurately reflect the energy resources of the tree. A mature tree divides its energy between growth and reproduction, and individuals vary considerably in their apportionment (Linhart and Mitton, 1985). The relationship between heterozygosity and growth should be most evident in the growth of young trees (Mitton, 1983), when all energy is invested in growth.

D. Resistance to Air Pollution

Stressful conditions appear to accentuate differences among genotypes, often producing overdominance or heterosis (Parsons, 1971, 1973, 1987; Mitton and Grant, 1984). Similarly, stress enhances correlations between enzyme heterozygosity and components of fitness (Koehn and Shumway, 1982; Diehl, 1989; Scott and Koehn, 1990; Teska et al., 1990). Deteriorating air quality is challenging the forests of Europe, causing either mortality or declining vigor in several species (Scholz et al., 1989; Giannini, 1991). Comparisons of resistant and susceptible trees from the same stands suggest that heterozygous genotypes are associated with resistance to airborne pollutants (Scholz and Bergmann, 1984; Muller-Stark, 1985; Bergmann and Scholz, 1984, 1987, 1989).

E. Resistance to Herbivores

Allozyme heterozygosity is associated with resistance to herbivores in pinyon pine (*P. edulis*) (Mopper et al., 1991b). Pinyon pines on the cinder soils around Sunset Crater, near Flagstaff, Arizona, experience chronic water and nutrient stress, and also sustain higher densities of herbivores than do pines living nearby on normal soils (Whitham and Mopper, 1985; Mopper and Whitham, 1986; Mopper et al., 1991a). However, the impact of the stress and herbivory is not uniform among the trees on the cinder soils. Some trees have little or no damage from herbivores, whereas others are trimmed so regularly by the stem moth, *Dioryctria albovittella*, that they assume a different growth form (Whitham and Mopper, 1985; Mopper and Whitham, 1986). By defoliating the tips of branches, the stem moths increase internal branching, producing an atypical, densely packed, closely trimmed growth form that records chronic herbivory. Resistant trees were significantly more heterozygous than susceptible trees at two of four allozyme loci (Mopper et al., 1991b) (Fig. 2). In addition, in both susceptible and resistant trees, older trees were more heterozygous than younger trees, revealing viability differentials favoring heterozygotes.

F. Oleoresin Pressure

The resin of conifers contains a variety of monoterpenes, diterpenes, and resin acids, and numerous studies and essays suggest that resin deters herbivores (Mason, 1969; Cates and Alexander, 1982; Sturgeon and Mitton, 1982, 1986; Raffa and Berryman 1982a,b, 1983; Snyder, 1992,

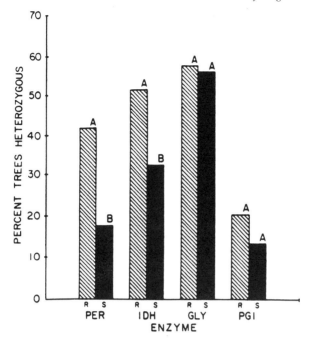

Figure 2 Heterozygosity of pinyon pine, *Pinus edulis*, resistant and susceptible to attack by the stem moth, *Dioryctria albovittella*. Genetic variation has been examined for peroxidase (PER), isocitrate dehydrogenase (IDH), glycerate dehydrogenase (GLY), and phosphoglucose isomerase (PGI). Letters above the bars designate statistically significant differences between groups. (From Mopper *et al.*, 1991b.)

1993; Snyder and Linhart, 1994; Linhart *et al.*, 1994). This point can be illustrated with two examples from ponderosa pine and an example from pinyon pine. Both oleoresin exudation pressure and flow rate were measured in ponderosa pine in the San Isabel National Forest, Colorado, immediately before the flight period of the mountain pine beetle, *Dendroctonus ponderosae*. Once the beetles had flown, attacked, and killed a portion of the trees, the data on resin pressure and flow were divided into two groups. Trees successfully attacked and killed by the beetles had significantly lower oleoresin pressures and flow rates than trees that were not attacked or that survived attack (Table II). In both years of this study, the flow rate of resistant trees was more than twice that of trees that died.

Abert's squirrel, *Sciurus aberti*, is a specialist on ponderosa pine, feeding heavily on the inner bark when other food sources are scarce. The flow rate of resin is the most effective variable for predicting which trees will be selected for feeding (Snyder, 1992, 1993). The resin flow rate, in

Table II Oleoresin Exudation Pressure and Resin Flow Rates of Nonattacked and Attacked *Pinus ponderosa* Trees in San Isabel National Forest, Colorado[a]

Attack status	Pressure (psi)	Flow rate (ml/24 hr)	N
1979			
Successfully attacked trees	4.9 ± 15.7	9.3 ± 16.0	18
Nonattacked trees	37.9 ± 47.0	25.7 ± 24.3	62
1980			
Successfully attacked trees	3.8 ± 9.1	11.1 ± 3.9	10
Nonattacked trees	46.8 ± 45.7	33.5 ± 27.8	34

[a] From Cates and Alexander (1982).

milliliters of resin per 24 hours, was 2.20 ± 0.45 for selected feeding trees, and 5.30 ± 0.72 for adjacent control trees ($P < 0.001$). To determine whether the differences in flow rate were caused by herbivory, Snyder simulated herbivory by clipping ponderosa pines for 2 years. Because this experimental manipulation had no effect on flow rate, Snyder concluded that Abert's squirrels select trees for feeding with low oleoresin pressure and flow rate. A study of the ponderosa pines utilized as nest trees revealed that nest trees differed from adjacent control trees for a variety of constituents in xylem, phloem, and oleoresin (Snyder and Linhart, 1994). In particular, the squirrels placed nests in trees with low levels of α-pinene.

The xylem pressures and resin flow rates of trees susceptible and resistant to the stem moth, *D. albovitella*, were compared in populations of trees growing in stressful conditions on cinder soils at Sunset Crater, Arizona (Mopper *et al.*, 1991a). Xylem pressures did not differ with herbivory, but resin flow rates were distinctly different ($P < 0.001$); resistant trees produced more than twice as much resin from wounds as did susceptible trees. Similar results were reported by Vite (1961), Rudinsky (1966), and Stark *et al.* (1968).

The study of resistance to herbivory in pinyon pine suggests a link between allozyme heterozygosity and resin pressures and flow rates (Mopper *et al.*, 1991b). Because resistant trees are more heterozygous than susceptible trees, and because resistant trees have greater resin flow rates than susceptible trees, it is reasonable to predict that resin pressure and flow rate will increase with allozyme heterozygosity.

Evidence for an association between allozyme genotypes and resin flow is seen in a study of monoterpene diversity in ponderosa pine (Table III). Resin flow was estimated for trees on a south-facing slope in Boulder Canyon, Colorado, and the data are arranged by genotype at the shikimate dehydrogenase locus. This locus segregates three alleles in a Mendelian manner, and the enzyme is in a metabolic pathway that

Table III Resin Production of Ponderosa Pine for Genotypes at the Shikimate Dehydrogenase Locus[a]

Measure	Genotype					
	11	12	22	13	23	33
Mean	10.1	16.1	24.2	12.0	19.7	16.2
SE	3.6	3.2	2.7	3.1	2.7	3.4
N	7	24	38	12	42	17

[a]SE is the standard error of the mean, and N is the sample size. Production is given as milliliters/hour. The most abundant genotypes (22, 23) have the highest resin production, and the rare genotypes (11, 13) have the lowest resin production. Data from Linhart and Mitton (1994).

leads to some of the plant secondary compounds that are carried in the resin (Linhart et al., 1981a). The resin flow rates are heterogeneous among the genotypes, with more than a twofold difference among genotypes. The magnitude of differences among genotypes is comparable to the differences between the victims and the survivors of a bark beetle attack (compare Tables II and III). Furthermore, the resin flow rates of the genotypes are related to their frequencies in the population; the rare genotypes (11 and 13) have the lowest flow rates (10.1 and 12.0) and the most abundant genotypes (22 and 23) have the highest flow rates (24.2 and 19.7).

G. Respiration

Many of the enzymes used by population biologists to estimate the heterozygosity of populations and of individuals are in the main corridor of metabolism: glycolysis, the pentose shunt, and the citric acid cycle. If these metabolic loci are directly responsible for the associations summarized above for growth rate, then we would predict that the enzyme loci would also be associated with some aspects of metabolism, such as those measured by the physiological ecologist.

The first clear demonstration of an association between individual heterozygosity and some aspect of metabolism was the study of oxygen consumption and heterozygosity in the American oyster, *C. virginica* (Koehn and Shumway, 1982). The reports of increases in growth rate with protein heterozygosity in the American oyster (Singh and Zouros, 1978; Zouros et al., 1980; Singh, 1982) prompted a more physiological study of the phenomenon. Under both control and stress conditions, oxygen consumption decreased dramatically with the number of loci heterozygous. Individuals in the most homozygous class consumed approximately twice as much oxygen as the most heterozygous class. These results are consistent with the observations on the higher growth rate of highly heterozygous individuals—individuals consuming less oxygen, and therefore respiring at a lower rate, would have greater amounts of energy to invest in growth.

A study of energetics and growth in the coot clam, *Mulinia lateralis*, revealed a very tight association between individual heterozygosity and metabolism (Garton *et al.*, 1984). Individual heterozygosity was measured with six polymorphic loci, and energy budgets were constructed from estimates of growth rates, rates of oxygen consumption, ammonia excretion, and clearance rates. Individual heterozygosity was associated with observed growth rates, and this association was driven primarily by a tight relationship between individual heterozygosity and routine metabolic costs. Routine metabolic costs decrease with increasing enzyme heterozygosity, leaving more energy to be invested in growth. A subsequent study of enzyme heterozygosity and growth in the coot clam revealed that the association of genotype with growth rate differed among enzyme groups (Koehn *et al.*, 1988). Correlations were high with enzymes involved in protein cycling, intermediate with glycolytic enzymes, and essentially zero for enzymes in the citric acid cycle.

Studies reporting a negative relationship between enzyme heterozygosity and respiration at rest have now been reported for the American oyster (Koehn and Shumway, 1982), the coot clam (Garton *et al.*, 1984; Koehn *et al.*, 1988), the blue mussel (Diehl *et al.*, 1985, 1986), *Thais hemostomata* (Garton, 1984), salmonid fishes (Danzmann *et al.*, 1987, 1988), and tiger salamanders (Mitton *et al.*, 1986). The result appears to be general, at least for animals. Physiological studies by Hawkins and colleagues (1986, 1989) report that these relationships may be tied to protein turnover rates, which decline with enzyme heterozygosity and form a substantial proportion of the routine metabolic costs. There are no comparable studies for forest trees, but there is evidence that enzyme variation directly affects respiration and growth in at least one plant, perennial ryegrass, *Lolium perenne*.

Extensive studies by Wilson and colleagues at the Welsh Plant Breeding Station in Aberswyth have demonstrated substantial differences among individuals in their rates of dark respiration. Two estimates of the heritability of dark respiration have been reported, and these indicate that 50% or more of variation in the rate of dark respiration is attributable to genetic variation (Wilson, 1981). Further evidence of the genetic basis for variability in the rate of respiration is found in the response of respiration rate to artificial selection. Selection for high and low respiration rates differentiated lines in just two generations of selection (Wilson, 1975, 1981; Day *et al.*, 1985). In perennial ryegrass, dark respiration is inversely proportional to production, or growth; plants selected for low levels of dark respiration were more productive than a random sample of plants (Wilson and Jones, 1982). Wilson has suggested that differences seen in productivity are the result of differential maintenance costs between the lines. Empirical data have led Hawkins and colleagues to the same hypothesis to explain genetic differences in growth among mussels (Hawkins *et al.*, 1986, 1989). The control of this variation in respiration rate in ryegrass is not in the mitochondria, but appears to reside in glycolysis (Day *et al.*, 1985).

Studies of enzyme genotypes reveal associations between enzyme genotypes and respiration in perennial ryegrass (Rainey *et al.*, 1987, 1990). Data are most extensive for 6-phosphogluconic dehydrogenase (6PGD), which is in the pentose shunt.

Enzyme kinetic studies of 6PGD reveal biochemical differences among genotypes (Rainey-Foreman, 1991). At 35°C, both Michaelis constants and V_{max} differ among genotypes, producing dramatic differences in V_{max}/K_m among genotypes. The Q_{10} of dark respiration, measured between 20 and 35°C, differs dramatically among the 6PGD genotypes (Rainey *et al.*, 1987). Over this temperature range, the Q_{10} of the 11 homozygote is approximately 30% higher than the Q_{10} of the 22 homozygote. The biological relevance of this difference was tested by exposing ryegrass to a temperature stress for 5 days, and estimating the tolerance of the genotypes to the prolonged stress (Rainey *et al.*, 1987). As expected, the genotypes with the lower values of Q_{10} survived this specific stress in better condition.

Respiration rates differ among 6PGD genotypes at 35°C (Rainey *et al.*, 1990). The variation in respiration rate is consistent with differences in V_{max}/K_m, with the 22 homozygote having a dramatic advantage over the 11 homozygote. In addition, the flux through the pentose shunt differs among 6PGD genotypes (Rainey-Foreman, 1991). Just as in the enzyme kinetics, differences are seen at 35°C but not at 25°C. The flux through the pentose shunt is highest in the 22 homozygote, the genotype with the highest value of V_{max}/K_m.

It is becoming increasingly apparent that genetic variation of enzymes has a direct and profound impact on the physiology of animals (Koehn *et al.*, 1983). However, these genetic/physiological studies have not yet been conducted on forest trees. Give the wide range of growth rate and seed production in natural stands, and the high level of genetic variation of forest trees, it seems appropriate to determine the correspondence between genetic and physiological variation in forest trees.

H. Viability

Conifers have a mixed mating system, producing seeds by both outcrossing and selfing (Mitton, 1992a,b). Therefore, a sample of seeds or seedlings may have a value of F, the inbreeding coefficient, reflecting a mixture of inbred and outcrossed seeds. However, viability selection typically reduces the value of F toward zero (Tigerstedt et al., 1982; Shaw and Allard, 1982; Farris and Mitton, 1984; Yazdani et al., 1985; Muona et al., 1987). This form of selection is undoubtedly selection against selfed and otherwise inbred genotypes.

Significant excesses of heterozygotes (summarized in Mitton and Jeffers, 1989) have been reported in mature stands of ponderosa pine (Linhart et al., 1981b) jack pine (Cheliak et al., 1985), black spruce (*Picea mariana*) (Boyle and Morgenstern, 1986; Yeh et al., 1986), Monterey pine (Plessas and Strauss, 1986), Douglas fir (Shaw and Allard, 1982), balsam fir (*Abies balsamea*) (Neale and Adams, 1985), Polish larch (*Larix decidua*) (Lewandowski et al., 1991), and *Pinus leucodermis* (Morgante et al., 1993). Whereas the decline in values of F from initial positive values to zero is consistent with selection against inbred individuals, the production of excesses of heterozygotes must be a different process, in which heterozygotes are favored (Shaw and Allard, 1982; Mitton and Jeffers, 1989). If selection acts only to eliminate selfed genotypes, then all outcrossed genotypes would have equal fitnesses. But when selection produces excesses of heterozygotes, then fitness must increase with heterozygosity, even within the pool of genotypes produced exclusively by outcrossing.

Survival of loblolly pine (*Pinus taeda*) appears to be dependent on enzyme genotype (Bush and Smouse, 1992a,b). Survival of heterozygous genotypes and the common homozygous genotypes is good, but survival of the homozygotes bearing the rarer alleles is poor. This same result has been reported for natural populations of annual ryegrass (*Lolium multiflorum*) (Mitton, 1989), and for two species of fishes (Mitton, 1993). All of these data, as well as those describing growth rate in pitch pine (*Pinus rigida*), fit the adaptive distance model of Smouse (1986), implying that variation at these enzymes (or genes in linkage disequilibrium with them) directly contributes to the probability of survival.

A survey of studies of plant population biology revealed that fitness differentials in natural populations of plants generally favor enzyme heterozygotes (Lesica and Allendorf, 1992). This result certainly has implications for forest management and conservation biology, but as yet the physiological mechanisms underlying this generality are little more than a subject of speculation (Mitton, 1993).

I. Analyses of the Mating System

When genetic data organized into open-pollinated families are used to estimate the level of outcrossing in a population (Mitton, 1992a,b; Brown 1990; Ritland, 1983), the single locus estimates of outcrossing are often heterogeneous among loci (e.g., Mitton et al., 1981). One of the more likely explanations for this heterogeneity is that genotypes at a small minority of polymorphisms are associated with reproductive phenology. If one homozygote tends to reproduce early in the season, and the alternate homozygote tends to reproduce late, then there is effectively positive assortative mating in time, producing a deficiency of heterozygotes, which yields a low estimate of the rate of outcrossing. Evidence for this assortative mating in time is strongest for several loci in corn (Bijlsma *et al.*, 1986; Allard, 1990), but similar data have been reported for Table Mountain pine (*Pinus pungens*) (Gibson and Hamrick, 1991) and Sitka spruce (Chaisurisri *et al.*, 1994). It is intriguing that biased estimates of outcrossing might identify loci that are related to, and perhaps influencing, reproductive phenology.

Although many conifers release seed as cones mature, trees with serotinous cones hold their seeds for many years. When seeds from serotinous cones are used to estimate the mating system, the rate of outcrossing changes over time (Cheliak *et al.*, 1985; Snyder *et al.*, 1985; Hamrick and Godt, 1990). The older the seeds, the higher the estimate of outcrossing obtained for the population. This directional change over time in separate populations is best explained by a general phenomenon, rather than coincidence of temporal trends. The temporal trends are consistent with higher viability of heterozygous seeds, perhaps due to lower routine metabolic costs (see Section IV,G). Measures of calorimetry or measures of respiration of seeds could directly test this hypothesis.

J. The Nature of cpDNA and mtDNA Variation

DNA markers are now employed widely by forest biologists, and for most applications, the variation is presumed to have no detectable effect on physiological variation. This may be a good assumption, but it is rarely tested. Data from other species indicate that the assumption is not always correct.

Various observations indicate that variation in mitochondrial DNA is not always neutral in its effect on physiology and components of fitness in animals. For example, King and Attardi (1989) removed the mitochondria from human tissue cells by chronic treatment with ethidium bromide. They then reintroduced isolated mitochondria into the cell culture, restoring the initial respiration rate and time required to double the number of cells. The experiment was then conducted again, but instead of reintroducing the mitochondria from the initial cell culture, mitochondria from other humans were injected. Although some of the new cultures were similar to the initial culture, the respiration rates of some cultures were dramatically higher, and some were dramatically lower. The investigators concluded that there was an interaction between the nuclear and mitochondrial genomes that profoundly influenced respiration.

Experiments with laboratory populations of *Drosophila* also suggest that variation in the mitochondrial genome influences components of fitness. A traditional selection experiment revealed that fitnesses differed among mitochondrial genotypes, altering the frequencies in population cages (MacRae and Anderson, 1990). A study of direct competition between mitochondria revealed differences between species (Miki et al., 1989). The mitochondria from *Drosophila mauritania* were injected into the eggs of *Drosophila melanogaster*, and four replicate populations were started with the eggs containing mitochondria from both species. Within 30 generations, three of the four populations were fixed for the *D. mauritania* mitochondria—the alien genome—and the fourth population had come to an equilibrium with 90% *D. mauritania* and 10% *D. melanogaster* mitochondria.

Studies of ongoing selection in composite crosses of barley (*Hordeum vulgare*) have revealed dramatic changes in fragments of DNA coding for ribosomal RNA that are consistent with strong selection (Saghai-Maroof et al., 1984, 1990; Allard et al., 1990; Zhang et al., 1990).

These diverse observations warn us against sweeping generalizations concerning the intensity of natural selection influencing genes, genomes, or fragments of DNA. Some sequences must be subject to natural selection, whereas others are certainly not. Statistically, it makes most sense to assume that markers are selectively neutral, for that assumption allows the use of neutral models in population genetics to make predictions concerning the distribution of genotypes in space and in time. But we know rather little about the physiological consequences of variation in cpDNA and mtDNA; the assumption should be made explicitly, and it deserves to be tested in the field.

V. Summary

Their large populations, immense geographic ranges, wide ecological amplitudes, high fecundities, long life-spans, and high potentials for gene flow lead us to expect conifers to maintain high levels of genetic variation. Data from numerous provenance studies and allozyme surveys are consistent with this expectation. Conifers are perhaps the most genetically variable group of species.

Provenance studies have revealed that the population structure of Rocky Mountain conifers is associated with environmental gradients. For Douglas fir, ponderosa pine, and lodgepole pine in the Rocky Mountains, populations differing by 300 m in elevation are likely to exhibit adaptation to different environments. These adaptations to environmental gradients and patchiness forced foresters to design seed collection zones for western conifers.

When population biologists adopted allozyme surveys to measure genetic variation, forest biologists used allozyme polymorphisms to describe mating systems, to estimate gene flow, and to describe geographic variation. However, allozyme surveys and provenance studies detect different magnitudes and patterns of geographic variation. In conifers, allozymes reveal populations with high levels of variation within populations, and only slight variation among populations. Most forest biologists have assumed that allozyme variation is adaptively neutral, but some empirical data are not consistent with this assumption.

Molecular techniques are now employed to examine variation in the nuclear, mitochondrial, and chloroplast genomes. In contrast to the nuclear genome, the mitochondrial and chloroplast genomes are not reshuffled during sexual reproduction, but are inherited without recombination. In the Pinaceae, mtDNA is inherited maternally and cpDNA is inherited paternally. In contrast to allozyme loci, mtDNA and cpDNA exhibit relatively little genetic variation within populations, but striking differentiation among populations.

Provenance studies reveal adaptation to local environments in most of the Rocky Mountain conifers that have been planted in common gardens. Where do the genes reside that produce the morphological, physiological, and phenological responses to environmental gradients? Although a few examples of microgeographic variation of allozyme loci have been documented, the geographic variation of allozyme frequencies appears to be too slight to cause the adaptive differentiation along elevational gradients. Do the mitochondrial and chloroplast genomes contribute to the adaptive variation? To answer these questions, we must examine the physiological and demographic consequences of allozyme loci, and variation of nuclear, mitochondrial, and chloroplast DNA.

References

Adams, W. T. (1992a). Gene dispersal within forest tree populations. *New For.* 6:217–240.
Adams, W. T. (1992b). *In* "Population Genetics of Forest Trees" (W. T. Adams, S. H. Strauss, D. L. Copes, and A. R. Griffin, eds.), pp. 217–240. Kluwer Academic Publishers, Dordrecht, The Netherlands.
Adams, W. T., and Birkes, D. S. (1991). Estimating mating patterns in forest tree populations. *In* "Biochemical Markers in the Population Genetics of Forest Trees" (S. Fineschi, M. E. Malvolti, F. Cannata, and H. H. Hattemer, eds.), pp. 157–172. SPB Academic Publishing, The Hague, The Netherlands.
Adams, W. T., Birkes, D. S., and Erickson, V. J. (1992). Using genetic markers to measure gene flow and pollen dispersal in forest tree seed orchards. *In* "Ecology and Evolution of Plant Reproduction: New Approaches" (R. Wyatt, ed.), pp. 37–61. Chapman & Hall, New York.
Allard, R. W. (1990). Future directions in plant population genetics, evolution, and breeding. *In* "Plant Population Genetics, Breeding, and Genetic Resources" (A. H. D. Brown, M. T. Clegg, A. L. Kahler, and B. S. Weir, eds.), pp. 1–19. Sinauer Associates, Sunderland, MA.

Allard, R. W., Saghai-Maroof, M. A., Zhang, Q., and Jorgensen, R. A. (1990). Genetic and molecular organization of ribosomal DNA (rDNA) variants in wild and cultivated barley. *Genetics* 126:743–751.

Baker, F. S. (1944). Mountain climates of the western United States. *Ecol. Monogr.* 14: 223–254.

Beckman, J. S. (1977). Adaptive peroxidase allozyme differentiation between colonizing and established populations of *Pinus ponderosa* Laws. on the Shanahan Mesa, Boulder, Colorado. Master's Thesis, University of Colorado, Boulder.

Beckman, J. S., and Mitton, J. B. (1984). Peroxidase allozyme differentiation among successional stands of ponderosa pine. *Am. Midl. Nat.* 112:43–49.

Bergmann, F., and Gregorius, H. R. (1992). Ecogeographical distribution and thermostability of isocitrate dehydrogenase (IDH) alloenzymes in European silver fir (*Abies alba*). *Biochem. Syst. Ecol.* 21:597–605.

Bergmann, F., and Scholz, F. (1984). Effects of selection pressure by SO_2 pollution on genetic structures of Norway spruce (*Picea abies*). *Lect. Notes Biomathe.* 60:267–275.

Bergmann, F., and Scholz, F. (1987). The impact of air pollution on the genetic structure of Norway spruce. *Silvae Genet.* 36:80–83.

Bergmann, F., and Scholz, F. (1989). Selection effects of air pollution in Norway spruce (*Picea abies*) populations. *In* "Genetics Effects of Air Pollutants in Forest Tree Populations" (F. Scholz, H.-R. Gregorius, and D. Rudin, eds.), Springer-Verlag, Berlin.

Bijlsma, R., Allard, R. W., and Kahler, A. L. (1986). Nonrandom mating in an open-pollinated maize population. *Genetics* 112:669–680.

Bongarten, B. C., Wheeler, N. C., and Jech, K. S. (1985). Isozyme heterozygosity as a selection criterion for yield improvement in Douglas-fir. *Proc. Can. Tree Improv. Assoc.*, pp. 121–128.

Boyle, T. J. B., and Morgenstern, E. K. (1986). Estimates of outcrossing rates in six populations of black spruce in central New Brunswick. *Silvae Genet.* 35:102–106.

Brown, A. H. D. (1990). Genetic characterization of plant mating systems. *In* "Plant Population Genetics, Breeding, and Genetic Resources" (A. H. D. Brown, M. T. Clegg, A. L. Kahler, and B. S. Weir, eds.), pp. 145–162. Sinauer Associates, Sunderland, MA.

Bush, R. M., and Smouse, P. E. (1992a). Evidence for the adaptive significance of allozymes in forest trees. *New For.* 6:179–196.

Bush, R. M., and Smouse, P. E. (1992b). *In* "Population Genetics of Forest Trees" (W. T. Adams, S. H. Strauss, D. L. Copes, and A. R. Griffin, eds.), pp. 179–196. Kluwer Academic Publishers, Dordrecht, The Netherlands.

Bush, R. M., Smouse, P. E., and Ledig, F. T. (1987). The fitness consequences of multiple-locus heterozygosity: The relationship between heterozygosity and growth rate in pitch pine (*Pinus rigida* Mill.). *Evolution (Lawrence, Kans.)* 41:787–798.

Campbell, R. K. (1979). Genecology of Douglas-fir in a watershed in the Oregon cascades. *Ecology* 60:1036–1050.

Cates, R. G., and Alexander, H. (1982). Host resistance and susceptibility. *In* "Bark Beetles in North American Conifers" (J. B. Mitton and K. B. Sturgeon, eds.), pp. 212–263. Univ. of Texas Press, Austin.

Ceulemans, R. J., and Saugier, B. (1991). Photosynthesis. *In* "Physiology of Trees" (S. S. Raghavendra, ed.), pp. 21–50. Wiley, New York.

Chaisurisri, K., El-Kassaby, Y. A., and Mitton, J. B. (1994). Variation in the mating system of Sitka spruce associated with genetic variation and crown level. *Am. J. Bot.* (in press).

Cheliak, W. M., Dancik, B. P., Morgan, K., Yeh, F. C. H., and Strobeck, C. (1985). Temporal variation and the mating system in a natural population of jack pine. *Genetics* 109: 569–584.

Clegg, M. T. (1990). Molecular diversity in plant populations. *In* "Plant Population Genetics, Breeding and Genetic Resources" (A. H. D. Brown, M. T. Clegg, A. L. Kahler, and B. S. Weir, eds.), pp. 98–115. Sinauer Assoc., Sunderland, MA.

Cobb, N., Mitton, J. B., and Whitham, T. (1994). Genetic variation associated with chronic water and nutrient stress in pinyon pine. *Am. J. Bot.* 81:936–940.

Conkle, M. T., and Critchfield, W. B. (1988). Genetic variation and hybridization of ponderosa pine. *In* "Ponderosa Pine: The Species and Its Management" (D. M. Baumgartner and J. E. Lotan, eds.), pp. 27–34. Washington State University, Pullman.

Danzmann, R. G., Ferguson, M. M., and Allendorf, F. W. (1987). Heterozygosity and oxygen-consumption rates as predictors of growth and developmental rate in rainbow trout. *Physiol. Zool.* 60:211–220.

Danzmann, R. G., Ferguson, M. M., and Allendorf, F. W. (1988). Heterozygosity and components of fitness in a strain of rainbow trout. *Biol. J. Linn. Soc.* 39:285–304.

Day, D., DeVos, O. C., Wilson, D., and Lambers, H. (1985). Regulation of respiration in the leaves and roots of two *Lolium perenne* populations with two contrasting mature leaf respiration rates and crop yields. *Plant Physiol.* 78:678–683.

Diehl, W. J. (1989). Genetics of carbohydrate metabolism and growth in *Eisemia foetida* (Oligochaeta: Lumbricidae). *Heredity* 61:379–387.

Diehl, W. J., Gaffney, P. M., McDonald, J. H., and Koehn, R. K. (1985). Relationship between weight standardized oxygen consumption and multiple-locus heterozygosity in the marine mussel. *Mytilus edulis* L. (Mollusca). *Proc. Eur. Mar. Biol. Symp., 19th, 1984*, 531–536.

Diehl, W. J., Gaffney, P. M., and Koehn, R. K. (1986). Physiological and genetic aspects of growth in the mussel *Mytilus edulis*. I. Oxygen consumption, growth, and weight loss. *Physiol. Zool.* 59:201–211.

Dong, J., and Wagner, D. B. (1993). Taxonomic and population differentiation of mitochondrial DNA diversity in *Pinus banksiana* and *Pinus contorta*. *Theor. Appl. Genet.* 86:573–578.

El-Kassaby, Y. A. (1991). Genetic variation within and among conifer populations: Review and evaluation of methods. *In* "Biochemical Markers in the Population Genetics of Forest Trees" (S. Fineschi, M. E. Malvolti, F. Cannata, and H. H. Hattemer, eds.), pp. 61–76. SPB Academic Publishing, The Hague, The Netherlands.

Farris, M. A., and Mitton, J. B. (1984). Population density, outcrossing rate, and heterozygote superiority in ponderosa pine. *Evolution (Lawrence, Kans.)* 38:1151–1154.

Furniss, R. L., and Carolin, V. M. (1977). Western forest insects. *Misc. Publ.—U.S. Dep. Agric.* 1339.

Garton, D. W. (1984). Relationship between multiple locus heterozygosity and physiological energetics of growth in the estuarine gastropod. Thais haemostoma. *Physiol. Zool.* 57:530–543.

Garton, D. W., Koehn, R. K., and Scott, T. M. (1984). Multiple-locus heterozygosity and the physiological energetics of growth in the coot clam, *Mulinia lateralis*, from a natural population. *Genetics* 108:445–455.

Giannini, R. (1991). "Effects of Pollution on the Genetic Structure of Forest Tree Populations," Proc. Meet., Rome, April 3, 1990. Consiglio Nazionale delle Ricerche (C.N.R.), Rome.

Gibson, J. P., and Hamrick, J. L. (1991). Heterogeneity in pollen allele frequencies among cones, whorls, and trees of table mountain pine (*Pinus pungens*). *Am. J. Bot.* 78:1244–1251.

Gillespie, H. H., and Turelli, M. (1989). Genotype-environment interactions and the maintenance of polygenic variation. *Genetics* 121:129–138.

Govindaraju, D. R., and Dancik, B. P. (1986). Relationship between allozyme heterozygosity and biomass production in jack pine (*Pinus banksiana* Lamb.) under different environmental conditions. *Heredity* 57:145–148.

Govindaraju, D. R., and Dancik, B. P. (1987). Allozyme heterozygosity and homeostasis in germinating seeds of jack pine. *Heredity* 59:279–283.

Grant, M. C., and Mitton, J. B. (1977). Genetic differentiation among growth forms of Engelmann spruce and subalpine fir at tree line. *Arct. Alp. Res.* 9:259–263.

Guries, R. P., and Ledig, F. T. (1982). Genetic diversity and population structure in pitch pine (*Pinus rigida* Mill.). *Evolution (Lawrence, Kans.)* 36:387–399.

Hamrick, J. L. (1976). Variation and selection in western montane species. II. Variation within and between populations of white fir on an elevational transect. *Theor. Appl. Genet.* 47:27–34.

Hamrick, J. L., and Godt, M. J. W. (1990). Allozyme diversity in plant species. In "Plant Population Genetics, Breeding and Genetic Resources" (A. H. D. Brown, M. T. Clegg, A. L. Kahler, and B. S. Weir, eds.), pp. 43–63. Sinauer Associates, Sunderland, MA.

Hamrick, J. L., and Libby, W. J. (1972). Variation and selection in western montane species. I. white fir. *Silvae Genet.* 21:29–35.

Hamrick, J. L, Linhart, Y. B., and Mitton, J. B. (1979). Relationships between life history characteristics and electrophoretically-detectable genetic variation in plants. *Annu. Rev. Syst. Ecol.* 10:173–200.

Hamrick, J. L., Blanton, H. M., and Hamrick, K. J. (1989). Genetic structure of geographically marginal populations of ponderosa pine. *Am. J. Bot.* 76:1559–1568.

Hamrick, J. L., Godt, M. J. W., and Sherman-Broyles, S. L. (1992a). Factors influencing levels of genetic diversity in woody plant species. *New For.* 6:95–124.

Hamrick, J. L., Godt, M. J. W., and Sherman-Broyles, S. L. (1992b). In "Population Genetics of Forest Trees" (W. T. Adams, S. H. Strauss, D. L. Copes, and A. R. Griffin, eds.), pp. 95–124. Kluwer Academic Publishers, Dordrecht, The Netherlands.

Hanover, J. R. (1963). Geographic variation in ponderosa pine leader growth. *For. Sci.* 9:86–95.

Hartl, D. L., and Clark, A. G. (1989). "Principles of Population Genetics." Sinauer Associates, Sunderland, MA.

Hawkins, A. J. S., Bayne, B. L., and Day, A. J. (1986). Protein turnover, physiological energetics and heterozygosity in the blue mussel, *Mytilus edulis*: The basis of variable age-specific growth. *Proc. R. Soc. London, Ser. B* 229:161–176.

Hawkins, A. J. S., Bayne, B. L., Day, A. J., Rusin, J., and Worrall, C. M. (1989). Genotype-dependent interrelations between energy metabolism, protein metabolism and fitness. In "Reproduction, Genetics and Distributions of Marine Organisms" (J. S. Ryland and P. A. Tyler, eds.), pp. 283–292. Olsen & Olsen, Fredensborg, Denmark.

Hedrick, P. W. (1986). Genetic polymorphism in heterogeneous environments: A decade later. *Annu. Rev. Ecol. Syst.* 17:535–566.

Hedrick, P. W., Ginevan, M. E., and Ewing, E. P. (1976). Genetic polymorphism in heterogeneous environments. *Annu. Rev. Ecol. Syst.* 7:1–32.

Hiebert, R. D., and Hamrick, J. L. (1983). Patterns and level of genetic variation in Great Basin bristlecone pine, *Pinus longaeva*. *Evolution (Lawrence, Kans.)* 37:302–310.

Hong, Y.-P., Hipkins, V. D., and Strauss, S. H. (1993). Chloroplast DNA diversity among trees, populations and species in the California closed-cone pines (*Pinus radiata, Pinus muricata*, and *Pinus attenuata*). *Genetics* 135:1187–1196.

Jelinski, D. E. (1993). Associations between environmental heterogeneity, heterozygosity and growth rates of trembling aspen in a cordilleran landscape. *Arct. Alp. Res.* 25:183–188.

Karl, S. A., and Avise, J. C. (1992). Balancing selection at allozyme loci in oysters: Implications from nuclear rflp's. *Science* 256:100–102.

King, M. P., and Attardi, G. (1989). Human cells lacking mtDNA: Repopulation with exogenous mitochondria by complementation. *Science* 246:500–503.

Knowles, P., and Grant, M. C. (1981). Genetic patterns associated with growth variability in ponderosa pine. *Am. J. Bot.* 68:942–946.

Knowles, P., and Mitton, J. B. (1980). Genetic heterozygosity and radial growth variability in *Pinus contorta*. *Silvae Genet.* 29:114–117.
Koehn, R. K., and Shumway, S. E. (1982). A genetic/physiological explanation for differential growth rate among individuals of the American oyster, *Crassostrea virginica* (Gmelin). *Mar. Biol. Lett.* 3:35–42.
Koehn, R. K., Zera, A. J., and Hall, J. G. (1983). Enzyme polymorphism and natural selection. *In* "Evolution of Genes and Proteins" (M. Nei and R. K. Koehn, eds.), pp. 115–136. Sinauer Assoc., Sunderland, MA.
Koehn, R. K., Diehl, W. J., and Scott, T. M. (1988). The differential contribution to individual enzymes of glycolysis and protein catabolism to the relationship between heterozygosity and growth rate in the coot clam, *Mulinia lateralis*. *Genetics* 118:121–130.
Ledig, F. T., Guries, R. P., and Bonefield, B. A. (1983). The relation of growth to heterozygosity in pitch pine. *Evolution (Lawrence, Kans.)* 37:1227–1238.
Lesica, P., and Allendorf, F. W. (1992). Are small populations of plants worth preserving? *Conserv. Biol.* 6:135–139.
Lewandowski, A., Burczyk, J., and Meinartowicz, L. (1991). Genetic structure and the mating system in an old stand of Polish larch. *Silvae Genet.* 40:75–79.
Lewontin, R. C. (1984). Detecting population differences in quantitative characters as opposed to gene frequencies. *Am. Nat.* 123:115–124.
Linhart, Y. B. (1988). Ecological and evolutionary studies of ponderosa pine in the Rocky Mountains. *In* "Ponderosa Pine: The Species and Its Management" (D. M. Baumgartner and J. E. Lotan, eds.), pp. 77–89. Washington State University, Pullman.
Linhart, Y. B. (1991). Disease, parasitism and herbivory: Multidimensional challenges in plant evolution. *TREE* 6:392–396.
Linhart, Y. B., and Mitton, J. B. (1985). Relationships among reproduction, growth rate, and protein heterozygosity in ponderosa pine. *Am. J. Bot.* 72:181–184.
Linhart, Y. B., and Mitton, J. B. (1994). In preparation.
Linhart, Y. B., Davis, M. L., and Mitton. J. B. (1981a). Genetic control of allozymes of shikimate dehydrogenase in ponderosa pine. *Biochem. Genet.* 19:641–646.
Linhart, Y. B., Mitton, J. B., Sturgeon, K. B., and Davis, M. L. (1981b). Genetic variation in space and time in a population of ponderosa pine. *Heredity* 46:407–426.
Linhart, Y. B., Snyder, M. A., and Habeck, S. A. (1989). The influence of animals on genetic variability within ponderosa pine stands, illustrated by the effect of Abert's squirrel and porcupine. *USDA For. Serv. Gen. Tech. Rep. RM* RM-185.
Linhart, Y. B., Snyder, M. A., and Gibson, J. P. (1994). Differential host utilization by two parasites in a population of ponderosa pine. *Oecologia* 98:117–120.
Loveless, M. D., and Hamrick, J. L. (1984). Ecological determinants of genetic structure in plant populations. *Annu. Rev. Ecol. Syst.* 15:65–95.
MacRae, A. F., and Anderson, W. W. (1990). Can mating preferences explain changes in mtDNA haplotype frequencies? *Genetics* 124:999–1001.
Madesen, J. L., and Blake, G. M. (1977). Ecological genetics of pondersoa pine in the Northern Rocky Mountains. *Silvae Genet.* 26:1.
Marr, J. W. (1961). Ecosystems of the east slope of the Front Range in Colorado. *Univ. Colo. Stud., Ser. Biol.* 8.
Mason, R. (1969). A simple technique for measuring oleoresin exudation flow in pines. *For. Sci.* 15:56–57.
Matos, J. (1992). Evolution within the *Pinus montezumae* complex of Mexico: Population subdivision, hybridization, and taxonomy. Ph.D. Thesis, Washington University, St. Louis, MO.
Miki, Y., Chigusa, S. I., and Matsuura, E. T. (1989). Complete replacement of mitochondrial DNA in *Drosophila*. *Nature (London)* 341:551–552.
Millar, C. J., and Westfall, R. D. (1992a). Allozyme markers in forest genetic conservation. *New For.* 6:347–371.

Millar, C. J., and Westfall, R. D. (1992b). *In* "Population Genetics of Forest Trees" (W. T. Adams, S. H. Strauss, D. L. Copes, and A. R. Griffin, eds.), pp. 347–371. Kluwer Academic Publishers, Dordrecht, The Netherlands.

Millar, C. I., Strauss, S. H., Conkle, M. T., and Westfall, R. (1988). Allozyme differentiation and biosystematics of the Californian closed-cone pines. *Syst. Bot.* 13:351–370.

Mitton, J. B. (1983). Conifers. *In* "Isozymes in Plant Genetics and Breeding" (S. D. Tanksley and T. J. Orton, eds.), pp. 443–472. Elsevier, Amsterdam.

Mitton, J. B. (1989). Physiological and demographic variation associated with allozyme variation. *In* "Isozymes in Plants" (D. Soltis and P. Soltis, eds.), pp. 127–145. Dioscorides Press, Portland, OR.

Mitton, J. B. (1992a). The dynamic mating systems of conifers. *New For.* 6:197–216.

Mitton, J. B. (1992b). *In* "Population Genetics of Forest Trees" (W. T. Adams, S. H. Strauss, D. L. Copes, and A. R. Griffin, eds.), 197–216. Kluwer Academic Publishers, Dordrecht, The Netherlands.

Mitton, J. B. (1993). Theory and data pertinent to the relationship between heterozygosity and fitness. *In* "The Natural History of Inbreeding and Outbreeding" (N. Thornhill, ed.), pp. 17–41. Univ. of Chicago Press, Chicago.

Mitton, J. B., and Grant, M. C. (1980). Observations on the ecology and evolution of quaking aspen, Populus tremuloides, in the Colorado Front Range. *Am. J. Bot.* 67:202–209.

Mitton, J. B., and Grant, M. C. (1984). Associations among protein heterozygosity, growth rate, and developmental homeostasis. *Annu. Rev. Ecol. Syst.* 15:479–499.

Mitton, J. B., and Jeffers, R. M. (1989). The genetic consequences of mass selection for growth rate in Engelmann spruce. *Silvae Genet.* 38:6–12.

Mitton, J. B., Linhart, Y. B., Hamrick, J. L., and Beckman, J. S. (1977). Observations on the genetic structure and mating system of ponderosa pine in the Colorado Front Range. *Theor. Appl. Genet.* 7:5–13.

Mitton, J. B., Sturgeon, K. B., and Davis, M. L. (1980). Genetic differentiation in ponderosa pine along a steep elevational gradient. *Silvae Genet.* 29:100–103.

Mitton, J. B., Linhart, Y. B., Davis, M. L., and Sturgeon, K. B. (1981). Estimation of outcrossing in ponderosa pine, *Pinus ponderosa* Laws, from patterns of segregation of protein polymorphisms and from frequencies of albino seedlings. *Silvae Genet.* 30:117–121.

Mitton, J. B., Carey, C., and Kocher, T. D. (1986). The relation of enzyme heterozygosity to standard and active oxygen consumption and body size of tiger salamanders, *Ambystoma tigrinum*. *Physiol. Zool.* 59:574–582.

Mitton, J. B., Stutz, H. P., Schuster, W. S., and Shea, K. L. (1989). Genotypic differentiation at PGM in Engelmann spruce from wet and dry sites. *Silvae Genet.* 38:217–221.

Monson, R. K., and Grant, M. C.. 1989. Experimental studies of ponderosa pine. III. Differences in photosynthesis, stomatal conductance, and water-use efficiency between two genetic lines. *Am. J. Bot.* 76:1041–1047.

Mopper, S., and Whitham, T. G. (1986). Natural bonsai of Sunset Crater. *Nat. Hist.* 95:42–47.

Mopper, S., Maschinski, J., Cobb, N., and Whitham, T. G. (1991a). A new look at habitat structure: Consequences of herbivore-modified plant architecture. *In* "Habitat Complexity: The Physical Arrangement of Objects in Space" (S. Bell, E. McCoy, and H. Mushinsky, eds.), pp, 260–280. Chapman & Hall, New York.

Mopper, S., Mitton, J. B., Whitham, T. G., Cobb, N. S., and Christensen, K. M. (1991b). Genetic differentiation and heterozygosity in pinyon pine associated with resistance to herbivory and environmental stress. *Evolution (Lawrence, Kans.)* 45:989–999.

Morgante, M., Vendramin, G. G., Rossi, P., and Olivieri, A. M. (1993). Selection against inbreds in early life-cycle phases in *Pinus leucodermis*. Ant. *Heredity* 70:622–627.

Muller-Stark, G. (1985). Genetic differences between "tolerant" and "sensitive" beeches

(*Fagus sylvatica* L.) in an environmentally stressed adult forest stand. *Silvae Genet.* 34: 241–247.
Muona, O., Yazdani, R., and Rudin, D. (1987). Genetic change between life stages in *Pinus sylvestris*: Allozyme variation in seeds and planted seedlings. *Silvae Genet.* 36:39–42.
Namkoong, G., Kang, H. C., and Brouard, J. S. (1988). "Tree Breeding: Principles and Strategies." Springer-Verlag, Berlin.
Neale, D. B., and Adams, W. T. (1985). Allozyme and mating-system variation in balsam fir (*Abies balsamea*) across a continuous elevational transect. *Can. J. Bot.* 63:2448–2453.
Neale, D. B., Wheeler, N. C., and Allard, R. W. (1986). Paternal inheritance of chloroplast DNA in Douglas-fir. *Can. J. For. Res.* 16:1152–1154.
Neale, D. B., Marshall, K. A., and Sederoff, R. R. (1989). Chloroplast and mitochondrial DNA are paternally inherited in *Sequoia sempervirens*. D. Don Endl. *Proc. Natl. Acad. Sci. U.S.A.* 86:9347–9349.
Neale, D. B., Marshall, K. A., and Harry, D. E. (1991). Inheritance of chloroplast and mitochondrial DNA in incense-cedar (*Calocedrus decurrens*). *Can. J. For. Res.* 21:717–720.
Nei, M. (1973). Analysis of gene diversity in subdivided populations. *Proc. Natl. Acad. Sci. U.S.A.* 70:3321–3323.
O'Malley, D. M., Allendorf, F. W., and Blake, G. M. (1979). Inheritance of isozyme variation and heterozygosity in *Pinus ponderosa*. *Biochem. Genet.* 17:233–250.
Parsons, P. A. (1971). Extreme environment heterosis and genetic loads. *Heredity* 26: 479–483.
Parsons, P. A. (1973). Genetics of resistance to environmental stresses in *Drosophila* populations. *Annu. Rev. Genet.* 7:239–265.
Parsons, P. A. (1987). Evolutionary rates under environmental stress. *Evolu. Biol.* 21: 311–347.
Plessas, M. E., and Strauss, S. H. (1986). Allozyme differentiation among populations, stands and cohorts in Monterey pine. *Can. J. For. Res.* 16:1155–1164.
Prager, E. M., Fowler, D. P., and Wilson, A. C. (1976). Rates of evolution in conifers (Pinaceae). *Evolution (Lawrence, Kans.)* 30:637–649.
Raffa, K. F., and Berryman, A. A. (1982a). Accumulation of monoterpenes and associated volatiles following inoculation of grand fir with a fungus transmitted by the fir engraver, *Scolytus ventralis* (Coleoptera: Scolytidae). *Can. Entomol.* 114:797–810.
Raffa, K. F., and Berryman, A. A. (1982b). Physiological difference between lodgepole pines resistant and susceptible to the mountain pine beetle and associated organisms. *Environ. Entomol.* 11:486–492.
Raffa, K. F., and Berryman, A. A. (1983). The role of host plant resistance in the colonization behavior and ecology of bark beetles (Coleoptera: Scolytidae). *Ecol. Monogr.* 53:27–49.
Rainey, D. Y., Mitton, J. B., and Monson, R. K. (1987). Associations between enzyme genotypes and dark respiration in perennial ryegrass, *Lolium perenne* L. *Oecologia* 74: 335–338.
Rainey, D. Y., Mitton, J. B., Monson, R. K., and Wilson, D. (1990). Effects of selection for dark respiration rate on enzyme genotypes in *Lolium perenne* L. *Ann. Bot. (London)* 66:649–654.
Rainey-Foreman, D. (1991). Physiological adaptations of enzyme polymorphisms in *Lolium perenne*. Ph.D. Thesis, University of Colorado, Boulder.
Read, R. A. (1980). Genetic variation in seedling progeny of ponderosa pine provenances. *For. Sci. Monogr.* 23:1–59.
Rehfeldt, G. E. (1979). Ecotypic differentiation in *Pinus monticola*—Myth or reality? *Am. Nat.* 114:627–636.
Rehfeldt, G. E. (1982). Differentiation of *Larix occidentalis* populations from the northern Rocky Mountains. *Silvae Genet.* 31:13–19.

Rehfeldt, G. E. (1988). Ecological genetics of *Pinus contorta* from the Rocky Mountains (USA): A synthesis. *Silvae Genet.* 37:131–135.

Rehfeldt, G. E. (1989). Ecological adaptations in Douglas-fir (*Pseudotsuga menziesii* var. *glauca*): A synthesis. *For. Ecol. Manage.* 28:203–215.

Rehfeldt, G. E. (1990). Genetic differentiation among populations of *Pinus ponderosa* from the upper Colorado River Basin. *Bot. Gaz. (Chicago)* 151:125–137.

Rehfeldt, G. E. (1991). Models of genetic variation for *Pinus ponderosa* in the Inland Northwest (U.S.A.). *Can. J. For. Res.* 21:1491–1500.

Rehfeldt, G. E. (1993). Genetic variation in the *ponderosae* of the southwest. *Am. J. Bot.* 80:330–343.

Rehfeldt, G. E., Hoff, R. J., and Steinhoff, R. J. (1984). Geographic patterns of genetic variation in *Pinus monticola*. *Bot. Gaz.* 145:229–239.

Ritland, K. (1983). Estimation of mating systems. *In* "Isozymes in Plant Genetics and Breeding, Part A" (S. D. Tanksley and T. J. Orton, eds.), pp. 289–302. Elsevier, Amsterdam.

Rudinsky, J. A. (1966). Scolytid beetles associated with Douglas-fir: Response to terpenes. *Science* 152:218–219.

Saghai-Maroof, M. A., Soliman, K. M., Jorgensen, R. A., and Allard, R. W. (1984). Ribosomal DNA spacer-length polymorphisms in barley: Mendelian inheritance, chromosomal location, and population dynamics. *Proc. Natl. Acad. Sci. U.S.A.* 18:8014–8018.

Saghai-Maroof, M. A., Allard, R. W., and Zhang, Q. (1990). Genetic diversity and ecogeographical differentiation among ribosomal DNA alleles in wild and cultivated barley. *Proc. Natl. Acad. Sci. U.S.A.* 87:8486–8490.

Scholz, F., and Bergmann, F. (1984). Selection pressure by air pollution as studied by isozyme-gene-systems in Norway spruce exposed to sulphur dioxide. *Silvae Genet.* 33:238–241.

Scholz, F., Gregorius, H.-R., and Rudin, D. eds. (1989). "Genetic Effects of Air Pollutants in Forest Tree Populations." Springer-Verlag, Berlin.

Schubert, G. H. (1974). Silviculture of ponderosa pine in the Southwest *U.S., For. Serv., Res. Pap. RM* RM-123.

Schuster, W. S. F., and Mitton, J. B. (1994). In preparation.

Schuster, W. S. F., Alles, D. L., and Mitton, J. B. (1989). Gene flow in limber pine: Evidence from pollination phenology and genetic differentiation along an elevational transect. *Am. J. Bot.* 76:1395–1403.

Scott, T. M., and Koehn, R. K. (1990). The effect of environmental stress on the relationship of heterozygosity to growth rate in the coot clam *Mulinia lateralis* (Say). *J. Exp. Mar. Biol. Ecol.*, 135:109–116.

Shaw, D. V., and Allard, R. W. (1982). Isozyme heterozygosity in adult and open-pollinated embryo samples of Douglas-fir. *Silva Fenn.* 16:115–121.

Shea, K. L. (1990). Genetic variation between and within populations of Engelmann spruce and subalpine fir. *Genome* 33:1–8.

Singh, S. M. (1982). Enzyme heterozygosity associated with growth at different developmental stages in oysters. *Can. J. Genet. Cytol.* 24:451–458.

Singh, S. M., and Zouros, E. (1978). Genetic variation associated with growth rate in the American oyster (*Crassostrea virginica*). *Evolution (Lawrence, Kans.)* 32:342–353.

Smith, D. E., and Klein, A. S. (1994). Phylogenetic inferences on the relationship of North American and European *Picea* species based on nuclear ribosomal 18S sequences and the internal transcribed spacer 1 region. *Mol. Phyl. Evol.* 3:17–26.

Smouse, P. E. (1986). The fitness consequences of multiple-locus heterozygosity under the multiplicative overdominance and inbreeding depression models. *Evolution (Lawrence, Kans.)* 40:946–957.

Snyder, M. A. (1992). Selective herbivory by Abert's squirrel mediated by chemical variability in ponderosa pine. *Ecology* 73:1730–1741.
Snyder, M. A. (1993). Interactions between Abert's squirrel and ponderosa pine: the relationship between selective herbivory and host plant fitness. *Am. Nat.* 41:866–879.
Snyder, M. A., and Linhart. Y. B. (1994). Nest-site selection by Abert's squirrel: Chemical characteristics of nest trees. *J. Mammal.* 75:136–141.
Snyder, T. P., Steward, D. A., and Strickler, A. F. (1985). Temporal analysis of breeding structure in jack pine (*Pinus banksiana* Lamb.). *Can. J. For. Res.* 15:1159–1166.
Somero, G. N. (1992). Adapting to water stress: Convergence on common solutions. *In* "Water and Life: Comparative Analysis of Water Relationships at the Organismic, Cellular, and Molecular Levels" (G. N. Somero, C. B. Osmond, and C. L. Bolis, eds.), pp. 3–18. Springer-Verlag, Berlin.
Squillace, A. E., and Silen, R. R. (1962). Racial variation in ponderosa pine. *For. Sci. Monogr.* 2:1–26.
Stark, R. W., Miller, P. R., Cobb, F. W., Wood, D. L., and Parmeter, J. R. (1968). Photochemical oxidant injury and bark beetle (Coleoptera: Scolytidae) infestation of ponderosa pine. I. Incidence of bark beetle infestation in injured trees. *Hilgardia* 39:121–126.
Steinhoff, R. J., Joyce, D. G., and Fins, L. (1983). Isozyme variation in *Pinus monticola*. *Can. J. For. Res.* 113:1122–1131.
Strauss, S. H. (1986). Heterosis at allozyme loci under inbreeding and crossbreeding in *Pinus attenuata*. *Genetics* 113:115–134.
Strauss, S. H., and Doerksen, A. H. (1990). Restriction fragment analysis of pine phylogeny. *Evolution (Lawrence, Kans.)* 44:1081–1096.
Strauss, S. H., Neale, D. B., and Wagner, D. B. (1989). Genetics of the chloroplast in conifers. *J. For.* 87:11–17.
Strauss, S. H., Hong, Y.-P., and Hipkins, V. D. (1993). High levels of population differentiation for mitochondrial DNA haplotypes in *Pinus radiata*, *muricata*, and *attenuata*. *Theor. Appl. Genet.* 85:605–611.
Sturgeon, K. B., and Mitton, J. B. (1982). Evolution of bark beetle communities. *In* "Bark Beetles in North American Conifers" (J. B. Mitton and K. B. Sturgeon, eds.), pp. 350–384. Univ. of Texas Press, Austin.
Sturgeon, K. B., and Mitton, J. B. (1986). Biochemical diversity of *Pinus ponderosa* Laws. and predation by bark beetles, *Dendroctonus* spp. (Coleoptera: Scolytidae). *J. Econ. Entomol.* 79:1064–1068.
Stutz, H. P., and Mitton, J. B. (1988). Variation in Engelmann spruce associated with variation in soil moisture. *Arct. Alp. Res.* 20:461–465.
Teska, W. R., Smith, M. H., and Novak, J. M. (1990). Food quality, heterozygosity, and fitness correlates in *Peromyscus polionotus*. *Evolution (Lawrence, Kans.)* 44:1318–1325.
Tigerstedt, P. M. A., Rudin, D., Niemela, T., and Tammisola, J. (1982). Competition and neighbouring effect in a naturally regenerating population of Scot pine. *Silva Fenn.* 16:22–129.
Vite, J. P. (1961). The influence of water supply on oleoresin exudation pressure and resistance to bark beetle attack in *Pinus ponderosa*. *Contrib. Boyce Thompson Inst.* 21:37–66.
von Wuehlisch, G., and Krusche, D. (1991). Single and multilocus genetic effects on diameter growth in *Picea abies* (L.) Karst. *In* "Biochemical Markers in the Population Genetics of Forest Trees" (S. Fineschi, M. E. Malvolti, F. Cannata, and H. H. Hattemer, eds.), pp. 77–86. SPB Academic Publishing, The Hague, The Netherlands.
Wagner, D. B. (1992a). Nuclear, chloroplast, and mitochondrial DNA polymorphisms as biochemical markers in population genetic analyses of forest trees. *New For.* 6:373–390.
Wagner, D. B. (1992b). *In* "Population Genetics of Forest Trees" (W. T. Adams, S. H.

Strauss, D. L. Copes, and A. R. Griffin, eds.), pp. 373–390. Kluwer Academic Publishers, Dordrecht, The Netherlands.
Wagner, D. B., Furnier, G. R., Saghai-Maroof, M. A., Williams, S. M., Dancik, B. P., and Allard, R. W. (1987). Chloroplast DNA polymorphisms in lodgepole and jack pines and their hybrids. *Proc. Natl. Acad. Sci. U.S.A.* 84:2097–200.
Wells, O. O. (1964). Geographic variation in ponderosa pine. I. The ecotypes and their distribution. *Silvae Genet.* 13:89–103.
Westfall, R. D., and Conkle, M. T. (1992a). Allozyme markers in breeding zone designation. *New For.* 6:279–309.
Westfall, R. D., and Conkle, M. T. (1992b). *In* "Population Genetics of Forest Trees" (W. T. Adams, S. H. Strauss, D. L. Copes, and A. R. Griffin, eds.), pp. 279–309. Kluwer Academic Publishers, Dordrecht, The Netherlands.
Wheeler, N. C., and Guries, R. P. (1982). Population structure, genic diversity and morphological variation in *Pinus contorta* Dougl. *Can. J. For. Res.* 12:595–606.
Wheeler, N. C., Adams, W. T., and Hamrick, J. L. (1992). Pollen distribution in wind-pollinated seed orchards. *In* "Advances in Pollen Management" (D. L. Bramlett, G. R. Askew, T. D. Blush, F. E. Bridgewater, and J. B. Jett, eds.), pp. 24–31. USDA For. Ser. Agric. Handbook 698.
Whitham, T. G., and Mopper, S. (1985). Chronic herbivory: Impacts on architecture and sex expression of pinyon pine. *Science* 228:1089–1091.
Wilson, D. (1975). Variation in leaf respiration in relation to growth and photosynthesis of *Lolium. Ann. Appl. Biol.* 80:323–338.
Wilson, D. (1981). Response to selection for dark respiration rate of mature leaves in *Lolium perenne* and its effects on growth of young plants and similar stimulated swards. *Ann. Bot. (London)* 49:303–312.
Wilson, D., and Jones, J. G. (1982). Effect of selection for dark respiration rate of mature leaves on crop yields of *Lolium perenne* L. *Ann. Bot. (London)* 49:313–320.
Wright, J. W. (1976). "Introduction to Forest Genetics." Academic Press, New York.
Wright, S. (1931). Evolution in Mendelian populations. *Genetics* 16:97–159.
Yazdani, R., Muona, O., Rudin, D., and Szmidt, A. E. (1985). Genetic structure of a *Pinus sylvestris* L. seed tree stand and naturally regenerated understory. *For. Sci.* 31:430–436.
Yeh, F. C. (1988). Isozyme variation of *Thuja plicata* (Cupressaceae) in British Columbia. *Biochem. Syst. Ecol.* 16:373–377.
Yeh, F. C., and El-Kassaby, Y. A. (1980). Enzyme genetic variation in natural populations of Sitka spruce (*Picea sitchensis* (Bong.) Carr.). I. Genetic variations patterns among trees from ten IUFRO provenances. *Can. J. For. Res.* 10:415–422.
Yeh, F. C., and Layton, C. (1979). The organization of genetic variability in central and marginal populations of lodgepole pine, *Pinus contorta* ssp. *latifolia. Can. J. Genet. Cytol.* 21:487–503.
Yeh, F. C., and O'Malley, D. (1980). Enzyme variations in natural populations of Douglas-fir *Pseudotsuga menziesii* (Mirb.) I. Genetic variation patterns in coastal populations. *Silvae Genet.* 29:83–92.
Yeh, F. C., Khalil, M. A. K., El-Kassaby, Y. A., and Trust, D. C. (1986). Allozyme variation in *Picea mariana* from Newfoundland: Genetic diversity, populations structure, and analysis of differentiation. *Can. J. For. Res.* 16:713–720.
Zhang, Q., Saghi-Maroof, M. A., and Allard, R. W. (1990). Effects on adaptedness of variations in ribosomal DNA copy number in populations of wild barley (*Hordeum vulgare* ssp. *spontaneum*). *Proc. Natl. Acad. Sci. U.S.A.* 87:8741–8745.
Zouros, E., Singh, S. M., and Miles, H. E. (1980). Growth rate in oysters: An overdominant phenotype and possible explanations. *Evolution (Lawrence, Kans.)* 34:856–867.

2
Long-Term Records of Growth and Distribution of Conifers: Integration of Paleoecology and Physiological Ecology

Lisa J. Graumlich and Linda B. Brubaker

I. Introduction

The specter of human-induced alteration of atmospheric composition, and the associated changes in climate, have focused attention on how species, communities, and ecosystems respond to climate change (Solomon and Shugart, 1993). Such concern has prompted a wide range of research, from short-term experiments that examine specific controls over the responses of individual tissues or plant parts to the modeling of a whole system that predicts ecological responses at different levels of biological organization and spatiotemporal scales (Ehleringer and Field, 1993). Although long-term, whole-system experimental programs in a broad range of biomes would be the ideal complement to efforts to understand physiological responses of individual plants, such idealism is constrained by funding, logistics, and time. An alternative source of information about how species, communities, and ecosystems respond to environmental change is the paleoecological record. Paleoecology offers insights in the nature of climate–vegetation interactions that derive from the well-documented response of plant communities to environmental changes of the past—a record that is equivalent to natural, if unplanned, experiments (Davis, 1989). Because these natural experiments typically include conditions not observed in the twentieth century, the paleorecord documents a substantially broader range of biotic responses to environmental variations than can be obtained from observational data.

The spatial and temporal resolution of paleoecological data sets has increased in recent decades, so that relatively detailed histories of conifer forests are available for much of North America and Europe. In addition, comparisons of records of past vegetation dynamics to paleoclimatic simulations by general circulation models have improved the understanding of the role of climate in governing past vegetation change. Several major findings of paleoresearch have importance to investigations of the effects of future climate change on the Earth's biota. These include the findings (1) that changing seasonality may result in unexpected vegetation patterns, (2) that climatic and vegetation changes can be rapid, with ecosystem-wide implications, and (3) that short-term, extreme events can have long-term effects on tree population structures.

In this chapter, we discuss patterns of coniferous forest response to climatic variation at two temporal scales: the Late Quaternary and the last millennium. Our examples illustrate the wide range of potential responses of coniferous forests to climatic variation, and emphasize opportunities for applying paleoecological findings to questions of ecophysiological research. Although we rely largely on examples from North America, our conclusions are well-supported by parallel research results in Europe and Asia.

II. Climatic Variation

Weather and climate vary on time scales from seconds to millions of years (Fig. 1). The response of forest vegetation to climate is scale dependent, ranging from physiological responses to relatively short-term (e.g., diurnal) weather events, to population responses to long-term (e.g., 10^3 to 10^5 years) climatic oscillations, leading to altered species distributions and community structure (Davis, 1986; Webb, 1988). We focus on two illustrative periods: (1) the last 18,000 years, a period that encompasses a global warming of 6°C from the end of the last glacial maximum to the warmest portion of the current interglacial cycle (Fig. 1c), and (2) the last 1000 years, a period of relatively small-magnitude (~1.5°C) temperature oscillations (Fig. 1b).

A. Climatic Change over the Last Glacial/Interglacial Cycle

Over the last 2 million years the Earth's climate has oscillated between glacial and interglacial conditions. Oxygen isotope records from deep sea sediments (Fig. 1e) indicate that these oscillations have a quasiperiodicity of ~100,000 years and that periods when ice sheets dominate high latitudes are relatively long (~90,000 to 120,000 years) compared

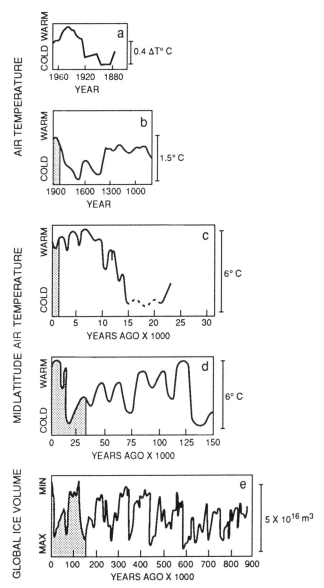

Figure 1 Estimated variability of the global mean temperature for various time scales from decades to hundreds of millennia (after Webb *et al.*, 1985). Times series are based on (a) averages from instrumental data, (b) general estimates from historical documents, with emphasis on the North Atlantic region, (c) general estimates from pollen data and alpine glaciers, with emphasis on midlatitudes from eastern North America and Europe, (d) the generalized oxygen isotope curve from deep sea sediments, with support from marine plankton and sea-level terraces, and (e) oxygen isotope fluctuations in deep sea sediments as indicators of global ice volume. From Webb *et al.* (1985), copyright 1985. Reprinted with the permission of Cambridge University Press.

to periods with little ice cover (~10,000 years) (Shackleton and Opdyke, 1973). The present interglacial cycle (the Holocene) began ~10,000 years ago, although ice sheets did not decrease to their current size until ~6000 years ago. These long-term climate oscillations are governed by periodic changes in the Earth's orbit, which modify the latitudinal and seasonal distributions of solar radiation at periods ranging from 10^3 to 10^5 years (Hays et al., 1976). Long-term changes in concentration of atmospheric trace gases (i.e., carbon dioxide, methane) mimic long-term changes in ice volume. Atmospheric concentrations (by volume) of carbon dioxide and methane are substantially higher during interglacial than during glacial periods (~100 ppmv for carbon dioxide and ~200 ppbv for methane). This observation has led to speculation as to the mechanisms by which trace gas variations may reinforce the control of glaciation by changes in solar radiation (Raynaud et al., 1993).

The factors that govern the regional climate (i.e., solar insolation, atmospheric trace gas and aerosol concentration, sea surface temperatures, ice volume) vary at different time scales and, as a result, their combined effects generate unique climatic states through time (COHMAP Project Members, 1988). These unique climatic states, often referred to as "no-analog" climates, have combinations of the seasonal temperature and precipitation with no counterparts in twentieth century observational climatic data. An oft-cited example of no-analog climate conditions is a rather long period encompassing the transition between the last glaciation and the present interglacial, ~12,000 to 6000 years ago. During this period the seasonal cycle of insolation was amplified due to the greater tilt of the Earth's axis and the occurrence in July of the perihelion (i.e., the point on the Earth's orbit that is nearest to the sun). These orbital changes increased summer insolation in middle to high latitudes by about 8%, as compared to the present, and decreased winter insolation by about 8%. Increased summer insolation contributed to the wastage of the northern hemispheric ice sheet, which had decreased to 25% of its maximum extent by ~9000 years ago. The expression of the increased seasonal cycle differed regionally. For example, summers in Alaska, western Canada, and the Pacific Northwest were warmer than present, but summer temperatures in eastern North America remained cold due to the continued influence of the Laurentide ice sheet on regional circulation. The most prominent aspect of summer climate in the southwestern United States was increased precipitation due to the intensification of monsoonal circulation (Spaulding and Graumlich, 1986). Periods of no-analog climate such as these serve as natural experiments in which the response of vegetation to altered climatic regimes can be studied. Such "experiments" are of particular importance given the likelihood that no-analog climates will arise from an enhanced greenhouse effect.

B. Climatic Change over the Last Millennium

Instrumental climatic observations as well as proxy climatic data from tree rings, high-resolution ice cores, and glacial deposits indicate decadal and longer term temperature variations over the last millennium. These variations are roughly synchronous over parts of the northern middle to high latitudes, although the magnitude, seasonality, and timing of warm and cold intervals vary from region to region (Grove, 1988; Hughes and Diaz, 1994). Such temperature records indicate a general pattern of more frequent warm episodes from ca. A.D. 1000 to 1300, more frequent cool episodes from ca. A.D. 1300 to 1850, and a trend toward higher temperatures from 1850 to the present (Fig. 1b). The causes of short-term (i.e., decadal to centennial scale) climatic variation differ from those of long-term climatic variation and are surprisingly poorly understood. Three external factors are hypothesized to contribute to climatic variation on these time scales (Wigley, 1989). First, variations in solar radiation, associated with variations in the sun's radius and sunspot activity, affect the amount of incoming radiation to the Earth's atmosphere. Second, sulfur-rich aerosols, derived from volcanic, industrial, and biological sources, are thought to reduce the Earth's temperature by increasing atmospheric reflectivity. Finally, increased carbon dioxide and other trace gases affect the Earth's climate by absorbing longwave radiation. Ascertaining the relative importance of each of these external factors in governing climate is complicated by the internal variability of the coupled atmosphere–ocean system. Simple global climate models demonstrate that the thermal inertia of the ocean as well as changes in the vertical (thermohaline) circulation of the ocean are sufficient to induce multidecadal trends in an otherwise unforced climate system (Wigley and Raper, 1990). A more complete understanding of the patterns and causes of decadal-scale climatic variation is likely in the near future with the increasing global coverage of various paleoclimatic indicators (Mosley-Thompson *et al.*, 1990).

III. Vegetation Responses

A. Changes in Coniferous Forests over the Last Glacial/Interglacial Cycle

The climatic changes marking the transition from the last glaciation to the current interglacial period caused major reorganization of coniferous forests worldwide. Counterintuitively, boreal forests covered substantially less area during full glacial times as compared to today. Conversely, the coniferous forest cover in southwestern North America expanded during full glacial times. A summary of coniferous forest response to climatic changes, emphasizing results from North America, serves to illustrate the linkages between global-scale climatic forcings and regional vegetation response.

1. Boreal Forest At the last glacial maximum, massive ice sheets covered much of mid- to high-latitude North America. Surface temperatures on and adjacent to the ice sheets were much lower than over equivalent modern land surfaces due to the high elevation and highly reflective surfaces of the ice sheets (COHMAP Project Members, 1988). Simulations by global climate models as well as paleoenvironmental evidence indicate that strong anticyclonic circulation (i.e., high pressure) over the continental ice masses displaced the westerly jet stream to the south and strengthened easterly winds along the southern flank of the ice sheet (COHMAP Project Members, 1988). As a result, conditions were much colder at middle to high latitudes as compared to the present, and boreal forests were restricted to relatively small areas south of the ice margin (Anderson and Brubaker, 1993; Ager and Brubaker, 1985; Ritchie, 1987). Compositionally, these forests were less diverse than modern boreal forests, with abundant sedge (Cyperaceae) but only minor representation of broad-leafed boreal taxa such as birch (*Betula*) and alder (*Alnus*). Several conifers with broadly overlapping modern ranges [e.g., spruce (*Picea*) and jack pine (*Pinus banksiana*)] seldom cooccurred in glacial-aged forests. The closest modern analog for the glacial-age boreal forest of North America is the spruce woodland in the western Ungava Peninsula of eastern Canada, a region characterized by relatively moderate temperatures and mesic conditions (Jacobson *et al.*, 1987). Structurally, these stands probably had lower tree densities, lower leaf areas, and more open ground as compared to modern boreal forests. As a result, competition among tree species may have been a less important process in structuring ancient forests.

As the ice sheets retreated, their influence on mid- to high-latitude climates diminished and the influence of seasonal variation in insolation receipt grew (Webb *et al.*, 1993). From 12,000 to 9000 years ago, for example, the anticyclonic circulation associated with the Lauentide ice sheet weakened, and westerly airflow replaced easterly airflow in northwestern North America (COHMAP Project Members, 1988).

Simultaneously, increased summer solar insolation potentially increased both surface heating and rates of evapotranspiration. During this period, boreal species moved northward, but at varying rates and in different community associations than at present. Variation in rates of northward migration can be explained, in part, by the relative dominance of increased summer solar insolation versus the effects of the lingering ice massess in governing regional climate. Rates of migration for eastern boreal forest species were considerably slower than those for the western boreal forest (see below), largely due to climatic effects associated with the Laurentide ice sheet (Ritchie, 1987; Webb, 1987, 1988). For example, spruce and associated taxa moved only ~500 km northward between 18,000 and 9000 years ago. From 12,000 to 9000 years ago, spruce grew in open woodlands with strong components of thermophilous hardwoods such as elm (*Ulmus*), ash (*Fraxinus*), and ironwood/hornbeam (*Ostrya/Carpinus*). Although this was the most widespread forest type in eastern North America at the time, the lack of a modern counterpart has made its interpretation in terms of climatic inferences difficult (Solomon, 1992). By 9000 years ago, when boreal forest taxa were expanding rapidly in northwestern North America, the eastern boreal forest had lost its thermophilous hardwood component and declined in geographical extent, becoming a minor forest type restricted to a few areas south of Hudson Bay. Although similar to modern forests in composition, these forests remained a minor component of eastern North American vegetation until ~3000 years ago. The extreme reduction in boreal forest cover from 9000 to 3000 years ago was associated with enhanced summer insolation, an intensification of midlatitude surface westerly winds, and an eastward expansion of prairie vegetation (Webb *et al.*, 1993). As modern climates were established 3000 years ago, eastern boreal forests expanded southward and reached their modern range limits.

Northwestern North America, located upstream from the atmospheric anomalies associated with the Laurentide ice sheet, experienced higher summer temperatures during the early Holocene than during any subsequent period of the Holocene (COHMAP Project Members, 1988, Barnosky *et al.*, 1987). As a result, many sites in northwestern North America show great sensitivity to early Holocene variations in solar insolation, including rapid migration rates across the postglacial landscape. For example, between 10,000 and 9500 years ago, white spruce (*Picea glauca*) spread northward from southern Alberta into central Alaska and to beyond its modern range limit in northwest Canada, encompassing distances of nearly 2000 km (Anderson and Brubaker, 1993; Ritchie *et al.*, 1983). Its range limits remained relatively stable until ~6000 years ago, when it expanded further west and south in Alaska, but retreated from its northern limit in Canada. The history of black spruce (*Picea mariana*) contrasts sharply with that of white spruce, even though both species have similar large-scale distribution patterns today. Black spruce spread from southern Alberta at the same time as white spruce, but its range expansion halted in the lower Mackenzie ~9000 years ago. By ~7000 years ago, its populations expanded rapidly again, catching up with a second expansion of white spruce in Alaska and meeting retreating populations of white spruce in northern Canada.

2. Western Conifer Forests The coniferous forests of western North America experienced similar levels of vegetation reorganization in the transition from late glacial to modern climates. Vegetation changes in western North America were driven both by changes in thermal regimes and by changes in soil moisture supplies as influenced by changes in the position and strength of westerly surface winds (COHMAP Project Members, 1988). In the Pacific Northwest and northern Rocky Mountains, conditions were colder and dryer during the late glacial and, as a result, coniferous forest species retreated to lower elevations and were, in general, far less abundant than during the ensuing interglacial period (Barnosky et al., 1987; Thompson et al., 1993). In the southwest (south of ~34°N latitude), conditions during the late glacial were colder and wetter than present due to a weakening of the Pacific subtropical high-pressure system and a southward displacement and strengthening of the westerly surface winds (COHMAP Project Members, 1988). Under these more mesic conditions, pinyon pine and juniper woodlands were widespread in lowland areas that today support only desert scrub vegetation (Spaulding et al., 1983). As in the case of the migrational histories of boreal forest taxa, species of the coniferous forests of western North America demonstrated highly individualistic behavior with respect to elevational displacement (Spaulding et al., 1983; Barnosky et al., 1987). Certain tree species [e.g., bristlecone pine (*Pinus longaeva*) and limber pine (*Pinus flexilis*) in the southwest; Engelmann spruce (*Picea engelmanii*) and lodgepole pine (*Pinus contorta*) in the Northwest] had broader altitudinal and latitudinal ranges in glacial times as compared to today. Other species [e.g., ponderosa pine (*Pinus ponderosa*), Colorado pinyon (*Pinus edulis*)] appear to have been restricted to small ranges in what is today southern Arizona and New Mexico (Van Devender et al., 1987).

During the early Holocene (~9000 to 6000 years ago), the regional contrasts in biotic response to large-scale climatic forcings continued. In the Pacific Northwest and the northern Rocky Mountains, the increased summer solar insolation was associated with dryer climatic conditions associated with an intensification of the eastern Pacific subtropical high (Thompson et al., 1993). Effectively dryer conditions and an associated increase in fire frequency resulted in an increase in drought-tolerant and fire-adapted species [lodgepole pine, Douglas fir (*Pseudotsuga menziesii*), alder (*Alnus*)] and the occurrence of open parklands in areas of currently closed forest. Conversely, in the southwest, increased summer solar insolation was associated with an intensification of summer, monsoonal rainfall (Thompson et al., 1993). Where summer soil moisture was adequate, montane conifers either expanded their lower elevational ranges or maintained their late-glacial range. In central Colorado, for example, increased summer temperatures combined with increased summer moisture to cause both a downward expansion of lower tree line (Markgraf and Scott, 1981) as well as an upward expansion of upper tree line (Fall, 1988; Carrara *et al.*, 1984). Elsewhere in the southwest, evidence for the persistence of montane conifers at relatively low elevations is abundant, although species vary in the pattern and timing of their retreat to higher elevations. Variation in the timing of upslope movement by coniferous species reflects both the individuality of species response to climatic trends and the heterogeneity of climatic regimes in the region (Thompson *et al.*, 1993, and references therein).

B. Changes in Coniferous Forests over the Last Millennium

Two broad classes of response have occurred in coniferous species in response to the temperature fluctuations of the last millennium. First, at several arctic and alpine tree lines, range limits have shifted due to altered reproductive and establishment rates. Second, in areas where climatic limitations to growth have changed, established trees have undergone phenotypic adjustments to the altered climate.

Although climatic variables may subtly alter rates of reproduction and establishment in temperate coniferous ecosystems, the role of climate in governing population processes is clearly seen at arctic and alpine ecotones. Tree establishment has increased in subalpine and tree line stands and young trees have established at elevations or latitudes beyond current tree lines in a variety of settings worldwide (Table I). One must be cautious in inferring from these data a global upward movement of tree line for several reasons. First, the fate of the young trees in these stands is unclear in that a return to more severe climatic conditions may slow or halt recruitment and decrease survivorship. Second, ecotonal movement may result from several different climatic factors (e.g., increased summer temperature, decreased snowpack, increased soil moisture supplies, alteration of wind regime) as well as different anthropogenic factors (e.g., grazing, altered fire regime). Finally, ecotonal movement is not universally observed in all regions that have experienced a recent upward trend in temperatures. A particularly well-studied example of this is the lichen–spruce woodland of northern Quebec in which only sites that are well-protected from the damaging effects of wind exposure show growth responses to recent climatic warming (Payette *et al.*, 1985). Further, rather than tree line advance in northern Quebec, radiocarbon-dated spruce charcoal records indicate that lichen–spruce woodlands have fragmented and retracted to their present distribution due to catastrophic fires during the past several millennia (Payette and Gagnon, 1979). The work of Payette and colleagues clearly demonstrates the complexity of the problem of assessing ecosystem response or resiliency to

Table I Examples of Twentieth Century Increases in Tree Recruitment at or above Tree Line

Region	Species	Interpretation	Reference
N. Québec	*Picea glauca* *Larix laricina*	Altitudinal seed regeneration limit increased ~ 100 m since nineteenth century at protected sites; increase of tree density near forest limit; attributed to climatic warming	Payette and Filion, 1985 Marin and Payette, 1984
Coast Range, British Columbia	*Abies lasiocarpa* *Tsuga mertensiana*	Decreased snowfall and resulting longer growing seasons led to increased recruitment in subalpine meadows	Brink, 1959
Canadian Rockies	*Abies lasiocarpa* *Picea engelmannii*	Increased seedling establishment in meadows at timberline associated with warmer summer temperatures	Kearney, 1982
Swedish Scandes	*Picea abies*	Upright growth of previously stunted individuals in conjunction with increased regeneration above tree line is correlated with climatic warming and two periods of heavy snowfall	Kullman, 1986
Finland	*Pinus sylvestris*	Climate amelioration led to increased regeneration at and above tree line	Hustich, 1958
Siberia, Russia	Boreal forest spp.	Increased regeneration in response to climate amelioration	Gorchakovsky and Shiyatov, 1978
Khibin Mts., Russia	*Pinus* sp.	Increased regeneration in response to presumed climatic amelioration began 200 years ago, but accelerated significantly in the 1900s	Kozobov and Shaydurov, 1965 (in Bray, 1971)

Location	Species	Description	Reference
S. Island, New Zealand	*Nothofagus* spp.	Climate amelioration led to increased regeneration at and above tree line	Wardle, 1963 (in Bray, 1971)
Cascade Range, Washington and Oregon	*Abies lasiocarpa* *Tsuga mertensiana* *Abies amabalis*	Warmer growing seasons and reduced snowpack led to increased regeneration and invasion of subalpine meadows	Frank *et al.*, 1971; Heikkinen, 1984; Agee and Smith, 1984
Olympic Mts., Washington	*Abies lasiocarpa* *Tsuga mertensiana*	Tree invasion of meadows during periods of warm growing season and reduced snowpack	Fonda and Bliss, 1969; Kuramoto and Bliss, 1970
Lemhi Mts., Idaho	*Abies lasiocarpa* *Picea engelmannii* *Pinus contorta* *Pseudotsuga menziesii*	Tree invasion of meadows during warm, dry periods of reduced snowpack in the early twentieth century	Butler, 1986
Sierra Nevada, California	*Pinus balfouriana*	Warmer growing seasons led to increased recruitment at and above tree line	Scuderi, 1987
White Mts., California and Nevada	*Pinus longaeva*	Warmer growing seasons led to increased recruitment at and above tree line	La Marche, 1973
Uinta Mts., Utah	*Picea engelmannii*	Young trees advancing toward relict tree line	J. Major, pers. comm. (in Bray, 1971)
La Sal Mts., Utah	*Picea* sp.	Climate amelioration led to increased recruitment	Richmond, 1962
Rocky Mountains, Colorado	*Pinus aristata* *Picea engelmannii*	Warmer growing seasons led to establishment of seedlings above current tree line	D. K. Yamaguchi, pers. comm., 1993

climatic change and emphasizes the importance of multiple climatic factors and exogenous factors such as disturbance.

Trees respond to decadal-scale climatic variation by systemic changes in carbon balance and nutrient status, which, in turn, lead to changes in either absolute growth rates or altered patterns of carbon allocation. Sustained changes in growth and productivity have been observed during the twentieth century in trees growing near latitudinal and altitudinal tree lines in western North America (Fig. 2b, d, and f) (Garfinkel and Brubaker, 1980; LaMarche et al., 1984; Graumlich and Brubaker, 1986; Graumlich, 1991; Graybill and Idso, 1993; Luckman, 1994). Such long-term growth trends often mirror similar trends in growing season temperature (Fig. 2a, c, and e). Although growing season temperature is generally the most important climatic variable governing growth at latitudinal and altitudinal tree lines, detailed analyses of year-to-year variability indicate that other climatic factors, such as soil moisture and depth of snowpack, interact with temperature in controlling growth at tree line sites in the Cascade Mountains and Sierra Nevada (Graumlich and Brubaker, 1986; Graumlich, 1991). Nonclimatic factors, specifically direct CO_2 fertilization, have been linked to the observed increasing trends in the growth of upper tree line bristlecone pines (LaMarche et al., 1984). However, a recent comparison of bristlecone pine growth in individual trees with strip bark (i.e., partial bark and cambium) versus full bark (i.e., entire bark and cambium) indicates that the growth enhancement is confined to strip bark individuals (Graybill and Idso, 1993). Further, analyses of growth trends of other subalpine conifers have failed to detect evidence of direct CO_2 fertilization (Kienast and Luxmoore, 1988; Graumlich, 1991). Direct enhancement of forest growth by CO_2 thus appears to be highly specific to strip bark bristlecone pine.

Figure 2 Summer (June through August) temperature trends over the period of observational climatic records and tree-growth indices over the last several hundred years for selected sites in coniferous forests of North America. These series demonstrate the degree to which the twentieth century growth rates at temperature-sensitive sites are anomalous relative to the previous several hundred years. For comparative purposes, all series are expressed as standard, normal deviates. In addition, the tree-ring data have been filtered to remove age-associated growth trends. The series represent (a) temperature data recorded at Fairbanks, Alaska, (b) white spruce growth in the Brooks Range of Alaska (Garfinkel and Brubaker, 1980), (c) grid point average temperature data for Washington (45°N, 120°W) (Jones et al., 1985), (d) subalpine larch (*Larix lyallii*) growth in the Cascade Range of Washington (Graumlich and Brubaker, 1986), (e) grid point average temperature for California (35°N, 120°W) (Jones et al., 1985), and (f) foxtail pine growth in the Sierra Nevada of California (Graumlich, 1991).

2. Paleoecology and Physiological Ecology   49

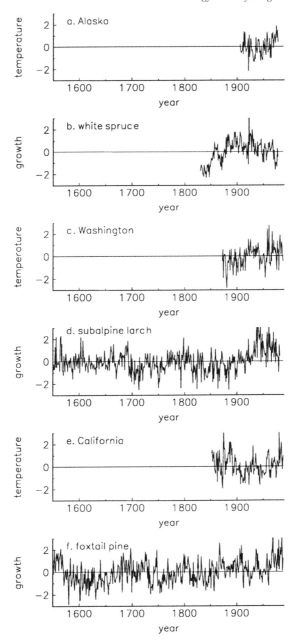

Table II Evidence of Altitudinal Tree Line Fluctuations over the Last Several Hundred Years, with Emphasis on Locations Where Relict Stands Are Preserved above Current Tree Line

Location	Interpretation	Dating	Reference
Siberia	Tree line higher 800–900 ^{14}C yr BP; tree line depressed 270–300 m after 600 ^{14}C yr BP; tree line advanced 200–250 ^{14}C yr BP	Radiocarbon	Kozubov and Shaydurov, 1965 (in Bray, 1971)
Europe	Tree line descended ~100–200 m in the 14th–17th century	Pollen Stratigraphy	Firbas and Losert, 1949 (in Bray, 1971)
Alps	Tree line descended ~70 m after AD 1300		Gams, 1937 (in Bray, 1971)
Canadian Rockies	Subfossil trees above current tree line date to late 1300s and early 1400s; dieback at these sites in late 1700s	Radiocarbon Dendrochronology	Luckman, 1994, and references therein
Yukon	*Picea* forest, 60 m higher than present tree line, buried under 1.5 m of volcanic ash	None	Rampton, 1969
Wyoming Range, Wyoming	Tree stumps 60 m above current tree line	None	Griggs, 1938
La Sal Mts., Utah	Relict forest 3 m above current tree line	None	Richmond, 1962
Big Snowy Range, Montana	Relict tree line 100 m higher than present	None	Bamberg and Major, 1968
Uinta Mts., Utah	*Picea engelmanii* relicts 80 m above current tree line	None	J. Major, pers. comm. (in Bray, 1971)
Sierra Nevada, California	Relict tree line up to 100 m above present tree line	Dendrochronology	Graumlich and Lloyd, 1994; Scuderi, 1987
Kirkliston Range, New Zealand	Logs of *Podocarpus totara* 155 m above present tree line dated 500–1450 ^{14}C yr BP	Radiocarbon	Burnett, 1926 (in Raeside, 1948)
New Zealand	*Nothofagus fusca* charcoal 213 m above present tree line	None	Molloy *et al.*, 1963

The two climatic responses documented above, increased regeneration rates and increased growth rates in tree line stands, appear unique when considered in the context of twentieth century observations. However, in the relatively rare circumstances in which particularly long-lived conifers provide growth or regeneration records that extend over the last millennium, the responses to recent climatic trends are not unique and have occurred in the past when temperatures were equivalent to late twentieth century values. In the Sierra Nevada, growth of foxtail pine exceeded that of late twentieth century values for several episodes in the past (1370–1440 and 1480–1580) (Graumlich, 1991). Similarly, evidence for tree establishment above current tree lines exists at several localities throughout the world (Table II), indicating that the climatic variables critical for regeneration and survival have fluctuated significantly on the time scales of hundreds of years to millennia. Therefore, whereas the warm temperatures of the twentieth century may be anomalous with respect to the life-span of most coniferous species, they may not be anomalous in the context of the recent history of a given forest stand.

IV. Implications for Ecophysiological Studies

A. Changing Seasonality May Result in Unexpected Vegetation Patterns

The natural experiments provided by variations in large-scale climatic controls demonstrate the potential for significant variation in the seasonality of temperature and precipitation. Long-term records of vegetation change indicate that altered seasonality, such as that associated with the glacial/interglacial transition, is commonly associated with vegetation types without analogs on the modern landscape. Two examples of no-analog forests extant at the glacial/interglacial transition, discussed below, illustrate how species-individualistic responses to separate elements of the climate system may combine with complex feedback processes to challenge our current understanding of vegetation–climate relationships.

1. Late Glacial Poplar Forests of Interior Alaska The modern vegetation of the interior plateau of Alaska is dominated by spruce forest with deciduous hardwoods, such as balsam poplar (*Populus balsamifera*), locally abundant along riparian corridors and on south-facing, well-drained hill slopes. Paleoecological evidence, including pollen accumulation rates and sedimentary chemistry, indicates that poplar, a minor element of the modern vegetation, formed extensive forests on upland sites during the glacial/interglacial transition ~10,500 years ago (Ager and Brubaker, 1985; Anderson et al., 1988; Anderson and Brubaker, 1993). Dendrochronological studies indicate that the growth of balsam poplar is favored by warm summer temperatures (Edwards and Dunwiddie, 1985), and thus increased population levels at a time of increasing summer solar insolation and increased potential evapotranspiration are consistent with known ecological requirements for growth. The nature of seedling establishment and long-term clonal growth on upland sites is more speculative. Establishment by seed during the glacial/interglacial transition may have been tied more closely to fire disturbances than to flooding, and root suckering may have allowed individuals to resprout after fire or to survive long periods without disturbance. The poplar expansion was relatively short-lived and pollen data indicate that populations declined ~9500 years ago, perhaps due to a loss of suitable establishment sites as open ground decreased and soil stabilized in the early Holocene (Hu et al., 1994).

The Alaskan record of poplar population fluctuations, therefore, indicates that warmer than present summer temperatures, potentially in concert with changes in soil moisture and disturbance regimes, favor boreal hardwood species rather than the expansion of boreal conifers. Although we are lacking a full explanation for the competitive dominance of a deciduous hardwood over evergreen conifers under warmer climatic conditions, this natural experiment is consistent with results from ecophysiological studies. Simulation experiments, which incorporate complex feedbacks between climate, vegetation processes, and soil thermal and nutrient regimes, suggest that warmer temperatures, particularly in the absence of increased precipitation, result in a decrease, rather than an increase, in the depth of soil thaw in upland stands and a concomitant decrease in conifer productivity (Bonan et al., 1990). Similarly, ecophysiological experiments indicate that spruce growth is limited not only by low-growing-season temperatures but also by soil moisture (Goldstein, 1981). These findings are particularly important in the context of predictions of the effects of future climatic change on the boreal forest. For example, models of the distribution of boreal forest life zones under climatic conditions forced by doubled CO_2 concentrations indicate that approximately 25% of the present boreal forests would be replaced by steppe and deciduous forest (Solomon, 1992). Taken together, the paleoecological and ecophysiological results confirm the necessity of specifying climatic scenarios in terms of both temperature and soil moisture regimes such that the complex feedbacks between climate, vegetation, and soil processes can be adequately addressed.

2. Boreal and Thermophilous Species Association of Eastern North America Like the poplar populations in Alaska, several thermophilous species, currently rare to absent in the eastern boreal forest, greatly increased their range during the late glacial. Between 12,000 and 9000 years ago, ash, ironwood, and elm populations were abundant in the unglaciated central United States, growing in association with pine and spruce (Webb, 1988). Indeed, these three hardwood taxa were more widespread during the late glacial than during any prior or subsequent period of the late Quaternary (Jacobson et al., 1987). The eastern boreal forest of the late glacial is thus another example of a widespread, ancient forest with no analog on the modern landscape.

Modern forests with strong mixtures of conifers and hardwoods are dominated by different genera (e.g., pine, hemlock, maple, beech). These modern mixed forests occur in the Great Lakes region, where the climate is characterized by generally subfreezing winter temperatures yet warm summers resulting from large seasonal shifts in the boundary between arctic and subtropical air masses. The late-glacial no-analog mixed forest probably experienced less seasonal variability in temperature. Summer insolation was more extreme than today, but increased insolation did not result in warmer summers because cool air flows off the Laurentide ice sheet decreased regional temperatures. Thus, summer temperatures were likely cooler than they are today. Although the physiological basis for the differences in species composition between late-glacial and modern mixed forests is not clear, the paleorecord suggests that responses to the magnitude of seasonal climatic variations may have been important.

These examples testify to the lack of continuity in forest makeup through time. Although most of the conifer species that dominate modern forests have been common throughout the Holocene, most modern forest types arose only 3000–6000 years ago. The unique histories of individual species result from their individual responses to climate. Thus the individualism of species physiological attributes (Chapin and Shaver, 1985) is also reflected in long-term vegetation dynamics. The fossil records suggest that individualistic responses to seasonal variations in temperature, precipitation, and insolation have played a particularly important role in structuring past forest composition. Seasonality effects are also relevant to considerations of future climate change because general circulation models predict independent changes in seasonal climates. Winter temperatures, for example, are predicted to warm more rapidly than summer temperatures, with potential implications to important components of western conifer forests. Leverenz and Lev (1988) have speculated that the chilling requirement of Douglas fir for bud development would not be met under a double CO_2 scenario, eliminating coastal Douglas fir from the majority of its range.

The realization that the characteristics of climate vary profoundly through time provides important lessons for thinking about vegetation responses to future climates and suggests guidelines for designing experiments to predict such responses. For example, we can expect climate factors to continue to behave in a quasi-independent manner such that current combinations of temperature, precipitation, and seasonality will not simply be shifted to different geographic locations in future landscapes. Instead, new climatic conditions should arise that have no present-day analogs. This realization enforces the already perceived need for a mechanistic understanding of vegetation processes. Given the wide range of potential future environments, it would be desirable to understand physiological responses to conditions not routinely encountered in field experiments in present environments. It is also reasonable to think that interactions of variables, nonlinearities in physiological responses, and interactions of different levels of biological organization will make it difficult to extrapolate results from current environments to predict responses to future conditions. Thus, we need models of physiological behavior that are based on the results of experimentation outside the normal range of variation and combinations of conditions in modern landscapes. This suggests that our experimental designs should not focus on selecting the norm for study. We should, for example, experiment with unusual microsites (e.g., slopes with odd combinations of exposures and microclimate) and unusual genotypes.

B. Climate and Vegetation Changes Can Be Rapid, with Ecosystem-Wide Implications

Climatic events or episodes that are infrequent can nevertheless have a strong influence on vegetation pattern if they are of sufficient magnitude to impact demographic processes. Paleoecological data can elucidate the imprint of extreme events on vegetation patterns and thus allow us to observe the integrated response of vegetation to events that are almost impossible to observe in the course of field studies.

In the context of the late Quaternary, short-term climatic variations are anomalies that persist over a few to several centuries. An example of such variations in the recent past is the increased frequency of warm episodes from A.D. 1000 to 1300 and more prevalent cold extremes from A.D. 1300 to 1850. Presumably similar short-term anomalies have occurred throughout the late Quaternary, but their effects on vegetation have been less pronounced than the effects of slowly changing boundary conditions. The existence of short-term vegetation variations, however, is becoming more widely recognized by the analysis of pollen records with fine-temporal and spatial resolution. Two examples illustrate that large population fluctuations of conifer species can occur on time scales of a few centuries.

1. Scotch Pine In a discrete layer in peatlands across much of northern Scotland, fossil stumps of Scotch pine are found 70–90 km beyond its modern range limit. Recent dendrochronological and ^{14}C dating studies show that these stumps represent a brief period of tree expansion ~4600–4200 years ago. Fine-resolution temporal studies of pollen in the associated peat also indicate that tree populations expanded locally in the peatlands. In addition, the charcoal content, pollen of light-demanding herb taxa, and degree of humification of the peat suggest that the establishment of pine stands was accompanied by an increase in fire frequencies and decomposition rates. Thus, major short-term changes probably occurred in a suite of ecosystem processes. The cause of these changes is thought to have been a shift in circulation patterns resulting from northward displacements of the Azores High and the jet stream during summer, which would have reduced rainfall over Great Britain and Europe. In terms of population dynamics, Scotch pines expanded at a rate of 400–800 m/year, similar to maximum rates of tree population spread during major Holocene expansions. The temporal distribution of the establishment and death dates of stumps confirms that seedling establishment was most sensitive to climatic shift. Population recruitment declined nearly 100 years before substantial loss in population by mortality. These findings emphasize the importance of (1) focusing on the physiology of seedling stages as most sensitive to climate change, (2) potential competitive and indirect effects of adult trees on seedlings that might dampen response times, based solely on seedling physiology, and (3) higher order ecosystem interactions, such as altered decomposition and disturbances that affect species responses to climate.

2. Black Spruce Pollen records from four small lakes near Great Slave Lake in north central Canada document a ~1000-year fluctuation of tundra to forest-tundra and return to tundra during the mid-Holocene. The initial shift from tundra to forest tundra encompassed only 150 years and coincided with a rapid change in diatom flora. The synchronous change of such different life forms (diatoms and trees) suggests a direct response to climatic controls. The authors suggest that the exceedingly rapid increase in black spruce may have been due to a climatic release and population expansion of krummholz trees already present in the vicinity of the lakes, rather than to wave-front advance from the tree line. Black spruce occurs in this area today (more than 50 km from the mapped forest–tundra boundary) as scattered krummholz individuals, and it is easy to imagine that a slight amelioration in climate might allow upright growth and seed production, allowing for rapid population increases from local seed sources. Other sediment evidence suggests that increases in total lake productivity lagged terrestrial vegetation changes by ~150–300 years. Presumably the increase in lake productivity was driven by vegetation and climate-induced changes in the terrestrial ecosystem. The temporal lag in lake productivity may have reflected the time required for increased soil leaching caused by increased production of acidic (conifer) litter, resulting in greater nutrient input to the lake. Thus, as in the Scotch pine example, ecosystem changes accompanied vegetation changes, suggesting that ecosystem feedbacks are important in determining the magnitude and rate of vegetation responses to climate. Sorting out such feedbacks from the paleorecord is difficult because these records are best at documenting the net result of such interactions. Ideally, we need independent evidence on the environmental stimulus and more specific evidence of ecosystem processes, which are typically lacking in paleorecords. Ecophysiological studies are faced with the opposite problem—such studies measure the environmental stimulus and physiological response, but can only extrapolate the net result at higher levels of organization.

C. Short-Term, Extreme Events Can Have Long-Term Implications

Age structure data from subalpine forest stands in the southern Sierra Nevada (Keifer, 1991) illustrate the potential role of anomalous climatic episodes in governing recruitment rates. Dendrochronologically determined dates of establishment for foxtail and lodgepole pine show pulses of recruitment that correspond to local fire events. Even in the absence of disturbance-related pulses of establishment, recruitment persists at low levels with several notable exceptions. Recruitment in these stands virtually ceased for three sustained periods (A.D. 1300 to 1360, 1530 to 1610, and 1870 to 1900). Each of these periods is associated with anomalous climatic events; however, each period differs in the nature of the anomaly in climate.

The most recent hiatus in regeneration, 1870 to 1900, is associated with widespread evidence of frost rings (i.e., ruptured xylem cells damaged when extracellular ice forms in the cambium during the growing season) (LaMarche and Hirschboeck, 1984). An unusually high frequency of frost events is recorded in Keifer's samples in the late 1800s as well as in regional frost-ring chronologies (LaMarche and Hirschboeck, 1984). The most recent hiatus in reproduction thus may result from cold damage to young trees.

An earlier and longer hiatus in recruitment, 1530 to 1610, is associated with a series of climatic anomalies. Regional climatic reconstructions indicate that the period 1549 to 1568 was characterized by unusually high winter precipitation values (the second wettest 20-year period in a 1000-year reconstruction) (Graumlich, 1993). High winter rainfalls are associated with deep and persistent snowpack, a condition resulting in lower than average tree growth (Graumlich, 1991). Spatial correlation between subalpine ecotones and snowpack distribution (Barbour *et al.*, 1991) supports the inference that episodes of anomalously high snowpack affect the temporal dynamics of recruitment near treeline. A subsequent precipitation anomaly, of the opposite direction, may have contributed to the hiatus in recruitment. An extremely severe drought in 1580 is reconstructed from independent data sets (Hughes and Brown, 1992; Graumlich, 1993). Most notably, 1580 is inferred to be the single most severe drought year since 101 B.C. (Hughes and Brown, 1992). Finally, the hiatus in recruitment encompasses the year 1601, which is characterized by extreme cold, as inferred from tree ring densities (Briffa *et al.*, 1992), tree ring widths (Scuderi, 1993; Graumlich, 1993), and the presence of frost rings both locally (Keifer, 1991) and regionally (LaMarche and Hirschboeck, 1984).

The earliest hiatus in recruitment, A.D. 1300 to 1360, is associated with more extended periods of anomalous temperature and precipitation. Climate reconstructions of summer temperature and winter precipitation (Graumlich, 1993) indicate that the period from ~A.D. 1250 to 1300 was the warmest and driest 50-year interval in the last 1000 years. Subsequently, from A.D. 1300 to 1350, precipitation values remained below twentieth century averages whereas temperatures approached twentieth century values.

Taken together, these results imply that recruitment processes in subalpine forest stands are sensitive to temperature and precipitation anomalies of varying magnitude and duration. Perhaps most surprisingly, anomalies of opposite directions (i.e., warm and cold, wet and dry) appear important in governing recruitment. The complexity of the climatic response of these stands arises from the varying climatic sensitivity of the multiple processes necessary for successful recruitment of individuals. Although the results presented above are entirely inferential in nature, they point to a series of simple field experiments in which the effects of climatic variables on specific processes, such as cone production, seed viability, seed germination, and seedling growth, can be manipulated to test the paleoecological inferences (Graumlich and Lloyd, 1994).

V. Conclusions

The paleoecological record of climatic variation and vegetation response summarized here demonstrates that the individualism of the behavior of a species with respect to physiological attributes (Chapin and

Shaver, 1985) is mirrored in long-term and large-scale vegetation dynamics. On Quaternary time scales we see that the communities we study are relatively short-lived, temporary assemblages characterized by continuous flux (West, 1964; Davis, 1981). Over the last 1000 years, we see that the growth of individuals and the dynamics of populations reflect complex interactions among multiple climatic variables. Regardless of the assumptions one makes regarding future climatic scenarios, we can expect climate factors to continue to behave in a quasi-independent manner and, as a result, future climates will be characterized by combinations of temperature and precipitation that are not replicated in the modern landscape. Consequently, predictions of vegetation response to climatic changes must be based on a mechanistic understanding of the relationship between climatic variation and vegetation processes.

What challenges do the findings reviewed here pose for ecophysiological research in coniferous forest systems? First, given the range of climatic variability experienced during the life-span of a typical conifer, one might question whether short-term, field-based measurements of ecophysiological responses are adequate to characterize the climatic variability that individuals experience. If an ecosystem is strongly affected by climatic variables whose variation is random and independent, then sampling strategies must simply conform to statistical common sense. Conversely, if an ecosystem is strongly affected by one or more climatic variables characterized by decadal and longer term trends, then short-term measurements of physiological variables may be inadequate for predictions of past or future behavior. Second, is it realistic to anticipate that ecophysiological models can be aggregated in such a manner as to explain or predict long-term vegetation dynamics? Our likelihood of success is potentially less limited by the sophistication of our ecophysiological understanding than by our knowledge of the feedbacks between ecophysiological processes and higher levels of biological organization.

References

Agee, J. K., and Smith, L. (1984). Subalpine tree reestablishment after fire in the Olympic Mountains, Washington. *Ecology* 65:810–819.

Ager, T. A., and Brubaker, L. B. (1985). Quaternary palynology and vegetational history of Alaska. *In* "Pollen Records of Late Quaternary North American Sediments" (V. M. Bryant, Jr., and R. G. Holloway, eds.) pp. 353–384. American Association of Stratigraphic Palynologists Foundation, Dallas.

Anderson, P. M., and Brubaker, L. B. (1993). Holocene climate and vegetation histories of Alaska. *In* "Global Climates since the Last Glacial Maximum" (J. Kutzbach, A. Perrot, W. F. Ruddiman, T. Webb, III, and J. E. Wright, Jr., eds), pp. 386–400. Univ. of Minnesota Press, Minneapolis.

Anderson, P. M., Reanier, R. E., and Brubaker, L. B. (1988). Late Quaternary vegeta-

tional history of the Black River region in northeastern Alaska. *Can. J. Earth Sci.* 25: 84–94.
Bamberg, S. A., and Major, J. (1968). Ecology of the vegetation and soils associated with calcareous parent materials in three alpine regions of Montana. *Ecological Monographs.* 38:127–167.
Barbour, M. G., Berg, N. H., Kittel, T. G. F., and Kunz, M. E. (1991). Snowpack and the distribution of a major vegetation ecotone in the Sierra Nevada of California. *J. Biogeogr.* 18:141–149.
Barnosky, C. W., Anderson, P. M., and Bartlein, P. J. (1987). The northwestern U.S. during deglaciation; vegetational history and paleoclimatic implications. *In* "North America and Adjacent Oceans During the Last Deglaciation" (W. F. Ruddiman and H. E. Wright, Jr., eds.), Geol. North Am., Vol. K-3. pp. 289–321. Geol. Soc. Am., Boulder, CO.
Bonan, G. B., Shugart, H. H., and Urban, D. L. (1990). The sensitivity of some high-latitude boreal forests to climatic parameters. *Clim. Change* 16:9–29.
Bray, J. R. (1971). Vegetational distribution, tree growth, and crop success in relation to recent climate change. *Advances in Ecological Research* 7:177–223.
Briffa, K. R., Jones, P. D., and Schweingruber, F. H. (1992). Tree-ring density reconstructions of summer temperature patterns across western North America since A.D. 1600. *J. Clim.* 5:735–754.
Brink, V. C. (1959). A directional change in the subalpine forest-heath ecotone in Garibaldi Park, British Colombia. *Ecology* 40:10–16.
Butler, D. R. (1986). Conifer invasion of subalpine meadows, central Lemhi Mountains, Idaho. *Northwest Science* 60:166–173.
Carrara, P. E., Mode, W. N., Meyer, R., and Robinson, S. W. (1984). Deglaciation and postglacial timberline in the San Juan Mountains, Colorado. *Quat. Res. (N.Y.)* 21:42–56.
Chapin, F. S., III, and Shaver, G. R. (1985). Individualistic growth response of tundra plant species to environmental manipulations in the field. *Ecology* 66:564–576.
COHMAP Project Members (1988). Climatic changes of the last 18,000 years: Observations and model simulations. *Science* 241:1043–1051.
Davis, M. B. (1981). Quaternary history and the stability of forest communities. *In* "Forest Succession: Concepts and Applications" (D. C. West, H. H. Shugart, and D. B. Botkin, eds.), pp. 132–153. Springer-Verlag, New York.
Davis, M. B. (1986). Climatic instability, time lags, and community disequilibrium. *In* "Community Ecology" (J. Diamond and T. J. Case, eds.), pp. 269–284. Harper & Row, New York.
Davis, M. B. (1989). Insights from paleoecology on global change. *Bull. Ecol. Soc. Am.* 70:222–228.
Edwards, M. E., and Dunwiddie, P. W. (1985). Dendrochronological and palynological observations on *Populus balsamifera* in northern Alaska, U.S.A. *Arctic and Alpine Research* 17:271–278.
Ehleringer, J. R., and Field, C. B. (1993). "Scaling Physiological Processes: Leaf to Globe." Academic Press, San Diego.
Fall, P. L. (1988). Vegetation dynamics in the southern Rocky Mountains: Late Pleistocene and Holocene timberline fluctuations. Unpublished Ph.D. Dissertation, University of Arizona, Tucson.
Firbas, F., and Losert, H. (1949). Untersuchungen über die Entstehung der heutigen Waldstufen in der Sudeten. *Planta* 36:478–506.
Fonda, R. W., and Bliss, L. C. (1969). Forest vegetation of the montane and subalpine zones, Olympic Mountains, Washington. *Ecol. Monog.* 39:271–296.
Franklin, J. F., Moir, W. H., Douglas, G. W., and Wiberg, C. (1971). Invasion of subalpine meadows by trees in the Cascade Range, Washington and Oregon. *Arctic and Alpine Research* 3:215–224.

Gams, H. (1937). Aus der Geschichteder Alpenwälder. Z. dt. öst Alpenver. 68:157–170.
Garfinkel, H. L., and Brubaker, L. B. (1980). Modern climate—Tree-ring relations and climatic reconstruction in sub-arctic Alaska. *Nature (London)* 286:872–873.
Goldstein, G. H. (1981). Ecophysiological and demographic studies of white spruce (*Picea glauca* [Moench] Voss) at treeline in the central Brooks Range of Alaska. Ph.D. dissertation, University of Washington, Seattle.
Gorchakovsky, P. L., and Shiyatov, S. G. (1978). The upper forest limit in the mountains of the boreal zone of the USSR. *Arctic and Alpine Research* 10:349–363.
Griggs, R. F. (1938). Timberlines in the northern Rocky Mountains. *Ecology* 19:548–564.
Graumlich, L. J. (1991). Subalpine tree growth, climate, and increasing CO_2: An assessment of recent growth trends. *Ecology* 72:1–11.
Graumlich, L. J. (1993). A 1000-year record of temperature and precipitation in the Sierra Nevada. *Quat. Res. (N.Y.)* 39:249–255.
Graumlich, L. J., and Brubaker, L. B. (1986). Reconstruction of annual temperature (1590–1979) for Longmire, Washington, derived from tree rings. *Quat. Res. (N.Y.)* 25:223–234.
Graumlich, L. J., and Lloyd, A. H. (1994). Dendroclimate, ecological and geomorphological evidence for long-term climatic change in the Sierra Nevada, U.S.A. *Radiocarbon* (in press).
Graybill, D. A., and Idso, S. B. (1993). Detecting the aerial fertilization effect of atmospheric CO_2 enrichment in tree-ring chronologies. *Global Biogeochem. Cycles* 7:81–95.
Grove, J. M. (1988). "The Little Ice Age." Methuen, New York.
Hays, J. D., Imbrie, J., and Shackleton, N. J. (1976). Variations in the earth's orbit: Pacemaker of the ice ages. *Science* 194:1121–1132.
Heikkinen, O. (1984). Forest expansion in the subalpine zone during the past hundred years, Mount Baker, Washington, U.S.A., *Erdkunde* 38:194–202.
Heusser, C. J. (1956). Postglacial environments in the Canadian Rocky Mountains. Ecological Monographs. 26(4):263–302.
Hu, F. S., Brubaker, L. B., and Anderson, P. M. (1993). A 12,000 year record of vegetation and soil development from Wein Lake, central Alaska. *Can. J. Bot.* 71:1133–1142.
Hughes, M. K., and Brown, P. M. (1992). Drought frequency in central California since 101 B.C. recorded in giant sequoia tree rings. *Clim. Dyn.* 6:161–167.
Hughes, M. K., and Diaz, H. F. (1994). "The Medieval Warm Period." Kluwer Academic Publishers, Dordrecht, The Netherlands.
Hustich, L. (1958). On the recent expansion of the Scotch pine in northern Europe. Fennia 82:1–25.
Jacobson, G. L., Webb, T., III, and Grimm, E. C. (1987). Patterns and rates of change during the deglaciation of eastern North America. *In* "North America and Adjacent Oceans During the Last Deglaciation" (W. F. Ruddiman and H. E. Wright, Jr., eds.), Geol. North Am., Vol. K-3, pp. 277–288. Geol. Soc. Am., Boulder, CO.
Jones, P. D., Raper, S. C. B., Santer, B. D., Cherry, B. S. G., Goodess, C., Bradley, R. S., Diaz, H. F., Kelley, P. M., and Wigley, T. M. L. (1985). "A Gridpoint Surface Air Temperature Data Set for the Northern Hemisphere: 1851–1984," U.S. DOE Tech. Rep. TR022. U.S. Dept. of Energy, Carbon Dioxide Res. Div. Washington, DC.
Keifer, M. (1991). Age structure and fire disturbance in the southern Sierra Nevada. MS Thesis, University of Arizona, Tucson.
Kienast, F., and Luxmoore, R. J. (1988). Tree-ring analysis and conifer growth responses to increased atmospheric CO_2. *Oecologia* 76:487–495.
Kozubov, G. M., and Shaydurov, V. S. (1965). Vertical zonality in the Khibin mountains and fluctuations of the timberline. Akademiia Nauk SSSR Izvestiia Ser. Geog. 3:101–104. (In Russian).

Kullman, L. (1986). Recent tree-limit history of *Picea abies* in the southern Swedish Scandes. *Can. J. For. Res.* 16:761–777.
Kuramoto, R. T., and Bliss, L. C. (1970). Ecology of subalpine meadows in the Olympic Mountains, Washington. *Ecol. Monog.* 40:317–345.
LaMarche (1973). Holocene climatic variations inferred from tree line fluctuations in the White Mountains, California. *Quat. Res.* 3:632–660.
LaMarche, V. C., Jr., and Hirschboeck, K. K. (1984). Frost rings in trees as records of major volcanic eruptions. *Nature (London)* 307:121–126.
LaMarche, V. D., Jr., Graybill, D. A., Fritts, H. C., and Rose, M. R. (1984). Increasing atmospheric carbon dioxide: Tree-ring evidence for growth enhancement in natural vegetation. *Science* 225:1019–1021.
Leverenz, J., and Lev, D. (1988). Effects of CO_2 induced climate change on the natural ranges of six major commercial tree species in the western U.S. *In* "The Greenhouse Effect, Climate Change and U.S. Forests" (W. F. Shands and J. S. Hoffman, eds.), pp. 123–155. Conservation Foundation, Washington D.C.
Luckman, B. H. (1994). Evidence for climatic conditions between ca. AD 900–1300 in the southern Canadian Rockies. *In* "The Medieval Warm Period" (M. K. Hughes and H. F. Diaz, eds.), Kluwer Academic Publishers, Dordrecht, The Netherlands.
Markgraf, V., and Scott, L. (1981). Lower timberline in central Colorado during the past 15,000 yr. *Geology* 9:231–234.
Marin, A., and Payette, S. (1984). Expansion récente du mélèze à la limite des fôrets (Québec nordique). *Can. J. Bot.* 62:1404–1408.
Molloy, B. P. J., Burrows, C. J., Cox, J. E., Johnston, J. A., and Wardle, P. (1963). Distribution of subfossil forest remains, Eastern South Island, New Zealand. *New Zealand J. Bot.* 1:68–77.
Mosley-Thompson, E., Barron, E., Boyle, E., Burke, K., Crowley, T., Graumlich, L., Jacobson, G., Rind, D., Shen, G., and Stanley, S. (1990). Earth system history and modeling. *In* "Research Priorities for the U.S. Global Change Research Program," Committee on Global Change. National Academy Press, Washington, DC.
Payette, S., and Filion, L. (1985). White spruce expansion at the tree line and recent climate change. *Can. J. For. Res.* 15:241–251.
Payette, S., and Gagnon, R. (1979). Tree-line dynamics in Ungava peninsula, northern Quebec. *Holarctic Ecol.* 2:239–248.
Payette, S., Filion, L., Gauthier, L., and Boutin, Y. (1985). Secular climate change in old-growth tree-line vegetation of northern Quebec. *Nature (London)* 315:135–138.
Raeside, J. D. (1948). Some post-glacial climatic changes in Canterbury and their effects on soil formation. Transactions of the Royal Society of New Zealand 77:153–171.
Rampton, V. (1969). Pleistocene geology of the Snag-Kiutian area of southwestern Yukon, Canada. Ph.D. thesis. Univ. of Minnesota.
Raynaud, D., Jouzel, J., Barnola, J. M., Chappellaz, J., Delmas, R. J., and Lorius, C. (1993). The ice record of greenhouse gases. *Science* 259:926–934.
Richmond, G. L. (1962). Quaternary stratigraphy of the La Sal Mountains, Utah. US Geological Survey Professional Paper 324. 135 pp.
Ritchie, J. C. (1987). "Postglacial vegetation of Canada." Cambridge University Press, Cambridge.
Ritchie, J. C., Cwynar, L. C., and Spear, R. W. (1983). Evidence from northwest Canada for an early Holocene Milankovitch thermal maximum. *Nature* 305:126–128.
Scuderi, L. A. (1987). Late-Holocene upper timberline variation in the southern Sierra Nevada. *Nature* 325:242–244.
Scuderi, L. J. (1993). A 2000-year record of annual temperatures in the Sierra Nevada Mountains. *Science* 259:1433–1436.
Shackleton, N. J., and Opdyke, N. D. (1973). Oxygen isotope and paleomagnetic stratig-

raphy of equatorial Pacific core V28-238: Oxygen isotope temperatures and ice volumes on a 10^5 year and 10^6 year scale. *Quat. Res. (N.Y.)* 3:39–55.

Solomon, A. M. (1992). The nature and distribution of past, present, and future boreal forests: Lessons for a research and modeling agenda. *In* "A Systems Analysis of the Global Boreal Forest" (H. H. Shugart, R. Leemans, and G. B. Bonan, eds.), pp. 291–307. Cambridge Univ. Press, Cambridge, UK.

Solomon, A. M., and Shugart, H. H. (1993). "Vegetation Dynamics and Global Change." Chapman & Hall, New York.

Spaulding, W. G., and Graumlich, L. J. (1986). The last pluvial climatic episode in the deserts of southwestern North America. *Nature* 319:441–444.

Spaulding, W. G., Leopold, E. B., and Van Devender, T. R. (1983). Late Wisconsin paleoecology of the American Southwest. *In* "Late-Quaternary Environments of the United States" (S. C. Porter, ed.), Vol. 1, pp. 259–293. Univ. of Minnesota Press, Minneapolis.

Thompson, R. S., Whitlock, C., Bartlein, P. J., Harrison, S. P., and Spaulding, W. G. (1993). Climatic changes in the western United States since 18,000 yr B.P. *In* "Global Changes Since the Last Glacial Maximum" (H. E. Wright, Jr., J. E. Kutzbach, T. Webb, III, W. F. Ruddiman, F. A. Street-Perrot, and P. J. Bartlein, eds.), pp. 468–513. Univ. of Minnesota Press, Minneapolis (in press).

Van Devender, T. R., Thompson, R. S., and Betancourt, J. L. (1987). Vegetation history of the deserts of southwestern North America; the nature and timing of the Late Wisconsin–Holocene transition. *In* "North America and Adjacent Oceans During the Last Deglaciation" (W. F. Ruddiman and H. E. Wright, Jr., eds.), Geol. North Am., Vol. K-3, pp. 323–352. Geol. Soc. Am., Boulder, CO.

Wardle, P. (1963). The regeneration gap of New Zealand gymnosperms. *New Zealand J. Bot.* 1:301–315.

Webb, T., III (1987). The appearance and disappearance of major vegetational assemblages: Long-term vegetational dynamics in eastern North America. *Vegetatio* 69: 177–187.

Webb, T., III (1988). Eastern North America. *In* "Vegetation History" (B. Huntley and T. Webb, III, eds.), pp. 385–414. Kluwer Academic Publishers, Dordrecht, The Netherlands.

Webb, T., III, Kutzbach, J. E., and Street-Perrot, F. A. (1985). 20,000 years of global climate change: Paleoclimatic research plan. *In* "Global Change" (T. F. Malone and J. G. Roederer, eds.), pp. 182–218. Cambridge Univ. Press, Cambridge, UK.

Webb, T., III, Bartlein, P. J., Harrison, S. P., and Anderson, K. H. (1993). Vegetation, lake levels, and climate in Eastern North America for the past 18,000 years. *In* "Global Climates since the Last Glacial Maximum" (H. E. Wright, Jr., J. E. Kutzbach, T. Webb III, W. F. Ruddiman, F. A. Street-Perrott, and P. J. Bartlein, eds.), pp. 415–467. University of Minnesota Press, Minneapolis.

West, R. G. (1964). Inter-relations of ecology and Quaternary paleobotany. *J. Ecol.* 52(Suppl.):47–57.

Wigley, T. M. L. (1989). Climatic variability on the 10–100-year time scale: Observations and possible causes. *In* "Global Changes of the Past" (R. Bradley, ed.), pp. 83–101. University of Colorado, Office for Interdisciplinary Studies, Boulder.

Wigley, T. M. L., and Raper, S. C. (1990). Natural variability of the climate system and detection of the greenhouse effect. *Nature (London)* 344:324–327.

3

Plant Hormones and Ecophysiology of Conifers

W. J. Davies

I. Introduction

Over the past 30 years, there have been very substantial fluctuations in the interests of plant scientists in the involvement of plant growth regulators in the control of physiology, growth, and development of plants. In the years following the identification of the five major classes of growth regulators and identification of other groups of compounds of somewhat more restricted interest, an enormous number of papers reported the effects of hormones applied externally to a very wide range of plants. During this period, it became very fashionable to compare effects of hormones with the effects of the environment on developmental and physiological phenomena and to suggest a regulatory role for the hormone(s) in the processes under consideration.

Ross *et al.* (1983) have published a very comprehensive survey of the effects of growth regulators applied externally to conifers, and even 10 years later, it is difficult to improve on what they have done. Nevertheless, in the light of recent changes in our understanding of how growth regulators may work (Trewavas, 1981), it is necessary to reexamine this field and ask what we really know about the involvement of growth regulators in the ecophysiology of conifers. It may be that we conclude that there is little point in trying to quantify growth regulator concentrations and distributions in plants in certain kinds of experiment, but if we do need to do this, we need to be sure that we are asking the correct kinds of questions.

In 1981, Trewavas criticized the types of conclusions drawn from many application-type experiments. One of the grounds for criticism was that endogenous concentrations of naturally occurring regulators were often several orders of magnitude lower than those apparently promoting the same effects when applied externally. We will argue here that we now know enough about the natural variation in concentration and distribution of some compounds of known biological activity such that we can begin to ascribe ecophysiological roles to these compounds. These arguments are not based on extrapolation of effects of random compounds sprayed onto plants held under arbitrary conditions, but rather on the need to explain particular physiological and developmental responses to environmental variation.

II. Quantification of Plant Growth Regulators

Until comparatively recently, there have been very few reliable estimates of the concentrations of growth regulators in trees and still fewer attempts at defining the intraplant distribution of individual regulators. Early identifications and attempts at quantification were made with bioassays. These may be used to identify an activity of the required type, but they are notoriously unreliable for quantitative work. Even the much-used gas chromatograph (GC) and high-performance liquid chromatograph (HPLC) cannot provide unequivocal identification, but can provide reasonable quantification. Data obtained using these techniques must always be validated using a mass spectrometer attached to the gas chromatograph (GC/MS) (Saunders, 1978). All of the chromatographic techniques suffer from the major disadvantage that samples must be purified prior to injection into the chromatograph, and these purification steps can take a considerable amount of time. An even bigger problem is the potential for loss of the product during the clean-up procedure. Sample purification is a particular problem with extracts from woody plants.

Under most circumstances, growth regulators occur in plants in extremely low concentrations. For example, abscisic acid (ABA) is found in the xylem at nanomolar to micromolar (10^{-9} to $10^{-6} M$) concentrations. Similar concentrations can be found in cells in the leaf and in the root. To detect concentrations of this order of magnitude using the chromatographic techniques described above, we require relatively large samples of tissue. This can be a problem if, for example, we are looking for activity in a specific plant part of a restricted size (e.g., the shoot apex).

The development of immunoassays has provided rapid and sensitive methods for hormone measurement in minute amounts of plant material. Many authors have now reported monoclonal antibody methods developed to quantify hormones using enzyme-linked immunoassay (ELISA), radioimmunoassay (RIA), and fluoroimmunoassay (FIA). For example, Walker-Simmons and Abrams (1991) provide a useful protocol for the determination of ABA concentrations using an indirect ELISA technique.

Immunological methods have the advantage that samples generally need only a limited amount of purification, which can save a lot of time. In fact, ABA can often be measured directly in crude extracts (Walker-Simmons, 1987) or after simple clean-up using C_{18} cartridges (Weiler et al., 1986). Hot water extractions can reduce interference by organic solvents. Immunoassay techniques are relatively inexpensive and do not require much specialized laboratory equipment. Published methods using an enzyme-amplified ELISA report sensitivity to as little as 10^{-16} mol of hormone (e.g., Harris and Outlaw, 1990). Sensitivity of this kind allows quantification of hormones in single cells. It is also possible to localize hormones in particular tissues using immunology. For example, ABA can be localized using immunogold techniques (Sossountzov et al., 1986) or using peroxidase–antiperoxidase (Sotta et al., 1985).

It is extremely important that immunoassay methods are modified and verified for each new type of plant tissue. Yield losses during extraction should be monitored using internal standards. Interference in the immunoassay from substances in the plant tissue can be detected by adding a range of authentic hormone standards to the plant sample dilutions and then checking that the amounts of hormone measured increase in proportion to the amounts added. Any deviation from the slope of the hormone standards alone, caused by the addition of tissue samples, indicates interference (Jones, 1987). Some interfering substances can be removed by adding compounds such as polyvinylpyrrolidone (Quarrie et al., 1988).

Validation of immunoassay results by physicochemical methods is desirable for any new tissue assayed or when extraction techniques are changed. There are comparatively few reports of studies wherein hormone contents of conifers have been quantified and still fewer reports of successful use of immunological techniques. Nevertheless, Leroux et al. (1985) have reported good correlation between ABA contents of extracts from *Pseudotsuga menziesii* measured using GLC with electron capture and an ELISA technique.

III. A Role for Plant Growth Regulator Biologists in Physiological Plant Ecology?

Reference to the review by Ross *et al.* (1983) shows clearly the wide range of processes that may be affected by plant hormones. For most of these processes, we cannot provide a detailed assessment of exactly how the hormone might be involved in much of this regulation. Physiological plant ecologists would be hard pressed to identify critical measurements of plant growth regulator concentrations in particular tissues that would help explain aspects of plant growth and development that are of particular interest.

The measurement of water relations and gas exchange has dominated in much of the analytical work undertaken in physiological plant ecology and such measurements, presumably for technological reasons, have even dominated the measurement of plant growth. Due largely to the influential writings of Kramer and others in the 1950s and 1960s, measurements of water relations variables have become central to many studies of the effects of environmental stress on gas exchange and growth, and it is commonly held that many stress effects are mediated via changes in plant water relations variables. More recently, careful analysis of the literature has shown that many environmental stresses will modify plant growth and development, even though shoot water relations do not change, and it is even tempting to conclude that water relations are controlled by variation in growth and stomatal behavior, rather than the converse. Some of the best examples of such responses are found with woody plants (Jones, 1985; Pereira and Chaves, 1993).

It is now generally accepted that plants may have the capacity to regulate physiology, growth, and development of shoots in response to changes in their root environment via a change in the transport of chemical signals moving through the plant in the xylem stream. Much recent analysis has shown that plant hormones can play a central part in this kind of chemical signaling.

IV. Chemical Signaling in Woody Plants

Gowing *et al.* (1990) separated the roots of individual apple trees, placing them in two separate rooting containers. The soil in one of these containers was allowed to dry and the other was kept well supplied with water. This treatment had no effect on plant water status but caused a restriction in leaf growth and stomatal conductance. Removal of the roots in drying soil from the plants caused a reopening of stomata and an increase in leaf growth rate. This result cannot be explained by an increased supply of water to the shoots and it seems likely that an interaction between roots and drying soil has generated an inhibitor of leaf growth and stomatal opening. Removal of the roots removes the source of this inhibitor.

There are now many examples of this type of signaling operating both in plants in controlled environments (Davies and Zhang, 1991) and in the field. For example, there is a clear relationship between stomatal conductance and xylem ABA concentration for *Prunus dulcis* growing in the Negev desert in Israel (Wartinger *et al.*, 1990), at least for the first part of the growing season. In these same leaves there was no obvious relationship between stomatal conductance and leaf water relations. Similar results have been obtained by Khalil and Grace (1993) for sycamore. In this study, there is a clear relationship between the intensity of the chemical signal generated and the water status of the soil around the root.

There are no published field studies of root to shoot communication in conifers but several of these studies are now underway. It is clear that ABA does accumulate in conifers in drying soil (Johnson, 1987) and that the hormone can regulate the stomatal behavior of gymnosperm species (Blake and Ferrell, 1977; Roberts and Dumbroff, 1986). In addition, there are many examples of field studies of gas exchange by conifers where limitation of stomatal conductance in plants in drying soil is not easily explained by water deficit in the needles (e.g., Beadle *et al.*, 1985); this is particularly the case in the afternoon hours. In addition, many simulations of plant gas exchange that incorporate only the effects of climatic variables predict reopening of stomata of droughted plants in the afternoon hours, but this type of response is not commonly observed (see, e.g., Schulze *et al.*, 1974). It is important to know whether variation in the balance of plant growth regulators in the shoots may have some controlling influence on gas exchange under these circumstances.

It is well known that a mild degree of soil drying can have a substantial effect on plant growth and development, if the stress lasts for a substantial period of time. One of the best examples of an effect of this kind is the effect of rainfall variation from year to year on the length of needles of *Pinus radiata* growing on experimental plots near Canberra (Landsberg, 1986). Myers (1988) has related this variation to a water stress integral (the summation of predawn water potential on every day during the period of interest) (Fig. 1). It is of interest to determine how a plant might "measure" soil water potential (predawn water potential) and thereby fix its developmental pattern over an extended period.

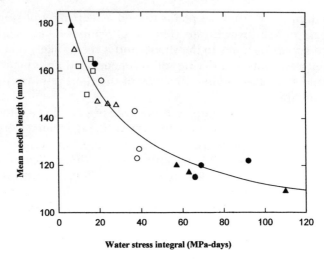

Figure 1 The relationship between mean final needle length of the current season's foliage on the 3-year-old whorl and water stress integral for *Pinus radiata* trees subjected to different irrigation and fertilizer treatments in each of 4 years. Each point is the mean length of between 120 and 240 needles. Modified from Myers (1988).

V. A General Model for Chemical Regulation of Stomatal Behavior, Water Relations, and Development of Plants in the Field

Tardieu and Davies (1993) have recently published a general model for the chemical regulation of stomatal behavior of plants growing in the field. This is based on a comprehensive data set collected for maize plants. Nevertheless, it seems likely that the principles will hold true for many plants. Because this is the only model of its kind available in the literature, it seems important to test some of the ideas contained therein for other species of interest. Individual components of the model can be varied from literature values for the species of interest.

In the model, stomatal conductance is a function of the concentration of ABA in the xylem, but the sensitivity of this function varies with the water relations of the shoot (Fig. 2), because Tardieu and Davies (1992) have reported that leaf water deficit increases stomatal responses to ABA, even though the water deficit has no effect on stomata. Xylem ABA concentration is a function of the water status of the root but also varies as a function of the flux of water through the plant (Fig. 2). Not surprisingly, when the vapor pressure deficit (VPD) is high, higher transpiration rates will dilute out the ABA signal even though the supply of ABA from the roots may increase due to greater root dehydration (Fig.

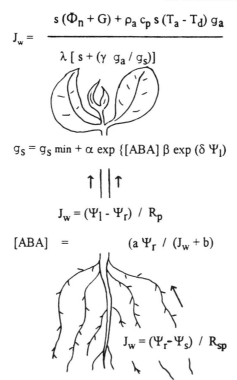

Figure 2 Representation of variables and equations of control in an interactive model describing water flux through maize plants [model described by Tardieu and Davies (1993)]. Input variables: ϕ, net radiation; T_a and T_d, air and dew point temperatures; and ψ_s, soil water potential. R_p and R_{sp} are the plant and the soil–plant resistance to water flux. Unknowns: g_s, stomatal conductance; ϕ_r and ψ_l, root and leaf water potentials; J_w, water flux; and [ABA], concentration of ABA in the xylem. Other symbols are constants (see Tardieu and Davies, 1993). Arrows symbolize transfers of water and/or ABA.

2) (Tardieu *et al.*, 1992b). Water fluxes in the model are calculated from the Penman–Monteith type of equation (Fig. 2), whereas the water relations of various plant parts are calculated from Ohm's law types of equations with appropriate resistances inserted (Fig. 2).

The solution to the model shows an adequate simulation of stomatal behavior in well-watered plants, with stomata opening in the morning, staying open all day, and closing late in the afternoon, largely as a function of a simple irradiance response, which is also included in the model. In plants with unfavorable conditions for water uptake (drying and compacted soil), however, stomata open in the morning and then close gradually throughout the day (Fig. 3), largely as a function of the sensitizing of the stomata to the ABA signal as the water potential falls. This

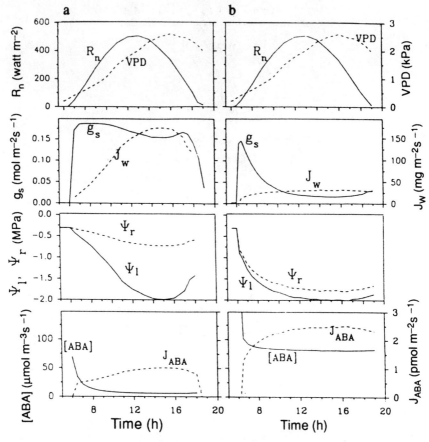

Figure 3 Simulations of the daily pattern of stomatal conductance (g_s), water flux in the soil–plant atmosphere continuum (J_w), leaf and root water potential (ψ_l, ψ_r), ABA concentration in the xylem ([ABA]), and ABA flux into the leaf (J_{ABA}). (a) Calculations with high evaporative demand for a root under favorable soil conditions; (b) calculations for soil conditions that are unfavorable for water uptake. For further details, see Tardieu and Davies (1993).

gradual closure of stomata is the response commonly shown by plants in drying soil in the field (see, e.g., Tenhunen et al., 1984). Interestingly, the model also predicts that shoot water relations of plants will be regulated and shows a relative stability of the ABA signal, even under conditions where the photosynthetically active radiation (PAR) is highly variable. It might be argued that a stable chemical signal, which increases in magnitude as the plant's access to soil water is restricted, might be a suitable regulator for the long-term development of the plant (Jones,

1990). It is clear from data collected in the field that the ABA concentration in the xylem at night provides a sensitive indicator of predawn leaf water potential or the water status of the soil (Tardieu *et al.*, 1992b). This might be the signal in the *P. radiata* plants described above, which links needle length with the water stress integral (see Fig. 1), although plants must show a very sensitive ABA response to soil drying because the reported differences in growth are the result of quite small changes in soil water availability (see Myers, 1988).

VI. Importance of Sensitivity Variation and Involvement of ABA in Stomatal Response to Climatic Variables

It is clear from our model that dynamic hour-by-hour regulation of stomatal behavior can be accomplished by the interaction between the chemical signal and the water status of the shoot (and perhaps other climatic variables). This places emphasis on the variation in sensitivity of plant processes to ABA, a feature of the hormone response emphasized in the writings of Trewavas (1981). We know that stomatal sensitivity to ABA can be greatly influenced by a large number of different variables. We have discussed elsewhere the possible basis for some of the sensitivity variation of this kind (Gowing *et al.*, 1993), which can be substantial. It is clear, however, that these interactions will contribute to a variety of stomatal responses that can be important in the control of gas exchange. For example, Gollan *et al.* (1992) and Schurr *et al.* (1992) have emphasized from their work with *Helianthus* that stomatal sensitivity to an ABA signal in the xylem can change by several orders of magnitude with quite small variation in xylem sap ion composition. Osonubi *et al.* (1988) show dramatic effects of soil drying on the ion concentrations in the xylem sap of spruce (Fig. 4), and it seems likely, therefore, that a proportion of the stomatal response to soil drying in this plant will be a function of changes of this kind.

The literature contains several examples of interactions between the effects of ABA and leaf temperature (Cornic and Ghashgaie, 1991; Rodriguez and Davies, 1982) and ABA and CO_2 (Snaith and Mansfield, 1982) on stomata. Reduction in temperature can completely suppress the effect of ABA whereas CO_2 effects can be accentuated by ABA treatment. Even in well-watered plants xylem sap can contain substantial amounts of ABA. A small amount of soil drying can increase this quantity. These observations raise the possibility that ABA could be involved in several of the well-documented stomatal responses to climatic fac-

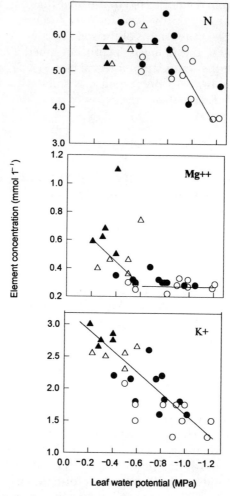

Figure 4 Changes in the xylem sap concentrations of nitrogen (N), magnesium (Mg^{2+}) and potassium (K^+) of *Picea abies* with decreasing leaf water potential. Modified from Osonubi et al. (1988).

tors. The model described above provides additional evidence for this assertion.

In the model we have simulated the combined effects of hydraulic and chemical changes on the behavior of stomata and the water relations of plants in soil where access to water is restricted. We refer to this as the interaction model. Tardieu (1993) has tested this model against a purely hydraulic model (no effect of a chemical signal) and a purely chemical model. Here, there is no sensitivity gain from an interaction between leaf water relations and the chemical signal. When the sensitivity of the stomata to the ABA signal is set at a comparable value to that shown by laboratory-grown plants (Zhang and Davies, 1990), the chemical-only model does not control stomatal conductance or water status of plants growing with a restricted supply of water. The simulation of the behavior of plants responding to a purely hydraulic model suggests reopening of stomata in the afternoon hours, which is not commonly observed in plants in drying soil (e.g., Schulze *et al.*, 1974).

Over the longer term of a soil-drying episode, the purely physical and combination models provided relatively similar predictions of variation in stomatal conductance, leaf water potential, and xylem ABA concentration on a day-to-day basis, provided that the variables were fitted to the same experimental data. This similarity has two important consequences: (1) The existence of tight relationships between xylem ABA and stomatal conductance such as those described by Tardieu *et al.* (1992a) and Khalil and Grace (1993) is not by itself conclusive evidence for stomatal control by root signals. (2) Control of water potential can be achieved by the interaction model, which does not involve a threshold leaf water potential.

Figure 5 shows simulations carried out using the interaction model with three constant evaporative demands (1.6, 4.2, and 7 mm/day). Simulations have been run over varying periods of time, and variations in gas exchange and water potential are presented as a function of the soil water reserve. It is clear that earlier stomatal closure and lower apparent leaf water potential for higher VPD, which are part of the "classical response" of plants to varying evaporative demand, do not necessarily correspond to a special mechanism and could be accounted for by a model of stomatal control taking into account root messages and water flux.

VII. Regulation of Growth and Development: A Conclusion

A role for ABA in the regulation of growth and development of a wide range of plants is inferred from the data collected in Table I. There is, however, comparatively little information on the mechanism of action of ABA in any of these processes. One system that has been investigated in detail is the regulation of root and mesocotyl growth of maize seedlings at low water potential. This is an important area because sustained root growth as the soil dries can lead to a maintenance of water uptake. Recent work by Saab *et al.* (1990; 1992) has shown that in maize seedlings grown in vermiculite at -1.6 MPa, ABA is necessary for continued root growth. If synthesis and accumulation of the hormone are restricted by the application of an inhibitor of carotenoid synthesis, then root growth

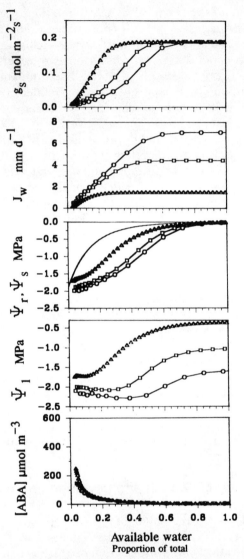

Figure 5 Simulated changes with time of stomatal conductance (g_s), water flux (J_w), soil and root water potentials (ψ_s and ψ_r), leaf water potential (ψ_l), and xylem [ABA] plotted against available soil water during a drying period with three contrasting evaporative demands. The model used for the simulation is that described in Fig. 3. The interval between two symbols represents 24 hours. Solid line, soil water potential. Symbols: △, 1.6 mm/day evaporative demand; □, 4.2 mm/day evaporative demand; ○, 7 mm/day evaporative demand. Modified from Tardieu (1993).

Table I Evidence for Involvement of ABA in Long-Term Plant Responses to Water Stress Based on the Similarity of Responses to Water Stress and to Exogenous ABA[a]

Response	Water stress	ABA	Correlation[b]
Biochemical and physiological			
Specific mRNA and protein synthesis	Increase	Increase	+ +
Proline and betaine accumulation	Increase	Increase	+ +
Osmotic adaptation	Yes	Yes	+
Photosynthetic enzyme activity	Decrease	Decrease	+
Desiccation tolerance	Increase	Increase	+
Salinity and cold tolerance	Induces	Induces	+ +
Wax production	Increase	Increase	+
Growth			
General growth inhibition	Yes	Yes	+ + +
Cell division	Decrease	Decrease	+ + +
Leaf initiation	Inhibits	Inhibits	+ +
Cell expansion	Decrease	Decrease	+ + +
Germination	Inhibits	Inhibits	+ +
Root growth	Increase	Increase	+ +
	Decrease	Decrease	
Suberization	?	Increase	
Morphology			
Production of trichomes	Increase	Increase	+ +
Production of spines	Increase	Increase	+ + +
Stomatal index	Decrease	Decrease	+
Tillering in grasses	Decrease	Decrease, increase	+
Conversion from aquatic to aerial leaf type	Yes	Yes	+ +
Induction of dormancy, terminal buds, or perennial organs	Yes	Yes	+ +
Apical dominance			
Reproductive			
Flowering in annuals	Often advanced	Often advanced	+ +
Flower induction in perennials	Inhibits	Inhibits	+ +
Flower abscission	Increase	Increase	+
Pollen viability	Decrease	Decrease	+
Seed set	Decrease	Decrease	+
Embryo maturation	?	Accelerate	

[a] From Trewavas and Jones (1991).
[b] The strength of correlation is indicated as ranging from weak (+) to strong (+ + +).

is significantly slowed. The same treatment will result in an increase in shoot growth and these responses therefore provide an explanation for the commonly observed increases in root:shoot ratio that are promoted by a soil-drying treatment (see Table I). More recent work has implicated ABA accumulation in osmotic regulation in root tips at low water potential and in the regulation of cell wall properties (Spollen et al., 1993).

It almost goes without saying that all of this detailed cell biology has been performed with relatively amenable, rapidly growing herbaceous species. Chemical control of the type described in this paper is no less important in woody plants than in herbaceous plants. Convenient, sensitive methods of growth regulator analysis are now readily available. It is important that in the next few years we see substantial increases in the understanding of the regulation of growth and physiology of conifers.

References

Beadle, C. L., Jarvis, P. G., Talbot, H., and Neilson, R. E. (1985). Stomatal conductance and photosynthesis in a mature Scots pine forest. II. Dependence on environmental variables of single shoots. *J. Appl. Ecol.* 22:573–586.

Blake, T. J., and Ferrell, W. K. (1977). The association between soil and xylem water potential, leaf resistance and abscisic acid content in droughted seedlings of Douglas fir (*Pseudotseuga menziesii*). *Physiol. Plant.* 39:106–109.

Cornic, G., and Ghashghaie, J. (1991). Effect of temperature on net CO_2 assimilation and photosystem II quantum yield of electron transfer of French bean (*Phaseolus vulgaris* L.) leaves. *Plant Physiol.* 185:255–260.

Davies, W. J., and Zhang, J. (1991). Root signals and the regulation of growth and development of plants in drying soil. *Annu. Rev. Plant Physiol. Plant Mol. Biol.* 42:55–76.

Gollan, T., Schurr, U., and Schulze, E.-D. (1992). Stomatal response to drying soil in relation to changes in the xylem sap composition of *Helianthus annuus*. 1. The concentration of cations, anions and amino acids in, and pH of the xylem sap. *Plant, Cell Environ.* 15:551–559.

Gowing, D. J. G., Davies, W. J., and Jones, H. G. (1990). A positive root-sourced signal as an indicator of soil drying in apple, *Malus* × *domestica* Borkh. *J. Exp. Bot.* 41:1535–1540.

Gowing, D. J. G., Davies, W. J., Trejo, C. L., and Jones, H. G. (1993). Xylem-transported chemical signals and the regulation of plant growth and physiology. *Philos. Trans. R. Soc. London* 341:41–47.

Harris, M. J., and Outlaw, W. H., Jr. (1990). Histochemical technique: A low volume enzyme-amplified immunoassay with sub-fmol sensitivity. Application to measurement of abscisic acid in stomatal guard cells. *Physiol. Plant.* 78:495–500.

Johnson, J. D. (1987). Stress physiology of forest trees: The role of plant growth regulators. *Plant Growth Regul.* 6:193–215.

Jones, H. G. (1985). Physiological mechanisms involved in the control of leaf water status: Implications for the measurement of tree water status. *Acta Hortic.* 171:291–296.

Jones, H. G. (1987). Correction for non-specific interference in competetive immunoassays. *Physiol. Plant.* 70:146–154.

Jones, H. G. (1990). Control of growth and stomatal behaviour at the whole plant level:

Effects of soil drying. *In* "Importance of Root to Shoot Communication in the Response to Soil Drying" (W. J. Davies and B. Jeffcoat, eds.), BSPGR Monogr. No. 21, pp. 81–93.

Khalil, A. A. M., and Grace, J. (1993). Does xylem ABA control the stomatal behaviour of water-stressed sycamore (*Acer pseudoplatanus* L.) seedlings? *J. Exp. Bot.* 44:1127–1134.

Landsberg, J. J. (1986). Experimental approaches to the study of the effects of nutrients and water on carbon assimilation by trees. *Tree Physiol.* 2:427–444.

Leroux, B., Maldiney, R., Miginiac, E., Sossountzov, L., and Sotta, B. (1985). Comparative quantitation of abscisic acid in plant extracts by gas–liquid chromatography and an enzyme-linked immunosorbent assay using the avidin–biotin system. *Planta* 166: 524–529.

Myers, B. J. (1988). Water stress integral—A link between short-term stress and long-term growth. *Tree Physiol.* 4:315–324.

Osonubi, O., Oren, R., Werk, K. S., Schulze, E.-D., and Heilmeier, H. (1988). Performance of two *Picea abies* (L.) Karst stands at two different stages of decline. IV. Xylem sap concentrations of magnesium, calcium, potassium and nitrogen. *Oecologia* 77:1–6.

Pereira, J. S., and Chaves, M. M. (1993). Plant water deficits in Mediterranean ecosystems. *In* "Water Deficits" (J. A. C. Smith and H. Griffiths, eds.), pp. 237–251. Bios Scientific Publishers, Oxford.

Quarrie, S. A., Whitford, P. N., Appleford, N. E. J., Wang, T. L., Cook, S. K., and Henson, I. E. (1988). A monoclonal antibody to (S0-abscisic acid: Its characterisation and use in a radioimmunoassay for measuring abscisic acid in crude extracts of cereal and lupin leaves. *Planta* 173:330–339.

Roberts, D. R., and Dumbroff, E. B. (1986). Drought resistance, transpiration rates and ABA levels in three northern conifers. *Tree Physiol.* 1:161–168.

Rodriguez, J. L., and Davies, W. J. (1982). The effects of temperature and ABA on stomata of *Zea mays* L. *J. Exp. Bot.* 33:977–987.

Ross, S. D., Pharis, R. P., and Binder, W. D. (1983). Growth regulators and conifers: Their physiology and potential uses in forestry. *In* "Plant Growth Regulating Chemicals" (L. G. Nickell, ed.), Vol. 2, pp. 35–78. CRC Press, Boca Raton, FL.

Saab, I., Sharp, R. E., Pritchard, J., and Voetberg, G. S. (1990). Increased endogenous ABA maintains primary root growth and inhibits shoot growth of maize seedlings at low water potential. *Plant Physiol.* 93:1329–1336.

Saab, I., Sharp, R. E., and Pritchard, J. (1992). Effect of inhibition of abscisic acid accumulation on the spatial distribution of elongation in the primary root and mesocotyl of maize at low water potentials. *Plant Physiol.* 99:26–33.

Saunders, P. F. (1978). The identification and quantitative analysis of abscisic acid in plant extracts. *In* "Isolation of Plant Growth Substances" (J. R. Hillman, ed.), pp. 115–134. Cambridge Univ. Press, Cambridge, UK.

Schulze, E.-D., Lange, O. L., Evanari, M., Kappen, L., and Buschbom, U. (1974). The role of air humidity and temperature in controlling stomatal resistance of *Prunus armeniaca* L. under desert conditions. I. A simulation of the daily course of stomatal resistance. *Oecologia* 17:159–170.

Schurr, U., Gollan, T., and Schulze, E.-D. (1992). Stomatal responses to drying soil in relation to changes in the xylem sap composition of *Helianthus annuus*. II. Stomatal sensitivity to abscisic acid imported from the xylem sap. *Plant, Cell Environ.* 15: 561–567.

Snaith, P. J., and Mansfield, T. A. (1982). Control of the CO_2 responses of stomata by indol-3-yl acetic acid and abscisic acid. *J. Exp. Bot.* 33:360–365.

Sossountzov, L., Sotta, B., Maldiney, R., Sabbagh, I., and Miginiac, E. (1986). Immunoelectron-microscopy localisation of abscisic acid with colloidal gold on Lowicryl-embedded tissues of *Chenopodium polyspermum* L. *Planta* 168:471–481.

Sotta, B., Sossountzov, L., Maldiney, R., Sabbagh, I., Tachon, P., and Miginiac, E. (1985). Abscisic acid localization by light microscopic immuno-histochemistry in *Chenopodium polyspermum* L. *J. Histochem. Cytochem.* 33:201–208.

Spollen, W. G., Sharp, R. E., Saab, I., and Wu, Y. (1993). Regulation of cell expansion in roots and shoots at low water potentials. *In* "Water Deficits" (J. A. C. Smith and H. Griffiths, eds.), pp. 37–52. Bios Scientific Publishers, Oxford.

Tardieu, F. (1993). Will increases in our understanding of soil–root relations and root signalling substantially alter water flux models? *Philos. Trans. R. Soc. London, Ser. B* 341:57–66.

Tardieu, F., and Davies, W. J. (1992). Stomatal response to abscisic acid is a function of current plant water status. *Plant Physiol.* 98:540–545.

Tardieu, F., and Davies, W. J. (1993). Integration of hydraulic and chemical signalling in the control of stomatal conductance and water status of droughted plants. *Plant, Cell Environ.* 16:341–349.

Tardieu, F., Zhang, J., Katerji, N., Bethenod, O., Palmer, S., and Davies, W. J. (1992a). Xylem ABA controls the stomatal conductance of field-grown maize subjected to soil compaction or soil drying. *Plant, Cell Environ.* 15:193–197.

Tardieu, F., Zhang, J., and Davies, W. J. (1992b). What information is conveyed by an ABA signal from maize roots in drying field soil? *Plant, Cell Environ.* 15:185–191.

Tenhunen, J. D., Lange, O. L., Gebel, J., Beyshlag, W., and Weber, J. A. (1984). Changes in photosynthetic capacity, carboxylation efficiency and CO_2 compensation point associated with midday depression of net CO_2 exchange of leaves of *Quercus suber*. *Planta* 193:193–203.

Trewavas, A. J. (1981). How do growth substances work? *Plant, Cell Environ.* 4:203–228.

Trewavas, A. J., and Jones, H. G. (1991). An assessment of the role of ABA in plant development. *In* "Abscisic Acid" (W. J. Davies and H. G. Jones, eds.), pp. 169–188. Bios Scientific Publishers, Oxford.

Walker-Simmons, K. (1987). ABA levels and sensitivity in developing wheat embryos of spouting and resistant and susceptible cultivars. *Plant Physiol.* 84:61–66.

Walker-Simmons, K., and Abrams, S. R. (1991). Use of ABA immunoassays. *In* "Abscisic Acid" (W. J. Davies and H. G. Jones, eds.), pp. 53–61. Bios Scientific Publishers, Oxford.

Wartinger, A., Heilmeier, H., Hartung, W., and Schulze, E.-D. (1990). Daily and seasonal courses of abscisic acid in the xylem sap of almond trees (*Prunus dulcis* M.) under desert conditions. *New Phytol.* 116:581–587.

Weiler, E. W., Eberle, J., Mertens, R., Atzorn, R., Feyerabend, M., Jourdan, P. S., Arnscheidt, A., and Wieczorek, U. (1986). Antisera- and monoclonal antibody-based immunoassay of plant hormones. *In* "Immunology in Plant Science" (T. L. Wang, ed.), pp. 27–58. Cambridge Univ. Press, Cambridge, UK.

Zhang, J., and Davies, W. J. (1990). Changes in the concentration of ABA in xylem sap as a function of changing soil water status can account for changes in leaf conductance and growth. *Plant, Cell Environ.* 13:271–285.

4

Ecophysiological Controls of Conifer Distributions

F. I. Woodward

I. Introduction

The boreal forest covers the most extensive worldwide area of conifer-dominated vegetation, with a total global area of about 12 million km^2 (Baumgartner, 1979). This large area is very species poor; in North America there are only nine widespread and dominant species of trees (Payette, 1992), of which six are conifers—*Picea mariana, Picea glauca, Abies balsamea, Larix laricina, Pinus contorta,* and *Pinus banksiana.* The remaining three angiosperms are *Betula papyrifera, Populus tremuloides,* and *Populus balsamifera.* In Fennoscandia and the former Soviet Union, 14 species dominate the boreal forest (Nikolov and Helmisaari, 1992), 10 of which are conifers—*Abies sibirica, Larix gmelinii, Larix sibirica, Larix sukaczewii, Picea abies, Picea ajanensis, Picea obovata, Pinus pumila, Pinus sibirica,* and *Pinus sylvestris.* The dominant angiosperm trees are *Betula pendula, Betula pubescens, Chosenia arbutifolia,* and *Populus tremula.* Such species paucity detracts from realizing the remarkable capacity of these species to endure the harshest forest climates of the world.

Both the short-term geological history and the current climate are major causes of the species paucity in the boreal forest. In general, the boreal forest has been present in its current distribution only since the Holocene era (Ritchie, 1987). In most cases, the dominant species of the boreal forest completed their postglacial expansion to their current distributions only over the past 2000 years. So the ecology of the forest is very young, in comparison with forests in warmer climates (Takhtajan, 1986). It might be expected that over subsequent millennia, with no climatic change, there could be a slow influx of new species to the boreal zone; however, the extreme climatic, edaphic, and disturbance characteristics of the area are likely to set insurmountable limits on this influx of diversity. The extremely cold winters (to nearly $-70°C$) (see Chapter 5, this volume) and short growing seasons strongly limit growth and maintain a constant permafrost in many areas (Larsen, 1980). The soils that melt during the short growing season are very nutrient poor, with very low rates of nitrogen mineralization, in some cases as low as 8 to 9 kg N/ha/year (Van Cleve *et al.*, 1981).

In combination with the poor climate and soils for growth, fire is a frequent occurrence in the boreal forest, with a return frequency in the range of 50 to 200 years (Bonan, 1992; Viereck, 1983). Defoliating insect outbreaks are also very extensive and relatively common (see Chapter 6) (Bonan, 1992). It is therefore apparent that the poor climate and extreme disturbance regimes severely limit the survival of species in the boreal forest biome. That species do survive and thrive indicates that nature can out-maneuver extreme exogenous conditions, and the chapters in this book all describe the different facets of these coniferous mechanisms. It is also important to note that conifers occur across the range of forests from the equator to beyond the Arctic circle (Walter, 1979) and they are very important sources of timber products worldwide (Mather, 1990). Although this chapter will be primarily concerned with the ecophysiological controls on conifer distribution in the boreal zone, there will be additional emphasis on the survival of conifers at the global scale.

II. Climatic Limits

The temperature extremes for meteorological stations in the boreal forest biome (Fig. 1) (Müller, 1982) indicate absolute minimum temperatures of around $-70°C$ and absolute maximum temperatures to about $40°C$. Boreal conifers demonstrate a capacity to endure temperatures well below $-70°C$ (Woodward, 1987; Sakai and Larcher, 1987). Only species with this capacity can survive the extreme cold of the boreal winter. In addition to coniferous species, the few angiosperm genera that occur in this region, e.g., *Betula* and *Populus*, also possess this extreme cold resistance.

The mechanisms of extreme cold tolerance are incompletely known; however there is evidence for both morphological and biochemical methods of survival, and the overwintering dormant bud and living xylem

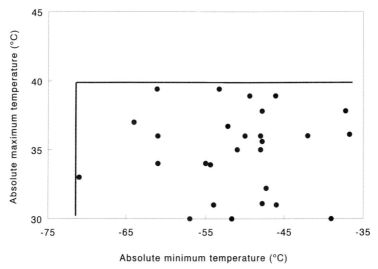

Figure 1 Absolute minimum and maximum temperatures for meteorological stations (●) located in the boreal forest biome. Boundary lines inserted by eye. (From Müller, 1982.)

appear to be the most sensitive structures wherein ice formation must be prevented (Sakai and Larcher, 1987; see also Chapter 5, this volume). In contrast, the cells in the evergreen needles of conifers appear to undergo extensive breakdown, particularly in the membranes of the chloroplasts and mitochondria. However, the degree of breakdown is reduced in cold-hardened leaves and repair occurs regularly in the spring (Senser and Beck, 1982).

There is no evidence that the absolute maximum temperatures in the boreal zone are currently limiting (Gauslaa, 1984). However, experimental studies in open-top chambers with CO_2 enrichment have shown increasingly high needle temperatures under full sun, as CO_2 is increased and stomatal conductance is reduced (Surano et al., 1986; Woodward et al., 1991). It appears likely, therefore, that increasing CO_2 concentrations may lead to close to lethal needle temperatures (~45 to 50°C) during periods of high air temperatures, low wind speeds, and high irradiance. However this is unlikely to be a problem in the present-day boreal forest.

Havranek and Tranquillini (Chapter 5) strongly emphasize the importance of a deep dormancy for enhanced winter survival. One feature of a lack of dormancy is that leaves may develop with an inadequate cuticle to survive abrasion by wind and ice crystals, leading to a subsequent tendency for desiccation during the winter (Tranquillini, 1979). The chance of desiccation is high, because during the winter period there will be little if any opportunity for water uptake from a frozen soil and through a frozen xylem (Hadley and Smith, 1990). A critical feature of the climate is therefore the length of the growing season, which will be sufficient for new needles to expand and reach morphological maturity. A crude estimate of the growing season has been taken as the number of months when the temperature is greater than 10°C (but see Chapter 5). This information and data on the number of months in which a frost is expected have been extracted from Müller (1982). The data, although rather coarse (Fig. 2), indicate that in the boreal forest biome the minimum growing season may be as little as 2 months, but more typically 3 to 4 months. Frosts are expected in the growing period, and for up to 8 months of the year. The frequent expectation of frost, even in the growing season, will select for plants with the capacity for short-term frost avoidance by supercooling and the necessity for rapid recovery from photosynthetic photoinhibition, when chilling and frosts occur during daylight (Chapter 5) (Ottander and Öquist, 1991).

The harsh and short growing season exerts a major limitation on the distribution of boreal conifers, particularly through low germination rates and high seedling mortality (Black and Bliss, 1980). The short growing season may limit the capacity of young seedlings to grow and develop a functional mycorrhizal network (Read, 1991). However, experimental work on red pine (*Pinus resinosa*) indicated germination and establishment of mycorrhizal associations at the northern limit of its dis-

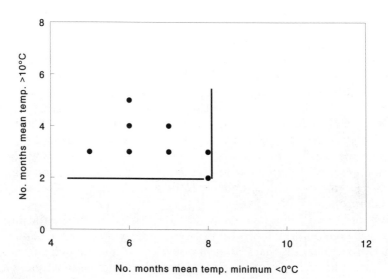

Figure 2 The number of months with mean minimum temperatures less than 0°C and when the mean temperature is greater than 10°C. (From Müller, 1982.)

tribution (Flannigan, 1993). For this species, at least, and unlike *Picea mariana* at its tree line (Black and Bliss, 1980), there is no evidence that restricted seed germination and establishment are major restrictions of northern spread.

III. The Xylem

As temperatures decline into winter, there is an increasing chance that water in the trunk xylem will freeze. This is inevitable in the boreal climate. As this water freezes, dissolved gases may bubble out of the liquid, and serve as potential sites for cavitation (Pallardy *et al.*, 1994). This cavitation may develop until xylem elements are completely air and vapor filled, when the now embolized vessels can no longer function for conducting water (Pallardy *et al.*, 1994). The cost for species in these cold regions is that, unless the emolism can be repaired, new vessels will have to be constructed each spring. If all of the vessels are embolized, then in the short growing season of the boreal zone (Fig. 2), the xylem must be either completely reconstructed or the embolisms ejected before significant leaf area development can occur (Sheriff *et al.*, 1994).

It follows, therefore, that in habitats where the winter climate is sufficiently cold to induce embolisms, but sufficiently warm to allow the occurrence of ring-porous species, leaf burst will occur in a definite series from the earliest for conifers, through diffuse-porous to ring-porous species last, based on the expectation and proportion of xylem that will need to be reconstructed. Similarly, it also follows that the frequently long-term evergreen habit of the conifer needles (Chapter 8, this volume) is possible because there is never a complete loss of xylem conductivity at the end of the winter and into spring. At this time leaf physiology will develop but with rather little carbohydrate available for constructing new xylem (Gower *et al.*, 1994), but with an immediate requirement for a continuous water supply from the xylem. At the same time of early spring, the extensive mycorrhizal compliment of the trees will also be developing a strong and perhaps competing sink activity for carbohydrates (Teskey *et al.*, 1994; Luxmoore *et al.*, 1994), in return for a new supply of nutrients (Read, 1991).

Experimental evidence clearly indicates that the tracheidal xylem of the conifer is superior to the vessel xylem of the angiosperms in its high resistance to and recovery potential from freeze-induced embolisms (Borghetti *et al.*, 1991; Sperry and Sullivan, 1992). The greater resistance is conferred by the smaller diameter of the vessel tracheids but is probably not due to the presence of pit membranes (Pallardy *et al.*, 1994).

The large ring-porous vessels of the angiosperms are the most sensitive to embolism, and as a consequence the conductive sapwood to heartwood area of these species is much lower than in the resistant conifers (Pallardy *et al.*, 1994), and with a significant annual production of sapwood and also a significant end-of-year loss in conductivity. It is interesting to note that the angiosperm species found in the boreal biome have diffuse-porous xylem, with specific water conductivities much closer to coniferous species than to the ring-porous species such as in oak (Pallardy *et al.*, 1994), and presumably with a greater resistance to embolism than the ring-porous species.

Whereas conifer tracheids are more resistant to freeze-induced embolisms than species with either diffuse- or ring-porous xylem, the same property does not follow in terms of a sensitivity to cavitation and embolism through drought under normal growth temperatures (Pallardy *et al.*, 1994; Sperry and Sullivan, 1992). Experimental and observational evidence (Pallardy *et al.*, 1994) indicates that the xylem pressures required for cavitation and the correlated change in photosynthetic rate with change in xylem pressure (Teskey *et al.*, 1994) are similar for species with tracheid, diffuse-porous and ring-porous xylem.

This response has significant impacts on the growth potential of the coniferous species in particular. With the same water potential drop across the top and bottom of the xylem, the water flux rate will be much lower for the coniferous species, because the tracheids have the lowest hydraulic conductivities (Tyree and Sperry, 1989; Sperry and Sullivan, 1992); if they were any higher, cavitation would occur. Therefore the transpiration rate and the stomatal conductances of coniferous species will be lower than for species with ring- and diffuse-porous xylem (Pallardy *et al.*, 1994). As a result, the photosynthetic rate will also be lower in the conifers (Stenberg *et al.*, 1994; Teskey *et al.*, 1994).

The obvious conclusion here, therefore, is that the high capacity of the coniferous tracheids to endure very low freezing temperatures has the negative effect that the species will have low potential rates of gross primary productivity. This effect will not be noticeable in the boreal biome, because no other xylem structure has the competitive edge over tracheids in this extreme environment. However, the difference will become more obvious as the climate warms into temperate and subtropical zones.

IV. Growth

Evidence for the slow growth potential of coniferous species can be most clearly seen for the growth of young seedlings (Fig. 3) (Jarvis and Jarvis, 1964; Bond, 1989). For four species that have overlapping geo-

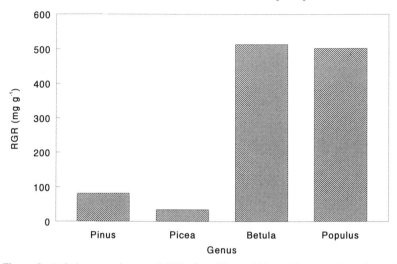

Figure 3 Relative growth rates (RGR) of seedlings of *Pinus*, *Picea*, *Betula*, and *Populus*. (From Jarvis and Jarvis, 1964.)

graphical distributions, the relative growth rates of the angiosperms are nearly an order of magnitude greater than those of the gymnosperms. It is not easy to recognize how the conifers could endure such powerful competition. As Bond (1989) clearly states, "Gymnosperms . . . may be excluded from environments where their seedlings encounter vigorous angiosperm competition." He goes on to suggest that gymnosperms will best persist in sites where the environment is poor for fast vegetative growth, or where the growing season is too brief. Therefore it is expected that the slow growth potential of seedling conifers will restrict their occurrence in sites with high nutrient and water status and high irradiance, sites where the high growth potential of the angiosperms will be realized.

However, the low growth potential of the conifers is not necessarily maintained as a life trait. Indeed the net primary productivity of coniferous forests may be as large or exceed some hardwood forests (Fig. 4) (Schulze, 1982). The reversal in conifer productivity from seedling to mature forest is most readily explained by the large leaf area indices observed for the coniferous forests (Fig. 5) (Schulze, 1982). The tracheidal xylem protects the sapwood from complete and irreversible embolism during winters with frost, therefore, as long as perennial needles develop a capacity for recovery after extreme winters (Chapter 5), then there may only be a relatively mild selection against survival of needles over many years. It is the accumulation, maintenance, and productivity of these long-lived leaves that maximize conifer productivity. This was

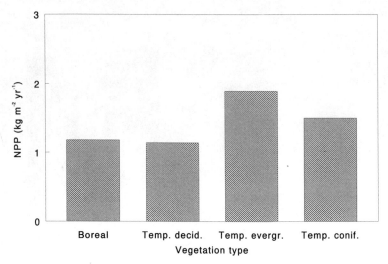

Figure 4 Net primary productivities (NPP) of four vegetation types: the boreal forest, temperate broad-leafed deciduous forest, temperate broad-leafed evergreen forest, and temperate evergreen coniferous forest. (From Schulze, 1982.)

clearly demonstrated for *Pinus radiata*, when, following the removal of 1- and 2-year-old leaves, growth was reduced by 51% (Rook and Whyte, 1976).

An important response for conifers to the environment is therefore to vary the period over which the canopy of leaves increases its leaf area

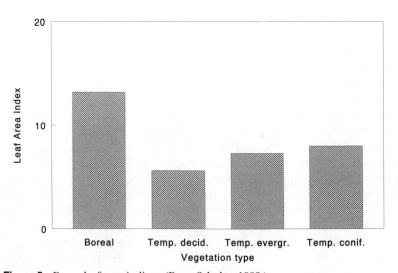

Figure 5 Forest leaf area indices. (From Schulze, 1982.)

index and the period over which leaves are retained on the plant. In the temperate zone, coniferous canopies may take 25 to 40 years to reach the maximum leaf area index, whereas the same response may be achieved within 5 years for a broadleaf deciduous forest (Bond, 1989). Across the full range of environments inhabited by conifers, leaf longevity can range from less than half a year (*Larix decidua*) to more than 40 years (*Pinus longaeva*); however there is significant overlap with broadleafed angiosperms in longevity (Chapter 8).

Increased proportions of lignin and decreased levels of leaf nitrogen have been found to be associated with increased longevity (Gower *et al.*, 1989; Reich *et al.*, 1992). Such leaves, when shed, will be very slow to decompose, and so it is expected that soil fertility will also decline as leaf longevity increases (Vitousek, 1982). The lower nitrogen levels are also associated with lower photosynthetic rates (Reich *et al.*, 1992). However, it appears that the decline in the maximum photosynthetic rate with longevity is not a *pro rata* decline because the multiple of leaf area index and canopy productivity can lead to increased forest productivity with leaf life-span (Chapter 8).

V. Beyond the Boreal Zone

The climatic limits of the boreal forest are most easily defined at the low temperature limits (Figs. 1 and 2) (Woodward, 1987). In contrast, it is not easy to recognize a regular high temperature limit and in particular it is not easy to recognize a high temperature limit that is associated with any specific ecophysiological processes (Woodward, 1988). The inevitable, but not generally demonstrable conclusion is that the high temperature limits of boreal conifers are a result of competition with more vigorous, thermophilous angiosperm species (Woodward, 1987). In some areas annual precipitation may be too low to support a forested vegetation; however, this limit applies equally to both coniferous and hardwood forests (Woodward, 1992).

Investigating the distribution of one genus, such as *Pinus*, clearly indicates that conifers are globally widespread (Fig. 6) (Critchfield and Little, 1966; Richardson and Bond, 1991). The species from low latitudes are not boreal species and presumably have appropriate ecophysiological responses to the local climate. However, the xylem limitation described earlier, through drought-induced cavitation, is likely to be more severe in warmer climates, and at first appearance it seems increasingly unlikely that conifers will occur with any great frequency because they will be out-competed, particularly during the early phase of seed regeneration by more vigorous angiosperm species (Fig. 3). Even regen-

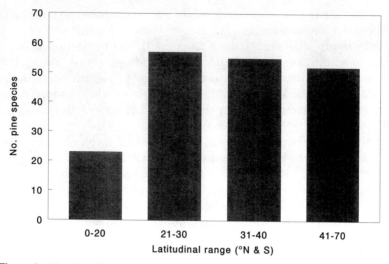

Figure 6 Number of pine species occurring within different latitudinal bands. (From Critchfield and Little, 1966; Richardson and Bond, 1991.)

eration of the very productive and drought-tolerant *P. radiata* is strongly influenced by grasses, particularly in semiarid regions such as in Australia (Squire, 1977; Sands and Nambiar, 1984). Grasses, in particular appear to out-compete conifers from habitats that in other respects are adequate for conifer survival (Sims and Mueller-Dombois, 1968; Richardson and Bond, 1991).

In their study of pine invasions, Richardson and Bond (1991) demonstrate that pine invades at all latitudes of the world, after disturbance (Fig. 7) (Richardson and Bond, 1991), and fire is the most universal disturbance to be associated with this invasion. This is no surprise because fire is a central factor in the ecology of pines and other conifers. For some species with serotinous cones, fire is a necessity both for releasing seeds and for optimizing the soil status for seed germination (Mirov, 1967; Richardson and Bond, 1991). Given a regular occurrence of natural and human-induced fires, pines will then be expected to be maintained over a wide range of climates. However, not all species of conifers are fire tolerant. In the boreal forest of the former Soviet Union, only species of *Larix* and *Pinus* are fire tolerant, a feature aided by the accumulation of a thick bark (Nikolov and Helmisaari, 1992). In contrast, species of *Abies* and *Picea* are fire intolerant, with relatively thin bark. These species, perhaps as a result of their poor fire tolerance, occur in wet and boggy sites, where fires are unlikely (Nikolov and Helmisaari, 1992).

If conifer seedlings survive beyond the initial stages of establishment,

through, for example, the continued effects of a previous fire, or though low-nutrient soils, low temperatures, or different and continued disturbances, then the conifer sapling will move to a different stage in which its growth becomes ever closer to that of angiosperm saplings (Bond, 1989). As the sapling becomes larger it will cast shade and, at least for shade-tolerant conifers, this shade may prevent the establishment of other species of lower shade tolerance (Woods, 1984).

In addition to aboveground effects, the conifer can exert very strong influences belowground. Conifers are strongly and prolifically ectomycorrhizal (Read, 1991). The ectomycorrhizas are a strong sink for plant photosynthate (Read, 1991; Teskey et al., 1994), in return for a supply of nitrogen and phosphorus from the soil. Both of these elements may be in a form that is unavailable to nonmycorrhizal species (Read, 1991), a feature that will favor the competitive balance toward the conifer. The presence of an active mycorrhizal sink may also stimulate the photosynthetic rate of the conifer (Dosskey et al., 1990). In addition, the poor litter quality of coniferous litter, with high levels of lignin (Read, 1991), will cause the development of increasingly nutrient-poor soils, as the conifer persists. This soil, with a very low rate of mineralization (Gower et al., 1989), will increasingly drift to a state where most of the nitrogen is organic and some is in a form that only mycorrhizas can extract (Read,

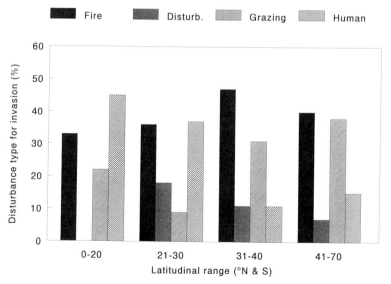

Figure 7 Types of disturbances favoring the invasion of pine, by latitudinal band. Disturbance types: fire, natural disturbance, animal grazing, and human-induced changes in the natural vegetation. (From Richardson and Bond, 1991.)

1991). At this stage the conifer, or angiosperm with a similar ectomycorrhizal association, dominates soil development.

Under continued CO_2 enrichment, mycorrhizal development is likely to be stimulated (O'Neill et al., 1987) through an increased rate of photosynthate supply to the mycorrhizas. It appears likely, therefore, that continued increases in atmospheric CO_2 may increasingly favor conifers through belowground effects. It remains to be seen, however, whether these belowground effects are more effective than aboveground stimulations of photosynthesis, a feature that will favor angiosperms (Woodward et al., 1991). However, observations of the growth of hardwood saplings on natural soil, in open-top chambers with CO_2 enrichment, have demonstrated that the aboveground stimulation of productivity by CO_2 fails to be sustained after as little as the first season of enrichment (Norby et al., 1992). This indicates a greater emphasis on belowground processes in determining future responses to CO_2 enrichment.

Although it might be possible that future CO_2 enrichment will favor conifers, particularly on nutrient-poor soils, it does not follow that any future global warming will have the same effect. The historical reconstruction of the history of red spruce, *Picea rubens*, in a mixed hardwood and softwood forest in the eastern United States, shows declining growth rates and increased mortality of the red spruce over the last 200 years (Hamburg and Cogbill, 1988). This effect is now so great that the formerly mixed forest is now a predominantly hardwood forest. The decline in the red spruce is correlated with a summer warming of about 2°C. Of course the decline in the red spruce may be due to other causes than warming. However, Hamburg and Cogbill (1988) suggest that in this area of forest, which has not been controlled by human activities, the sensitivity of the change in red spruce closely matches the predictions of stand models, based solely on climatic and edaphic responses, for the species in the forest (Davis and Botkin, 1985), adding weight to their premise that a small degree of warming favors the hardwood over the softwood species.

VI. Conclusions

A wide range of specialized mechanisms exist that conifers use for surviving the very coldest climates of the world. For winter survival in the boreal zone it is the capacity of the xylem that enables a species to avoid and recover from freeze-induced embolisms, a critical factor for both evergreen and deciduous species. In evergreen species, it is additionally important for the needles to be able to repair winter damage in the first part of the short growing season.

The mechanisms for enduring the impacts of frost do not generally translate to a capacity for conifers to endure extremely arid conditions, although the low xylem conductivities severely limit the effective rates of transpiration and photosynthesis, features that might favor survival in arid lands. A general characteristic that maximizes the potential for conifers to occur in warm climates but on very poor soils is the very extensive ectomycorrhizal association, which enables extraction of nutrients from sources unavailable to the roots of the host plants. The continued presence of conifers at a site will also slowly decrease the nutrient quality of the site, through the accumulation of high-lignin, low-nutrient, and slowly decomposing litter.

Seedling growth in conifers is generally much slower than in angiosperms, and in particular herbaceous species, which will be the general competitors during early regeneration. Strong competition may therefore limit the distribution of conifers, perhaps most strongly at this stage of the life cycle. The regeneration of the conifers may be favored, however, following extensive disturbance, particularly fire.

Although seedling conifers have slow growth rates, forest productivities are very similar, and sometimes higher than for hardwood species. This feature results from the capacity of coniferous forest canopies to reach very high leaf area indices, a feature that is in turn favored by the capacity to retain long-lived leaves.

References

Baumgartner, A. (1979). Climatic variability and forestry. *In* "Proceedings of the World Climate Conference," WMO-No. 537, pp. 581–607. World Meteorol. Organ., Geneva.

Black, R. A., and Bliss, L. C. (1980). Reproductive ecology of *Picea mariana* (Mill.) at the tree line near Inuvik, Northwest Territories, Canada. *Ecol. Monogr.* 50:331–354.

Bonan, G. B. (1992). Processes in boreal forests. *In* "A Systems Analysis of the Global Boreal Forest" (H. H. Shugart, R. Leemans, and G. B. Bonan, eds.), pp. 9–12. Cambridge Univ. Press, Cambridge, UK.

Bond, W. J. (1989). The tortoise and the hare: Ecology of angiosperm dominance and gymnosperm persistence. *Biol. J. Linn. Soc.* 36:227–249.

Borghetti, M., Edwards, W. R. N., Grace, J., Jarvis, P. G., and Raschi, A. (1991). The refilling of embolized xylem in *Pinus sylvestris* L. *Plant, Cell Environ.* 14:357–369.

Critchfield, W. B., and Little, E. L. (1966). Geographic distribution of the pines of the world. *Misc. Publ. U.S. Dep. Agric.* 991:1–97.

Davis, M. D., and Botkin, D. B. (1985). Sensitivity of the cool-temperate forests and their fossil pollen to rapid climatic change. *Quat. Res. (N.Y.)* 23:327–340.

Dosskey, M. G., Linderman, R. G., and Boersma, L. (1990). Carbon sink stimulation of photosynthesis in Douglas-fir seedlings by some ectomycorrhizas. *New Phytol.* 115: 269–274.

Flannigan, M. D. (1993). Environmental controls of red pine. (*Pinus resinosa* Ait.) distribution and abundance. Ph.D. Thesis, University of Cambridge, Cambridge, UK.

Gauslaa, Y. (1984). Heat resistance and energy budget in different Scandinavian plants. *Holarctic Ecol.* 7:1–78.

Gower, S. T., Grier, C. C., and Vogt, K. A. (1989). Aboveground production and N and P use by *Larix occidentalis* and *Pinus contorta* in the Washington Cascades, USA. *Tree Physiol.* 5:1–11.

Gower, S. T., Isebrands, J. G., and Sheriff, D. W. (1994). Carbon allocation and accumulation in conifers. *In* "Resource Physiology of Conifers: Acquisition, Allocation, and Utilization (W. K. Smith and T. M. Hinckley, eds.), pp. 217–254. Academic Press, San Diego.

Hadley, J. L., and Smith, W. K. (1990). Influence of leaf surface wax and leaf area to water content ratio on cuticular transpiration in western conifers, USA. *Can. J. For. Res.* 20: 1306–1311.

Hamburg, S. P., and Cogbill, C. V. (1988). Historical decline of red spruce populations and climatic warming. *Nature (London)* 331:428–431.

Jarvis, P. G., and Jarvis, M. S. (1964). Growth rates of woody plants. *Physiol. Plant.* 17: 654–666.

Larsen, J. A. (1980). "The Boreal Ecosystem." Academic Press, New York.

Luxmoore, R. J., Oren, R., Sheriff, D. W., and Thomas, R. B. (1994). Source–sink–storage relationships of conifers. *In* "Resource Physiology of Conifers: Acquisition, Allocation, and Utilization" (W. K. Smith and T. M. Hinckley, eds.), pp. 179–216. Academic Press, San Diego.

Mather, A. S. (1990). "Global Forest Resources." Belhaven Press, London.

Mirov, N. T. (1967). "The Genus Pinus." Ronald Press, New York.

Müller, M. J. (1982). "Selected Climatic Data for a Global Set of Standard Stations for Vegetation Science." Junk Publ., The Hague, The Netherlands.

Nikolov, N., and Helmisaari, H. (1992). Silvics of the circumpolar boreal forest tree species. *In* "A Systems Analysis of the Global Boreal Forest" (H. H. Shugart, R. Leemans, and G. B. Bonan, eds.), pp. 13–84. Cambridge Univ. Press, Cambridge, UK.

Norby, R. J., Gunderson, C. A., Wullschleger, S. D., O'Neill, E. G., and McCracken, M. K. (1992). Productivity and compensatory responses of yellow-poplar trees in elevated CO_2. *Nature (London)* 357:322–324.

O'Neill, E. G., Luxmoore, R. J., and Norby, R. J. (1987). Increases in the mycorrhizal colonization and seedling growth in *Pinus echinata* and *Quercus alba* in an enriched CO_2 atmosphere. *Can. J. For. Res.* 17:878–883.

Ottander, C., and Öquist, G. (1991). Recovery of photosynthesis in winter-stressed Scots pine. *Plant, Cell Environ.* 14:345–349.

Pallardy, S. G., Cermak, J., Ewers, F. W., Kaufmann, M. R., Parker, W. C., and Sperry, J. S. (1994). Water transport dynamics in trees and stands. *In* "Resource Physiology of Conifers: Acquisition, Allocation, and Utilization" (W. K. Smith and T. M. Hinckley, eds.), pp. 301–389. Academic Press, San Diego.

Payette, S. (1992). Fire as a controlling process in the North American boreal forest. *In* "A Systems Analysis of the Global Boreal Forest" (H. H. Shugart, R. Leemans, and G. B. Bonan, eds.), pp. 144–169. Cambridge Univ. Press, Cambridge, UK.

Read, D. J. (1991). Mycorrhizas in ecosystems. *Experientia* 47:376–391.

Reich, P. B., Walters, M. B., and Ellsworth, D. S. (1992). Leaf lifespan in relation to leaf, plant, and stand characteristics among diverse ecosystems. *Ecol. Monogr.* 62:365–392.

Richardson, D. M., and Bond, W. J. (1991). Determinants of plant distribution: Evidence from pine invasions. *Am. Nat.* 137:639–668.

Ritchie, J. C. (1987). "Postglacial Vegetation of Canada." Cambridge Univ. Press, Cambridge, UK.

Rook, D. A., and Whyte, A. G. D. (1976). Partial defoliation and growth of 5-year-old Radiata pine. *N.Z. J. For. Sci.* 6:40–56.

Sakai, A., and Larcher, W. (1987). "Frost Survival of Plants. Responses and Adaptation to Freezing Stress," *Ecol. Stud.*, Vol. 62. Springer-Verlag, Berlin.

Sands, R., and Nambiar, E. K. S. (1984). Water relations of *Pinus radiata* in competition with weeds. *Can. J. For. Res.* 14:233–237.

Schulze, E. D. (1982). Plant life forms and their carbon, water and nutrient relations. *Encycl. Plant Physiol., New Ser.* 12B:615–676.

Senser, M., and Beck, E. (1982). Frost resistance in spruce (*Picea abies* (L.) Karst.). V. Influence of photoperiod and temperature on the membrane lipids of the needles. *Z. Pflanzenphysiol.*, 108: 71–85.

Sheriff, D. W., Margolis, H. A., Kaufmann, M. R., and Reich, P. B. (1994). Resource use efficiency. *In* "Resource Physiology of Conifers: Acquisition, Allocation, and Utilization" (W. K. Smith and T. M. Hinckley, eds.), pp. 143–178. Academic Press, San Diego.

Sims, H. P., and Mueller-Dombois, D. (1968). Effect of grass competition and depth to water table on height growth of coniferous tree seedlings. *Ecology* 49: 597–603.

Sperry, J. S., and Sullivan, J. E. M. (1992). Xylem embolism in response to freeze-thaw cycles and water stress in ring-porous, diffuse-porous, and conifer species. *Plant Physiol.* 100:605–613.

Squire, R. O. (1977). Interacting effects of grass competition, fertilizing and cultivation on the early growth of *Pinus radiata* D. Don. *Aust. J. For. Res.* 7: 247–252.

Stenberg, P., DeLucia, E. H., Schoettle, A. W., and Smolander, H. (1994). Photosynthetic light capture and processing from the cell to canopy. *In* "Resource Physiology of Conifers: Acquisition, Allocation, and Utilization" (W. K. Smith and T. M. Hinckley, eds.), pp. 1–38. Academic Press, San Diego.

Surano, K. A., Daley, P. F., Houpis, J. L. J., Shinn, J. H., Helms, J. A., Palasson, R. J., and Costello, M. P. (1986). Growth and physiological responses of *Pinus ponderosa* Dougl. ex. P. Laws to long-term elevated CO_2 concentration. *Tree Physiol.* 2:243–259.

Takhtajan, A. (1986). "Floristic Regions of the World." Univ. of California Press, Berkeley.

Teskey, R. O., Sheriff, D. W., Hollinger, D. Y., and Thomas, R. B. (1994). External and internal factors regulating photosynthesis. *In* "Resource Physiology of Conifers: Acquisition, Allocation, and Utilization" (W. K. Smith and T. M. Hinckley, eds.), pp. 105–140. Academic Press, San Diego.

Tranquillini, W. (1979). "Physiological Ecology of the Alpine Timberline," *Ecol. Stud.*, Vol. 31. Springer-Verlag, Berlin.

Tyree, M. T., and Sperry, J. S. (1989). Vulnerability of xylem to cavitation and embolism. *Ann. Rev. of Plant Physiol. and Mol. Biol.* 40:19–38.

Van Cleve, K., Barney, R., and Schlentner, R. (1981). Evidence of temperature control of production and nutrient cycling in two interior Alaska black spruce ecosystems. *Can. J. For. Res.* 11:258–273.

Viereck, L. A. (1983). The effects of fire in black spruce ecosystems of Alaska and northern Canada. *In* "The Role of Fire in Northern Circumpolar Ecosystems" (R. W. Wein and D. A. MacLean, eds.), pp. 201–220. Wiley, New York.

Vitousek, P. M. (1982). Nutrient cycling and nutrient use efficiency. *Am. Nat.* 119: 553–572.

Walter, H. (1979). "Vegetation of the Earth and Ecological Systems of the Geobiosphere," 2nd ed. Springer-Verlag, New York.

Woods, K. D. (1984). Patterns of tree replacement: Canopy effects on understory pattern in hemlock-northern hardwood forests. *Vegetatio* 56:87–107.

Woodward, F. I. (1987). "Climate and Plant Distribution." Cambridge Univ. Press, Cambridge, UK.

Woodward, F. I. (1988). Temperature and the distribution of plant species. *Symp. Soc. Exp. Biol.* 42:59–75.

Woodward, F. I. (1992). A review of the effects of climate on vegetation: Ranges, competition, and composition. *In* "Global Warming and Biological Diversity" (R. L. Peters and T. E. Lovejoy, eds.), pp. 105–123. Yale Univ. Press, New Haven, CT.

Woodward, F. I., Thompson, G. B., and McKee, I. F. (1991). The effects of elevated concentrations of carbon dioxide on individual plants, populations, communities and ecosystems. *Ann. Bot. (London)* 67 (Suppl. 1):23–38.

5

Physiological Processes during Winter Dormancy and Their Ecological Significance

Wilhelm M. Havranek and Walter Tranquillini

I. Introduction

Coniferous forests are found predominantly in the boreal and temperate zones or at higher elevations in more southern latitudes, where strong seasonal changes in climate occur. In the annual rhythmicity of trees, summer growth phases alternate with periods of winter dormancy. In autumn, growth may come to an abrupt end with the onset of subfreezing temperatures. Before full winter dormancy is attained, the development of a fundamental winter hardness is required for winter survival. This winter hardness must be maintained throughout winter, despite the potential occurrence of relatively warm periods, and must persist until a prolonged dehardening period begins in spring. Full winter dormancy is characterized by the formation of resting buds, the suspension of growth processes (elongation and divisional growth), a reduction in metabolic activity, the enhancement of frost and desiccation resistance, and changes in cellular and cytoplasmic structures. These physiological and structural adaptations enable evergreen conifer trees to survive the cold and desiccation of winter.

A substantial transition period into winter dormancy is a prerequisite for acquiring sufficient winter hardness. This transition is characterized by a number of physiological and structural changes from the cell to the whole-plant level. Three stages of winter dormancy can be distinguished: pre-dormancy, true dormancy, and post-dormancy. The onset of dormancy does not occur in all organs simultaneously or with the same intensity. For instance, vegetative buds in boreal conifers are already dormant at the end of summer and cannot be induced to flush, whereas other tissues are still actively growing (e.g., roots, cambium). Even meristematic activity in the buds, leading to the formation of primordia, may occur until late autumn. Concomitantly, frost resistance may increase substantially (Aronsson *et al.*, 1976; Unterholzner, 1979) and high photosynthetic rates at low temperatures result in an accumulation of reserve carbohydrates. These seasonal changes in various organs and their physiological functions may also involve endogenous rhythmicity via a genetically based "physiological clock" that is controlled precisely by temperature and day length.

A strong "endogenous" or "autonomic" rhythmicity becomes apparent during dormancy, when buds are unable to flush under favorable conditions and cambium has lost its potential to grow (Mellerowicz *et al.*, 1992). During true dormancy warmer temperatures and a lengthening of the photoperiod may not individually alleviate dormancy. This phase may end only after a specific requirement of chilling and/or the occurrence of short days, especially in more northern conifers. Growth potential may be restored only after this combination of chilling and day length is experienced. During post-dormancy physiological activity is gradually restored, under the influence of phytohormones, in response to increased photoperiod and/or warmer temperatures (Lavender and Silim, 1987). The renewal of growth is then inhibited only by unfavorable weather conditions, a growth phase often referred to as imposed dormancy.

Survival during winter and recovery from winter dormancy may be the most important characteristics of conifer tree ecophysiology. In this chapter we focus on the most important physiological processes in conifers during winter dormancy (i.e., tolerance to frost and desiccation, water relations, and carbon metabolism) and discuss the regulatory influences of environmental and endogenous factors. These factors are fundamental to understanding conifer forest ecology and may be influenced strongly by anthropogenic effects.

II. The Coniferous Forest Zone

A. Distribution and Climate

The boreal coniferous forest in Eurasia extends from Scandinavia across northern Russia and Siberia to Kamchatka, and to the other side of the Bering Strait. In North America, conifer forests dominate from Alaska across Canada to Labrador, Newfoundland. The northern border of the boreal forest is characterized by a transitional zone, the "forest tundra," in which the forest gradually blends into the treeless tundra in a mosaic-like pattern. At the southern border, the boreal forest is replaced by deciduous forest in maritime climate and by a mixed boreal forest and steppe in more interior continental areas. Within the deciduous forest zone, conifer trees dominate on mountains from the upper montane zone to the highest portion of the subalpine zone where the upper timberline ecotone occurs. With increasing altitude, low temperatures and short growth periods probably limit the range of deciduous, broad-leafed tree species, as in the boreal zone. According to Walter (1968), the boreal zone begins where the climate becomes too unfavorable for broad-leafed trees, that is, with a vegetational growth period of less than 120 days (days with mean temperatures $>10°C$), cold periods longer than 6 mo, with monthly mean minimum temperatures $<0°C$. Although the northern boundary of the boreal zone is defined as having a growth period of 30 days and a dormancy period of over 8 mo, the climate within this zone is not uniform, but ranges from a cold oceanic climate with fairly constant temperatures to a cold continental climate with temperatures ranging from $+40$ to $-60°C$. The climate is also humid throughout the boreal zone, that is, precipitation (with summer maximum) exceeds potential evaporation, which leads to formation of peat bogs and limits the growth of trees.

Table I shows climatic data from several stations across the American and Eurasian boreal zones. Absolute temperature minima are generally lower than $-40°C$ but have been recorded as low as $-67°C$ in eastern Siberia Oimekon, the "frost pole" of the Northern Hemisphere. With mean annual temperatures $<-10°C$, the soil is typically frozen to depths of 250–400 cm in winter. During the relatively warm summers, at least the upper 10–50 cm of the permafrost thaws, although 100–150 cm depths may thaw in well-drained soils. This depth of soil thaw is enough to allow the growth of *Larix gmelinii* (syn. *L. dahurica*) forests, which in the "light taiga" of eastern Siberia covers about 2.5 million km^2. The dominant tree species of the "dark taiga" in western Siberia are *Pinus sibirica*, *Abies sibirica*, and *Picea obovata*. In North America, a similar but slightly less extreme cold continental climate occurs only in the area around Fort Vermillion and Fort Yukon (Alaska). The warming effect of the oceanic influence can be seen in the data from Sweden and Finland, and from Klutschevskoje, Petropavlowsk on Kamchatka (Table I).

The polar tree line in more maritime areas is formed by *Betula tortuosa* in northwestern Europe and by *Betula ermanii* and *Pinus pumila* in the Far East. In Eurasia, *Picea abies* and *P. obovata* extend farthest north. In the more continental East, both polar and mountain forest timberlines are dominated by *Larix sibirica* and *L. gmelinii* (Gorchakovsky and Shiyatov,

Table I Data of Climate Stations within the Boreal Zone[a]

Station	Annual mean temperature (°C)	Annual sum precipitation (mm)	>10°C[b] (days)	<0°C[c] (months)	T_{max} (°C)[d] Mean	T_{max} (°C)[d] Absolute	T_{min} (°C)[e] Mean	T_{min} (°C)[e] Absolute
North America/Canada[f]								
Sept-Iles	+1.1	1124	95	7	19.8	32.2	−19.3	−43.3
Wabush Lake	−3.8	894	77	8	18.5	31.1	−27.9	−47.8
Cameron Falls	+1.7	793	119	6	23.6	38.9	−22.3	−46.1
Big Trout Lake	−3.0	580	92	8	20.9	35.6	−29.5	−47.8
Fort Vermillion	−1.2	382	114	7	22.9	39.4	−28.5	−61.1
Norman Wells	−6.4	328	92	8	22.0	33.9	−32.9	−54.4
Fort Yukon	−6.7	172	89	8	—	—	−33.0	−57.2
Fairbanks	−3.4	287	94	7	22.2	33.9	−28.9	−54.4
Europe/northern Asia[g]								
Östersund (S)	+2.4	496	84	5	—	—	−14.0	−41.0
Oulu (SF)	+1.8	545	87	5	—	—	—	−37.9
Archangelsk	+0.4	466	94	7	—	—	−17.4	−44.8
Tomsk, Ob	−0.8	478	108	—	—	—	—	−51.1
Irkutsk	−1.3	369	105	—	—	—	—	−50.0
Nikolajewsk	−2.4	447	100	—	—	—	—	−46.1
Yakutsk, eastern Siberia	−10.4	187	96	—	—	—	—	−57.0
Oimekon, eastern Siberia	−16.3	131	78	—	—	—	—	−67.0
Klutschevskoje	−1.6	459	86	—	—	—	—	−48.3
Petropavlowsk	+0.2	767	68	—	—	—	—	−33.9

[a] After Walter (1968) and Loris (1991).
[b] Number of days with a mean temperature of >10°C.
[c] Number of months with a mean temperature minimum of <0°C.
[d] Temperature maximum, mean of warmest month.
[e] Temperature minimum, mean of coldest month.
[f] From eastern Canada to Alaska.
[g] From Scandinavia via Siberia to Kamchatka.

1978), whereas *Picea mariana* and *Picea glauca* dominate the polar timberline in North America. More detailed information about the distribution and silvics of circumpolar boreal forest tree species is given by Nikolov and Helmisaari (1992). Low temperature minimums during winter do not appear to limit the survival of the highly frost resistant trees of this zone; rather, this seems to result from the short duration of the growth season. Norway spruce in Central Europe need a period of at least 50 days without frost lower than $-3°C$ to avoid frost damage of the new flush, and another 40 days to develop adequate needle resistance to winter desiccation (Tranquillini, 1987).

Calculation of the length of the vegetative growth period is often an approximation based on the number of days with mean air temperatures $>10°C$ at a particular climate station. More precise information about the actual duration of summer growth periods can be obtained from plant phenological and physiological measurements. Such measurements at the Central Alpine timberline (2100 m elevation; Ötztal, Tirol) showed that the photosynthetic and vegetative period of *Larix decidua* averaged over 6 yr was about 128 days (Friedel, 1967). For comparison, the daily mean air temperature was $>10°C$ on 45 days and $>5°C$ on 125 days, the latter corresponding closely to the observed vegetation period. Conditions at the timberline near Innsbruck (1950 m elevation) appeared to be similar: 46 days $>10°C$ and 124 days $>5°C$. At this location the snow-free period lasted 169 days, and the period when warmer temperatures were uninterrupted (consecutive days $>-4°C$), averaged over 10 yr, was 163 days (Havranek, 1987). Both periods can be considered estimations for the time available for the net photosynthetic gain of evergreens, which is approximately 40 days more than the time available for the deciduous larch. After growth seasons with less than 130–140 days, winter damage occurred that may have been linked to insufficient needle maturation.

B. Temperature Relations in Winter

Low temperature is most often considered the main factor responsible for the duration and intensity of winter dormancy. Climatologically, winter is often defined as the number of days with air temperatures lower than $0°C$. However, meteorological frost is not always relevant for predicting influences on the physiological processes of plants. For example, water uptake from the soil and xylem water transport become impossible below $-1°C$, although the needle-like leaves of conifers appear not to freeze until approximately $-4°C$. Thus, needles do not freeze on every meteorological frost day. Moreover, strong radiation under windless conditions may raise needle temperature considerably above air temperature, and long-wave radiation in clear nights may decrease needle temperatures well below air temperature (Tranquillini and Turner, 1961; Hadley and Smith, 1987; Gross, 1989; Jordan and Smith, 1994b).

Soil surface temperatures in a closed forest can be several degrees lower than in the open (Aulitzky, 1961a; Hadley and Smith, 1987; Bonan, 1992) because of greater shading and a later melt-out of winter snow. The low heat conductivity of snow inhibits the penetration of subfreezing air temperatures into the soil, and early snowfall (before the soil is frozen) may result in unfrozen soils throughout winter. If the snow cover is missing or thin, soil water may freeze to a depth of 1 m, as was observed at the alpine timberline where soil frost lasted nearly 5 mo (Aulitzky, 1961b). Near Fairbanks, Alaska, soil temperature at a depth of 10 cm under a stand of *P. glauca* remained below $-1°C$ for approximately 4–5 mo from December to April (Viereck, 1970). Very little information is available concerning temperature effects on the ecophysiology of conifer tree species from most of the Asian and American portions of the boreal zone. Because of the similarity of the winter temperature conditions, some results of the more intensive research in the subalpine zone (e.g., Tranquillini, 1979; Lassoie *et al.*, 1985; Smith and Knapp, 1990, reviews) may be extended cautiously to the boreal region.

III. Frost Resistance

A. Biochemical and Structural Changes during Cold Acclimation

Intracellular freezing may cause ice crystallization in the cell, which can fracture the fine structures of the cytoplasm and organelle membranes, generating a potentially lethal condition. Thus, water in the cells at subzero temperatures must be capable of supercooling or must be segregated to extracellular spaces, where it can be deposited safely as ice. Alterations in lipid composition may act to stabilize membrane functions at low temperatures which, in conjunction with membrane augmentation, enables more rapid water translocation during rapid freezing. These factors also protect membranes and the cytoplasm from destructive freeze dehydration (DeYoe and Brown, 1979; Senser and Beck, 1982; Yoshida, 1984).

In a first step of frost hardening, intensive changes at the cellular level can occur in virtually all metabolic pathways and may include alterations in sugars and related compounds, amino acids, nucleic acids, proteins, lipids, abscisic acid, and cytological structures (Sakai and Larcher, 1987). Full frost hardening is achieved most effectively at subfreezing temperatures (-5 to $-10°C$) when many large invaginations of the chloroplast envelope occur, as well as an increase in the number of chloroplasts and mitochondria (Senser and Beck, 1984), the migration and conglomeration of chloroplasts in the cell, and the dissolution of a single large vacuole into numerous small ones (Holzer, 1958). In conjunction with frost hardiness, tolerance to desiccation also increases (Pisek and Larcher, 1954).

B. Frost Resistance of Organs Limiting Species Distribution

Cold acclimation may not advance in all organs of the conifer tree at the same time and with the same intensity. Needles and cortex are the first to acclimate in most northern conifers, often achieving a frost tolerance of $-70°C$ (or less) by "extracellular equilibrium freezing" (Sakai, 1983). This process has been defined by Olien (1981) as the transition from water to ice outside the protoplast at small displacements of temperature from the balanced state associated with low crystallization energy. However, the maximum frost resistance of shoot and flower primordia was found to be less (-40 to $-60°C$), except in species from the cold-continental boreal zone. For example, *P. glauca*, *P. mariana*, *P. obovata*, *Pinus sylvestris*, *Pinus banksiana*, *L. sibirica*, and *L. gmelinii* survived intensive freeze dehydration at $-60°C$ or below when cooled very slowly. Leaf primordia in Pinaceae also had extracellular equilibrium freezing, whereas "extra-organ freezing" in *Larix*, *Picea*, and *Abies* occurred via a containment of ice crystals beneath the "crown," a special tissue that allows dehydration of the whole primordial shoot but segregates the bud from the xylem during dormancy (Sakai, 1983).

In the xylem ray parenchyma of some conifers, deep supercooling occurred at a low temperature exotherm of approximately $-40°C$, indicating sudden freezing and, hence, cell death at this temperature (Becwar *et al.*, 1981). For many hardwoods as well as conifers of the subalpine zone and lower (e.g., *Abies lasiocarpa*, *Abies concolor*, *Picea engelmannii*, and *Pseudotsuga menziesii*), the winter isotherm of $-40°C$ is considered the limit of their latitudinal and altitudinal distribution (Becwar *et al.*, 1981). However, no stem low temperature exotherm was found in species of the continental boreal zone at minimum temperatures lower than $-60°C$ (George *et al.*, 1974).

Roots generally harden later in fall and become less hardened (-5 to $-20°C$) than aboveground parts (Sakai and Larcher, 1987). Coleman *et al.* (1992) found no relationship between cold hardiness of roots, in four subalpine tree species, and their native distribution in either maritime (*Abies amabilis*, *Tsuga mertensiana*) or continental climates (*A. lasiocarpa*, *Pinus contorta*). Root cold hardiness was equal in *A. amabilis* and *A. lasiocarpa* ($-11.5°C$) and was least in *P. contorta* ($-7.5°C$). However, in the boreal environment of St. Petersburg, root exposure of old trees, by removing the surrounding soil, showed great differences in the frost resistance of the root cambium. Exotic tree species (e.g., *Quercus rubra*) were less frost resistant (-3 to $-8°C$) than the indigenous broad-leafed species *Q. robur* and *Tilia cordata* (-7 to $-16°C$). The roots of boreal conifer species (*P. abies, L. sibirica, P. sylvestris*) and of birch (*Betula pendula*) were most resistant and showed the first damage only between -13 and $-29°C$. In the latter species, carbohydrates in the roots increased more quickly in autumn and remained at higher levels during the cold months than those in oak and linden (Korotaev, 1994).

In *P. abies*, marked increases in raffinose produced by the plant and trehalose produced by the mycorrhizal root fungus were observed in winter and correlated positively with tolerance to frost and desiccation (Niederer *et al.*, 1992). Also, Moser (1958) reported differences in the freezing tolerance of isolated mycorrhizal fungi of conifer trees, and in the growth of strains of the same fungus from mountain or valley habitats at near $0°C$.

C. Endogenous and Environmental Control of Frost Hardiness

Freezing tolerance in northern conifers appears strongly dependent on dormancy, the only effective survival mechanism in areas where frost is severe and seasonally prolonged (Glerum, 1973; Sakai and Larcher, 1987). The gradual changes during cold acclimation that occur at the cellular level seem to be partly due to a genetically based endogenous rhythm (Siminovitch, 1982; Mellerowicz *et al.*, 1992). Once changes have been induced through a decrease in photoperiod, the development of a relatively high frost resistance ($-30°C$) can be observed even at temperatures between 10 and $20°C$ (Schwarz, 1968; Kandler *et al.*, 1979), and without frost in the field (Repo, 1992). However, low but above-freezing temperatures alone often appear to be sufficient for the development of full winter hardness (Weiser, 1970). Also, different effects of short days and subzero temperatures on winter hardening were reported for cell membrane alterations, one of the key processes in frost hardening (Senser and Beck, 1982). In spruce needles, short days led to an augmentation of membrane lipids, particularly phospholipids, whereas exposure to subfreezing temperatures increased the degree of unsaturation of the fatty acid component of the lipids. Under natural conditions, both processes could take place concomitantly or in close succession (Senser and Beck, 1982). Perhaps the maintenance of a basic frost resistance throughout winter is primarily due to changes in endogenous rhythm. Further hardening and the fluctuation of actual hardiness during winter is modified by temperature and by the degree of dehydration of membranes and cytoplasm according to acclimational status.

The important influence of the endogenous rhythm becomes evident when the effects of developmental stage on frost hardening are compared (Sarvas, 1974; Fuchigami et al., 1982; Mellerowicz et al., 1992). In early winter, subzero temperatures created rapid hardening beyond that required in the field, whereas high temperatures caused only little dehardening (Pisek and Schiessl, 1947). The opposite trend occurred in later winter. Note that large daily fluctuations in needle temperature during midwinter were also associated with an increase in frost resistance (Gross, 1989). Moreover, needles in the shade or under snow were found to be less frost resistant.

D. Occurrence of Frost Damage in the Field

In general, native conifer trees possess the genetic potential to develop adequate frost resistance to survive most frost events expected in their natural environment during their lifetime. Frequent and severe frost damage can be expected only at the margin of a species' distribution (e.g., timberline), or if the regional climate changes (Kullman, 1989). Tree species with a wide latitudinal or altitudinal distribution can have ecotypes with genetically based differences in the timing (rhythmoecotypes) and degree of frost hardiness (Sakai and Weiser, 1973; Maronek and Flint, 1974; Rehfeldt, 1988, 1989). In forestry practice, the use of provenances whose photoperiodic responses are not matched to the environment can result in substantial damage, primarily due to a premature loss of hardiness in the spring or delayed cold hardiness in the autumn. Damage due to late frost occurs mainly on warm slopes or in lower elevation valleys in spring, but is rare near the upper timberline where frost episodes may persist into the summer and return in early autumn (Jordan and Smith, 1994a).

Measurements over four winters revealed that the difference between frost resistance (leaf temperatures at which 50% mortality occurred; TL_{50}) and the corresponding minimum temperature of needles in the field varied between 12°C and 40°C (Gross, 1989). However, initial needle damage (TL_i) occurred at temperatures up to 12°C higher than TL_{50} in the same study, indicating that needles of some individuals were approaching permanent damage. Similar differences between the curves of frost resistance and minimum air temperature were found in *Picea rubens* (Sheppard et al., 1989). In this case, TL_{50} (measured over one winter) was always less than the lowest daily minimum air temperatures that had occurred over the past 22 yr. However, comparing air temperatures instead of needle temperatures with frost resistance may result in a considerable underestimation of possible injury. Spruce needle temperatures up to 11°C lower than air temperatures were measured on clear, windless winter nights (Gross, 1989). In both studies, the measured variation in frost hardiness among trees of about 10°C was surprisingly high, probably because of both intraspecific genetic variability and acclimation according to twig exposure. Although frost resistance is persistent in midwinter, extreme climatic anomalies such as prolonged mild temperatures in December and January followed by a rapid return to colder temperatures caused substantial bud damage (Van Der Kamp and Worrall, 1990). Under these conditions, chilling requirements may be fulfilled too early for vegetative buds, which could result in premature loss of dormancy and dehardening (Sarvas, 1974).

Under extreme winter conditions, an intensive reddening of trees at specific elevations can be observed. In Switzerland, such an event was associated with extremely rapid freeze–thaw cycles caused by vertically oscillating upper fog limits and a warm airstream immediately above the cold, foggy air mass (Turner, 1988). A more frequent similar symptom, the "red belt" that occurs on south- and west-facing slopes in the Canadian Rocky Mountains, has also been associated with warm winds above a cold air pool in the valley (MacHattie, 1963). Field conditions that correspond to the appearance of winter injuries are complex and distinguishing between causes (White and Weiser, 1964; Kincaid and Lyons, 1981; Strimbeck et al., 1991) or their succession (Hadley and Amundson, 1992) is difficult. Solar radiation leads to frequent warming and thawing of needles even during winter. At the subalpine timberline, monthly maximum temperatures of spruce needles were 11 to 25°C higher than air temperatures between February and May (Gross, 1989), and mean maximum temperatures of *Pinus cembra* needles were 4.3°C and 11°C above air temperatures in April and May, respectively (Tranquillini, 1957). Similar large increases above air temperature have been reported for "krummholz" mats in the timberline ecotone of the Rocky Mountains in the United States (Hadley and Smith, 1987). During changes from sun to shade, temperature drops of up to 13°C hr^{-1} were recorded within the range of -4 to $-8°C$ (Gross, 1989). Since freezing of conifer needles begins at about $-4°C$ in winter, and most of the freezable water is frozen at $-12°C$ (Pisek and Kemnitzer, 1968), the transition from water to ice occurs rapidly within a relatively small temperature span. In needles with water deficits, proportionally less water is turned into ice than in saturated needles (Tranquillini and Holzer, 1958), making water reabsorption easier during high cooling and thawing rates. However, experiments have confirmed that natural freezing and thawing is rapid enough to cause tissue damage (Gross et al., 1991), especially if it occurs in a critical temperature range (Levitt, 1980).

IV. Winter Water Relations

A. Adjustment of Osmotic Potential

During cold acclimation, low temperatures were associated with increases in osmotically active substances, especially soluble carbohydrates (Kandler *et al.*, 1979; Levitt, 1980; Sakai and Larcher, 1987). By midwinter the osmotic and the water potentials were frequently at their annual minima in many boreal conifers. Low osmotic potentials can result from an augmentation of osmotically active substances and/or from a passive concentration of the cell sap through desiccation. Ritchie and Shula (1984) inferred a true seasonal osmotic adjustment in Douglas fir shoots from the fact that it occurred at full turgor, when passive cell dehydration is excluded. Winter osmotic adjustments to lower potentials may also be achieved by decreasing saturated water contents without increasing osmotically active substances (Little 1970; Teskey *et al.*, 1984a,b), in conjunction with a reduction in symplastic volume (Ritchie and Shula, 1984; Gross and Koch, 1991). The last condition appears to be influenced by the occurrence of short days during cold acclimation (Levitt, 1980). In most cases, an increase in sugars and other alterations in the cytoplasm and vacuole seem to be associated with the higher tolerance to desiccation. This possibility is also suggested by the fact that osmotic adjustments and changes in frost resistance are much smaller in roots than in needles. During the transition to spring activity, osmotic potential and water content at saturation increases again until it reaches the annual maximum during flushing (Ritchie and Shula, 1984).

B. Decrease of Stomatal Conductance and Gas Exchange in the Fall

Decreasing air temperatures and light intensities in autumn also result in decreased gas exchange rates. During mild and clear days, however, photosynthetic rates similar to summer rates are still possible as long as needles do not freeze. At that time, trees are already dormant and most of the cellular changes that increase frost resistance to about $-30°C$ have been completed. According to the decreased osmotic potential, freezing of needles occurs at temperatures below -3 to $-5°C$. On clear autumn nights, such needle temperatures may often be approached even if air temperatures near $0°C$ are recorded (Jordan and Smith, 1994a). Night frost occurs early in autumn at high altitudes and causes a stepwise decrease of photosynthesis and stomatal conductance, depending on its frequency and intensity (Cartellieri, 1935; Tranquillini, 1957; Smith *et al.*, 1984; Smith, 1985). The impact of minimum air temperatures on seasonal gas exchange was demonstrated for six conifer species of the Central Rocky Mountains. Smith *et al.* (1984) found a significant linear relationship between the daily maximum leaf conductance and the mean of the minimum air temperatures of three preceding nights, conductance becoming zero at $-9°C$. Körner and Perterer (1988) found a similar relationship for the gas exchange of *P. abies* and *P. sylvestris* near Innsbruck, Austria, if the minimum air temperatures were higher than $-4°C$, which is the freezing point of the needles. These researchers also found a close relationship between stomatal conductance and photosynthesis from November to March. Apparently, needle freezing is not required to initiate the strong decline in leaf conductance and transpiration that is typical of cold-hardened conifers. Continuously low but above-freezing temperatures (without intermittent warm periods) have also been observed to decrease stomatal conductance and photosynthesis substantially (Christersson, 1972; Andersson, 1980; Öquist *et al.*, 1980; Teskey *et al.*, 1984b, Bahn, 1988; Strand and Öquist, 1988).

The specific physiological mechanisms that may be responsible for the observed decrease in stomatal conductance in response to near-freezing temperatures are unknown at this time. During the transition to winter dormancy, water potentials at predawn are often remarkably negative but increase during the day, the converse of their daily course in summer (Benecke and Havranek, 1980; Smith *et al.*, 1984). Soil temperatures during this period, although decreasing, are still high enough that serious interference with water uptake can be excluded as a limiting factor (Havranek, 1972; Day *et al.*, 1990). A more probable explanation may be that xylem freezing of thin twigs, needles, or branches generates a low water potential. Rapid declines in the xylem flow (Teskey *et al.*, 1983), as a result of freezing, could also influence the release of abscisic acid (ABA), a known influencer of stomatal behavior. ABA may be a dominant factor in the conversion of environmental signals into the changes in gene expression involved in frost hardening (Qamaruddin *et al.*, 1993, and references therein), and in the acceleration of the transition into true winter dormancy. ABA was shown to increase stomatal sensitivity, both by promoting transpiration in unstressed seedlings and by some contribution to the decrease in transpiration and photosynthesis in water-stressed seedlings (Blake *et al.*, 1990). Thus, the role of ABA in the common down-regulation of stomatal conductance and photosynthesis observed during winter hardening must be investigated further.

C. Water Uptake and Movement during Winter

In boreal forests and in the timberline ecotone of mountain forests, soil temperature is often still above freezing with the first snowcover of autumn. As a result, soil surface temperatures often remain near zero because of the high insulating qualities of snow that inhibit heat loss from the soil or heat transfer to the colder air. This insulation and maintenance of above-freezing soil temperatures may be particularly important to numerous biological processes. At a soil temperature near zero, hyphae of mycorrhizal fungi are able to grow (Moser, 1958), whereas fine root growth is inhibited (Turner and Streule, 1983; Coleman *et al.*, 1992). Water uptake from the soil, although impeded by low temperatures, is still possible at about $-1°C$ (Larcher, 1980). Nonetheless, xylem transport through the frozen stem can be restricted severely (Hadley and Smith, 1987). Also, stems in a closed forest may remain frozen longer than those of isolated trees; certain portions of a stem may be more sun exposed and, thus, experience greater thaw (Michaelis, 1934). Stem thawing may enable xylem water movement of variable quantities from roots to stems and leaves, while some water uptake from melting snow, via the bark of twigs, may also be possible. Water absorption by bark occurs in *P. abies* (Katz *et al.*, 1989); considerable reductions of water deficits in spruce needles because of cuticular water absorption has been reported by Stalfelt (1944). The capacity for cuticular water uptake has been reported to be of the same magnitude as that for cuticular transpiration (Härtel and Eisenzopf, 1953), a potentially important factor for winter water balance (see also Stone, 1963, for review).

The physical process of freezing and thawing can cause significant water movement in the hydraulic system of trees (Havis, 1971; Zimmermann, 1983). Under freezing conditions in the field, the xylem may freeze centripetally, proceeding from thin to thicker twigs and from distal to proximal ones. As initial ice formation begins in the xylem, the increase in volume generates pressure (Robson and Petty, 1987) that can displace water downward or into the heartwood. Because of mechanical stiffness, freezing of the xylem has almost no effect on diameter change, whereas water transport from bark cells into intercellular spaces can cause a distinct shrinking of the stem diameter (Loris, 1981).

When xylem water freezes, released gases may form bubbles that can cause cavitation and blockage of conducting pathways (Robson *et al.*, 1988). Embolisms induced by low water potential were found to block xylem flow both in summer and in winter (Sperry and Tyree, 1990). However, several freeze–thaw cycles did not increase embolisms in *A. lasiocarpa* (Sperry and Sullivan, 1992) and no embolisms occurred in several conifers at field sites influenced by the milder Atlantic climate in late winter and spring (Cochard, 1992). Comparing conifers with ring-porous and diffuse-porous trees, Sperry and Sullivan (1992) concluded that small-volume conduits are adaptive for minimizing embolism formation by freeze–thaw cycles, and that the dominance of conifers in cold climates is consistent with the presence of tracheids. Xylem conductance can be re-established following embolism by new growth of conduits, as in ring-porous trees, or by positive xylem sap pressure, as in diffuse-porous trees. The mechanisms that lead to the reversal of embolism in conifers are still not well understood.

D. Cuticular Transpiration and Winter Desiccation

Periods with continuous frost, typical of the alpine timberline and boreal environment, are characterized by some period of complete stomatal closure during winter. However, "cuticular transpiration" may still occur when stomata are fully closed, although transpiration is restricted not only by the cuticle, but also by epicuticular waxes and cutinized epidermal layers. Also, some stomata may close only partially and, thus, make a major contribution to cuticular transpiration. In general, cuticular transpiration in conifers is less than 1/100 of summer transpiration, but may be crucial for winter water balance when water uptake is severely inhibited by frozen soils and/or stems. Data from tree species at the alpine timberline in Europe (Larcher 1957,1963; for a review see Tranquillini, 1979,1982) and in North America (Lindsay, 1971; Hansen and Klikoff, 1972; Richards and Bliss, 1986; Herrick and Friedland, 1991) show clearly that needle desiccation increases during winter and reaches a maximum in spring. In the timberline ecotone above the closed forest, needle water content often falls below the lethal level (Hadley and Smith, 1986). Moreover, needle desiccation is greatly accelerated by cuticle abrasion due to blowing snow, a dramatic feature of needle desiccation and death in timberlines of the Rocky Mountains in the United States (Hadley and Smith, 1987,1990). The "krummholz" mats and "flagged" tree forms of this ecotone are the result of this abrasion and needle death along wind-exposed shoots. Another reason for the frequent occurrence of winter desiccation in this ecotone may be the much smaller water reserve in smaller stems of stunted trees, compared with forest trees (Larcher, 1963). Winter desiccation may also be enhanced by a short and/or cool summer growth period when maturation of epidermal protective tissue is insufficient. Excessively high cuticular transpiration may deplete water reserves early and lead to desiccation damage (Baig et al., 1974; Tranquillini, 1979; Vanhinsberg and Colombo, 1990).

Because of small water reserves, conifer seedlings may desiccate after a short time if not protected by snow (Rossa and Larsen, 1980). In contrast, needles of mature trees in low-elevation forests of central Europe rarely desiccate to harmful levels (Michael, 1967). In the boreal zone, where frost and frozen soil persist for several months (Bonan, 1992), adequate maturation of tissues and lower evaporative demand during winter seem to prevent damage by frost drought. However, in contrast to alpine conditions, frequent polar storms are thought to have a strong desiccating effect, especially at the polar treeline (Holtmeier, 1971; Walter and Breckle, 1986). Thus, wind-protected and snow-covered locations, due to topography and growth form, may be important for the survival of evergreen species in both alpine and polar regions (also see Smith and Knapp, 1990).

Whether winter desiccation is an important selective factor in the most continental areas of the boreal zone where *Larix* species dominate is unknown. Here, the deciduous growth form may be necessary because of the severe winter desiccation potential in evergreens. Investigations on *Larix lyallii* at a Rocky Mountain timberline revealed that winter desiccation damage in needles and buds of the sympatric conifers *A. lasiocarpa*, *P. engelmannii*, and *P. contorta* was considerably greater than in *Larix*, in which buds were protected by hydraulic segregation from the twig xylem (Richards and Bliss, 1986). This adaptation allows buds to remain dehydrated in a more frost-resistant state of dormancy, while xylem water potentials may fluctuate widely.

V. Carbon Metabolism in Winter

A. What Causes the Winter Depression of Photosynthesis?

Short days and near-freezing temperatures can initiate the development of frost resistance via major cellular alterations (Sensor and Beck, 1979), some of which have been mentioned already. For example, chloroplasts of mesophyll cells become more clumped during midwinter in *A. balsamea* (Chabot and Chabot, 1975). Also, microscopic examinations of *P. cembra* and *P. abies* needles reveals that, after fall frosts to $-5°C$, most of the chloroplasts move into light-protected corners of cells (*P. cembra*) or become concentrated around cell nuclei (*P. abies*; Holzer, 1958). Also, after repeated freezing and accompanying cell dehydration, the central vacuole is split into numerous tiny vacuoles and the cytoplasm turns opaque (Holzer, 1958). Winter chloroplasts of *P. abies* have reductions in thylakoid membrane system and chlorophyll content, and increases in chloroplast volume due to stromal swelling (Senser and Beck, 1979). Near-freezing temperatures during frost hardening of pines also has been reported to alter chlorophyll organization (Öquist and Strand, 1986).

At the whole-leaf level, the photosynthetic capacity of spruce and cembra pines is not substantially less during natural hardening, as long as needles remain above freezing (Pisek and Winkler, 1958). Artificial frost hardening has no significant effect on the quantum yield of CO_2 uptake under light-limiting conditions, although light-saturated photosynthesis is lower in frost-hardened than in non-hardened *P. sylvestris* needles (Öquist *et al.*, 1980). Freezing of the needles stops CO_2 uptake completely in *Abies alba*, although photosynthesis recovers to pre-freezing values if temperatures remain above the level of frost resistance. Recovery at 15°C is dependent on the intensity and duration of the frost exposure (Pisek and Kemnitzer, 1968). In the field, several mild night frosts in sequence or a single strong night frost can cause substantial decreases in photosynthetic capacity that are accompanied by stomatal closure. Chlorophyll fluorescence measurements show that this inhibition of photosynthesis at subfreezing temperatures occurs at both the stomatal and the chloroplast level.

In *P. sylvestris*, winter inhibition of photosynthesis appears to be caused by the combined effects of freezing temperatures and light. Frost causes inactivation of enzymes in the photosynthetic carbon reduction cycle, whereas high light levels cause photoinhibition of photosystem II (PS II; Strand and Öquist, 1985,1988). Photoinhibition is often viewed as a stress response that is secondary to temperature inhibition of the Calvin cycle. In winter-stressed pine, the reversible photoinhibition of PS II is considered to be a mechanism for a long-term down-regulation of PS II related to low or inhibited consumption of NADPH and ATP (Ottander and Öquist, 1991). The photoinhibited state enables excitation energy to be dissipated thermally and prevents photodestruction of thylakoid membranes. Thus, the interaction of low temperatures and photoinhibition of photosynthesis could be important to the success of evergreens in cold climates.

B. Variability of Winter Depression of Photosynthesis

In the milder maritime winters of southwestern England and Scotland (temperature rarely below 5°C), only a small winter depression in photosynthesis during bud dormancy has been reported in conifers (Fry and Phillips, 1977). In Scotland, where frost hardiness of Sitka spruce increases rapidly in late October and early November, photosynthetic depressions are also accompanied by sharp declines in the optimum temperatures for net photosynthesis (Neilson *et al.*, 1972). In a comparative study of the effects of winter thermal history on photosynthetic capacity, Douglas fir, silver fir, and Norway spruce had threshold temperature values for initial photosynthetic depression of -1 to $-2°C$ (Guehl, 1985). However, in the two species from milder winter climates, photosynthesis in winter was affected by thermal history only, whereas in the more frost-resistant continental Norway spruce, thermal history was superimposed by a seasonal depression (Guehl, 1985).

In regions with moderately cold winters (e.g., New Zealand timberline), winter depression of the same European timberline species was much shorter and less pronounced than at the alpine timberline (Benecke and Havranek, 1980). In central alpine valleys, gas exchange is suspended completely only for short frost periods. However, gas exchange is significant in warm zones along mountain slopes where winter photosynthetic rates of up to 50% of summer values have been recorded in *P. abies* and *P. sylvestris* (Körner and Perterer, 1988). In central Sweden at 150 m elevation, positive net photosynthesis has been shown in *P. sylvestris* on many favorable days in midwinter (Troeng and Linder, 1982). At the alpine timberline, *P. abies* needles photosynthesize and transpire at a moderate level in February after days of intensive overheating of needles in calm air and only minor night frost (G. Wieser, personal communication). However, severe frost followed by shorter favorable periods does not result in CO_2 uptake at timberline (Tranquillini and Machl-Ebner, 1971) or in many parts of the boreal zone (Tranquillini, 1957,1979; Pisek and Winkler, 1958; Schulze *et al.*, 1967; Ungerson and Scherdin, 1968; Schwarz, 1971). Similarly, in young *P. cembra*, *P. sylvestris*, and *P. abies* saplings transferred from the field to favorable laboratory conditions during winter, photosynthetic capacity does not recover to summer values (Pisek and Winkler, 1958; Zelawski and Kucharska, 1967; Schwarz, 1971).

Photosynthetic capacity also decreases during winter in pine seedlings kept in climate chambers from fall through winter, even under conditions of a long-day photoperiod and a constant temperature of 15°C (Bamberg *et al.*, 1967); similar results have been obtained for several other conifer species kept in a heated glasshouse throughout winter (Bourdeau, 1959). This decline in photosynthetic capacity under favorable, experimental conditions is assumed to be due to the influence of endogenous rhythms, although induction could have occurred before transfer from the field. In *P. cembra*, measurements of chlorophyll fluorescence show a sustained winter inhibition of PS II that is less intense and of shorter duration in trees growing at low elevation valley sites than in those growing at the timberline (Nagele, 1989). Under favorable growth conditions in the laboratory (water saturation, 25°/18°C, 16/8 hr light/dark), full recovery of the F_v/F_m ratio to summer values can be obtained within 1–2 days in October and November, 2 wk in December, about 1 wk in January, and 3–5 days after February. These differences are thought to reflect not only the cumulative stress experienced before recovery, but also the influence of seasonal endogenous status (Nagele, 1989).

Species differences in recovery potential during winter also have been demonstrated by Schwarz (1971). Rapid and full recovery of photosynthesis in less than 1 wk occurred for shoots of *P. menziesii* and *P. glauca* placed in favorable conditions throughout the winter, but not in *P. contorta* from the same site. Slower chlorophyll regeneration and tissue rehydration in *P. contorta* needles were considered possible contributing mechanisms.

The results presented here suggest that photosynthetic capacity during winter is curtailed severely by both freezing temperatures and endogenous status. Desiccation also appears to be coupled with low temperature effects, and rehydration may be one of the repair mechanisms required in addition to the more rapid return of favorable temperatures. Also, both stomatal and nonstomatal factors contribute to photosynthetic inhibition. In particular, structural and biochemical changes in chloroplasts occur, and are part of a seasonally developed resistance to permanent frost damage.

C. Spring Recovery of Gas Exchange

With diminishing frost occurrence and warmer days, frost resistance diminishes in spring and the reorganization of cell structures can be observed. Small vacuoles rejoin into a single central vacuole that no longer disintegrates on the occasional nights with severe frost (Holzer, 1958). Similarly, spring chloroplasts become enriched in thylakoid membranes and starch content (Senser and Beck, 1979). By February or March, full recovery of photosynthesis is achieved experimentally within approximately 48 hr in *P. sylvestris* under favorable laboratory conditions, whereas photochemical efficiency of PS II recovers fully within 24 hr (Ottander and Öquist, 1991). Field measurements reveal that the F_v/F_m ratios of both shaded and exposed (i.e., less and more photoinhibited) pine needles begin to increase in April, more than 1 mo before the recovery of photosynthesis (Ottander and Öquist, 1991). When tested in May, F_v/F_m rises to its maximum within a few days.

High temperatures are well known to have a strong activation potential during the spring phase of post-dormancy. In *P. contorta*, *P. sylvestris*, and *P. abies*, the F_v/F_m ratio is still low (close to 0.1) at the end of March, but recovers almost completely within 3 days when shoots are kept at 20°C (Lundmark *et al.*, 1988). At 5°C, F_v/F_m increases less than half as much within a full week. Under colder field conditions in northern Sweden, recovery in F_v/F_m ratio takes much longer—from the end of March until June—and shows a similar exponential recovery in all three species. Although high daytime temperatures promote recovery, strong solar radiation following night frosts also creates substantial photoinhibition (Lundmark *et al.*, 1988). Part of the slow recovery of F_v/F_m is attributed to the *de novo* synthesis of photosynthetic pigments and of photodamaged proteins of PS II. Pigment synthesis in Norway spruce also is reported to occur only at temperatures above $-2°C$ (Godnev and Hodasevic, 1965; cited by Andersen *et al.*, 1991).

Photoinhibition and the increase of photoprotective agents in winter (Grill and Pfeifhofer, 1985; Gillies and Vidaver, 1990; Anderson *et al.*, 1992; Demmig-Adams and Adams, 1992) appear, under certain circumstances, to be insufficient for protection from photo-oxidative damage.

Chlorophyll decomposition often takes place in needles exposed to the strongest solar radiation, both in the valley and at the timberline, especially in late winter (Tranquillini, 1957; Holzer, 1959; Benecke, 1972). The degree of needle yellowing varies from tree to tree according to the level of radiation exposure, even among needles on the same branch. Reflection from snow (albedo > 95%; Budyko, 1974) may also be an important factor involved in photoinhibition, especially for lower branches (Day *et al.*, 1990). Biologically effective UV radiation, which increases with altitude, may also increase due to reflection from snow (DeLucia *et al.*, 1992). At the timberline, photo-oxidative yellowing of needles delays the recovery of photosynthesis in sun-exposed needles in *P. cembra* for several weeks (Tranquillini, 1957); shade needles remain greener with higher photosynthetic capacity during winter than sun needles (Pisek and Winkler, 1958).

Needles of young *P. cembra* that first emerged from the snow cover in May after almost 5 mo of burial appear to have changed from the true winter dormancy to spring post-dormancy (Tranquillini, 1957). These needles have lower frost resistance, increased osmotic potential and water saturation, newly synthesized chlorophyll, and photosynthesis rates higher than those measured in December before permanent snow burial (Tranquillini, 1957). These differences in needles buried beneath snow for most of winter could be due to a host of factors, including endogenous differences, reduced photodamage, warmer and more constant temperatures, and less desiccation. The critical importance of snow burial for alpine timberline conifers (e.g., "krummholz" mats) has also been reported by Hadley and Smith (1990).

As mentioned earlier, both temperature and water status of needles may be decisive factors in the recovery process of photosynthesis. As long as the water supply from frozen soil is insufficient, no active reaction centers of PS II and only reduced activity in PS I are found in needles of *P. sylvestris* and *P. abies* near Moscow (Tsel'niker and Chetverikov, 1988). Some recovery of gas exchange in conifers is observed when soils are still frozen, but only when stem water can be utilized (Troeng and Linder, 1982). However, full recovery of gas exchange in spring appears to be strongly dependent on soil thawing and adequate water uptake capacity (Tranquillini, 1957; Havranek; 1972; Smith, 1985; Jurik *et al.*, 1988; Day *et al.*, 1990). Ultimate recovery from winter dormancy may also involve other rhizosphere factors such as mycorrhizal activity and root hormones (Doumas and Zaerr, 1988; Smith and Knapp, 1990).

D. Respiration and Carbon Balance during Winter

With the transition to full winter dormancy, total respiration may be reduced to the level of maintenance respiration. For example, dark respiration of spruce and pine twigs at a given temperature is substantially lower in winter than in summer (Pisek and Winkler, 1958). The data collected by these researchers have also demonstrated that spruce twigs from the timberline have higher respiration rates at the same temperatures than those from the low-elevation valley sites, especially at the end of winter dormancy and during the summer growth period. This feature could reflect both genotypic and acclimational adaptations to the low temperatures characteristic of higher elevations (Larcher, 1980). Genotypic adaptations in respiratory activity also have been found in a spruce provenance trial under the relatively mild winter conditions of western Norway. There, dark respiration of German provenances decreased moderately during early winter and reached a minimum in January or February. In provenances from northern Sweden and eastern Norway, dark respiration decreased suddenly in late autumn, but rose to higher values earlier in spring than in German provenances (Saetersdal, 1956). These differences may express genotypic adaptations in the timing of dormancy in response to both the photo- and the thermoperiod of the site. Respiration rates during the summer growth period are also higher in alpine and polar timberline provenances than in low-latitude, low-elevation provenances (Pelkonen and Luukkanen, 1974). However, no detailed investigations have been done of the respiration rates of higher plants at very low subfreezing temperatures (Larcher, 1981); the low temperature for inactivation of mitochrondria in conifers, or when alternative respiration pathways are used, remains unknown.

When twigs thaw after frost exposure, a temporary rise in respiratory rates above prefreezing rates is frequently observed that appears to be dependent on the duration and minimum temperatures of the preceding frost period (Pisek and Winkler, 1958; Pisek and Kemnitzer, 1968). Experimental freezing of frost-hardened needles of *A. alba* increased respiration rates up to 2.5-fold within hours, before gradually decreasing to normal levels after a few days (Bauer *et al.*, 1969). This transient overshoot of dark respiration has been interpreted as having a repair effect on pathological disorders that occur during freezing (Larcher, 1981). A similar effect has been measured in *L. decidua* at the timberline when frozen stems or branches began to thaw at a temperature of approximately $-2°C$ (W. M. Havranek, unpublished data). In February, the daily sum of stem respiration increases 10 times in a day when temperature measured at 1–2 cm into the stem increases from -4 to $-0.5°C$. During the following 4 days, respiration declines gradually although stem temperatures remain constant at about $0°C$. Whether this initial burst in respirational CO_2 production, which is accompanied by a sudden increase in stem (bark) diameter, is caused by repair processes only or by a sudden release of accumulated CO_2 is not clear. Within 5 days, CO_2 emission equals the amount lost during the previous 5 wk in the frozen state. Relatively small absolute values of CO_2 release and a low frequency of freeze–thaw episodes appear to minimize the input of respiration during true dormancy. For a mature subalpine *L. decidua* tree at the timberline, respiration of the whole tree during the leafless period from October to April is calculated to be only 2.3% of the annual net photosynthetic carbon gain (W. M. Havranek, unpublished data).

Continuous gas-exchange measurements in Switzerland during winter demonstrate that all monthly CO_2 balances of shoots of mature spruce are positive in the valley (685 m elevation) and only slightly negative at 1600 m elevation (Häsler, 1991). A net carbon gain was recorded in young *P. abies* and *P. sylvestris* during a relatively mild winter in Innsbruck (Bahn, 1988). Even in the boreal zone in central Sweden, CO_2 balance is only slightly negative from December to February (Troeng and Linder, 1982); small carbon losses also occur at the alpine timberline, where cold and warm period fluctuations occur more often than at high latitudes. On cold days (minima -15 to $-19°C$), both photosynthesis and respiration are immeasurable in needles of *P. cembra* (Tranquillini, 1957) and *P. sylvestris* (Ungerson and Scherdin, 1968). Depending on the degree of inactivation caused by preceding frost days, little or no respiration occurs during night and day, and no CO_2 exchange is measured during the day, even if needle temperatures are above freezing and photosynthesis is compensating respiration. It may certainly be adaptive that respiration during dormancy is reduced to very low values, minimizing carbon losses during the long boreal winters. However, more data are needed to confirm this hypothesis.

During winter dormancy, incorporation of ^{14}C into cellulose occurs in unfrozen stems of *L. decidua*, *P. abies*, *P. sylvestris*, *P. cembra*, and *A. alba* (Christmann, 1982). ^{14}C has been found in both ray parenchyma and dormant cambium. In *P. abies* needles, raffinose and stachyose increase in midwinter and appear to act as substrates for respiration in late winter (Senser *et al.*, 1971). In *P. sylvestris* needles, soluble sugar and fat content also rise dramatically during winter (Fischer and Höll, 1991), whereas changes in storage material in living xylem cells of the stem are comparatively small throughout the year (Fischer and Höll, 1992). In comparison, a high winter maximum of both soluble carbohydrates and free glycerols occurs in the outer 5–7 xylem rings in *P. abies* trunks (Höll, 1985). In conclusion, significant levels of metabolic activity in needles, xylem, and bark can occur during winter dormancy in temperate zone conifers, based on measured changes in the quality and quantity of stored carbon compounds.

Storage material accumulated during autumn appears to be little affected by winter respiration in *Abies veitchii* at the timberline in Japan and is thought to be mainly used for new spring growth (Kimura, 1969). In the mild winter climate of Bayreuth (Germany) ^{14}C labeling of *P. sylvestris* reveals that soluble carbohydrates produced in autumn are utilized, together with recent photosynthates, for root growth and incorporated into cell walls of the latest xylem and phloem elements during winter. Budbreak and new growth in spring, however, are not dependent on reserve materials but are exclusively supplied by recent photosynthates of the previous year's needles (Hansen and Beck, 1994). Therefore any restriction on carbohydrate production during spring before budbreak could result in a major inhibition of new flush. Although stunted thin needles can frequently be observed in *P. cembra* and *P. abies* of the kampfzone, measurements of carbohydrate contents are not available.

VI. Conclusions

Lengthy and severe winters require that trees in the forests of boreal and mountain zones undergo winter dormancy. Physiologically, a high resistance to subfreezing temperatures and concomitant dehydration are necessary. To accomplish this dormancy, both physiological and structural changes are needed at the cellular level that require induction by endogenous and photoperiodic control early in autumn. Endogenous rhythmicity promotes cold hardening in early autumn and the persistence of hardiness throughout the winter. Numerous physiological functions are maintained at a reduced level, or become completely inhibited during true winter dormancy. Although gas exchange is strongly reduced at low positive temperatures, transpiration and photosynthesis can occur at low levels as long as needles do not freeze, thus maximizing annual carbon gain. Frost occurrence can inhibit photosynthesis completely, both by stomatal closure and at the chloroplast level. Although photoinhibition and low temperatures can restrict photosynthesis severely, the former is considered a mechanism to protect chloroplasts from photo-oxidative damage. Although winter respiration may not impose a major quantitative stress on carbon balance, the long winter periods of 5–8 mo may result in significant cumulative effects.

Winter hardiness also includes the capability to minimize water loss effectively when water uptake is severely impeded or impossible. Anatomical features such as tracheids act to minimize xylem embolism during frequent freeze–thaw cycles, and "crown" tissues enable buds to stay in a dehydrated and, thus, more resistant state during winter. Both these structural features are adaptations that contribute to the dominance of conifers in cold climates. Interestingly, deciduous tree species rather than evergreen conifers dominate in the most severe winter climates, although it is not clear whether limitations during winter, during the summer growth period, or during both are most limiting to conifer tree ecology. Additional work that evaluates the importance of winter and summer growth restriction, and their interaction, is needed before a comprehensive understanding of conifer tree ecophysiology will be possible. This information could be crucial to understanding future changes in conifer distribution patterns due to natural or anthropogenic change.

References

Andersen, C. P., McLaughlin, S. B., and Roy, W. K. (1991). Foliar injury symptoms and pigment concentrations in red spruce saplings in the southern Appalachians. *Can. J. For. Res.* 21:1119–1123.
Anderson, J. V., Chevone, B. I., and Hess, L. J. (1992). Seasonal variation in the antioxidant system of eastern white pine needles. *Plant Physiol.* 98:501–508.
Andersson, L.-A. (1980). Water transport in hardened and non-hardened seedlings of Scots pine. *Ecol. Bull.* 32:215–218.
Aronsson, A., Ingestad, T., and Lööf, L.-G. (1976). Carbohydrate metabolism and frost hardiness in pine and spruce seedlings grown at different photoperiods and thermoperiods. *Physiol. Plant.* 36:127–132.
Aulitzky, H. (1961a). Die Bodentemperaturen in der Kampfzone oberhalb der Waldgrenze und im subalpinen Zirben-Lärchenwald. *Mitt. Forstl. Bundesversuchsanst. Mariabrunn* 59:153–208.
Aulitzky, H. (1961b). Die Bodentemperaturverhältnisse an einer zentralalpinen Hanglage beiderseits der Waldgrenze. *Arch. Meteorol. Bioklimatol., Ser. B* 10:445–532.
Bahn, M. (1988). Der winterliche Gaswechsel frostexponierter und frostgeschützter Fichten und Kiefern. Diplomarbeit Botanik, Univ. Innsbruck.
Baig, M. N., Tranquillini, W., and Havranek, W. M. (1974). Cuticuläre Transpiration von *Picea-abies-* und *Pinus-cembra-*Zweigen aus verschiedener Seehöhe und ihre Bedeutung für die winterliche Austrocknung der Bäume an der alpinen Waldgrenze. *Centralbl. Gesamte Forstwes.* 91:195–211.
Bamberg, S., Schwarz, W., and Tranquillini, W. (1967). Influence of daylength on the photosynthetic capacity of stone pine (*Pinus cembra* L.). *Ecology* 48:264–269.
Bauer, H., Huter, M., and Larcher, W. (1969). Der Einfluss und die Nachwirkung von Hitze- und Kältestress auf den CO_2-Gaswechsel von Tanne und Ahorn. *Ber. Dtsch. Bot. Ges.* 82:65–70.
Becwar, M. R., Rajashekar, C., Hansen Bristow, K. J., and Burke, M. J. (1981). Deep undercooling of tissue water and winter hardiness limitations in timberline flora. *Plant Physiol.* 68:111–114.
Benecke, U. (1972). Wachstum, CO_2-Gaswechsel und Pigmentgehalt einiger Baumarten nach Ausbringung in verschiedene Höhenlagen. *Angew. Bot.* 46:117–135.
Benecke, U., and Havranek, W. M. (1980). Gas exchange of trees at altitudes up to timberline, Craigieburn Range, New Zealand. *N.Z. For. Serv., Tech. Pap.* 70:195–212.
Blake, T. J., Bevilacqua, E., Hunt, G. A., and Abrams, S. R. (1990). Effects of abscisic acid and its acetylenic alcohol on dormancy, root development and transpiration in three conifer species. *Physiol. Plant.* 80:371–378.

Bonan, G. B. (1992). Soil temperature as an ecological factor in boreal forests. *In* "A Systems Analysis of the Global Boreal Forest" (H. H. Shugart *et al.*, eds.), pp. 126–143. Cambridge Univ. Press, Cambridge, UK.

Bourdeau, P. F. (1959). Seasonal variations of the photosynthetic efficiency of evergreen conifers. *Ecology* 40:63–67.

Budyko, M. I. (1974). "Climate and Life," Int. Geophys. Ser., Vol. 18. Academic Press, New York and London.

Cartellieri, E. (1935). Jahresgang von osmotischem Wert, Transpiration und Assimilation einiger Ericaceen der alpinen Zwergstrauchheide und von *Pinus cembra*. *Jahrb. Wiss. Bot.* 82:460–506.

Chabot, J. F., and Chabot, B. F. (1975). Developmental and seasonal patterns of mesophyll ultrastructure in *Abies balsamea*. *Can. J. Bot.* 53:295–304.

Christersson, L. (1972). The transpiration rate of unhardened, hardened and dehardened seedlings of spruce and pine. *Physiol. Plant.* 26:258–263.

Christmann, A. (1982). Über Stofftransport und Zellulosebau einiger Nadelhölzer während des Winterhalbjahres. Ph.D. Dissertation, Bot. Inst., Univ. Hohenheim, Stuttgart.

Cochard, H. (1992). Vulnerability of several conifers to air embolism. *Tree Physiol.* 11: 73–83.

Coleman, M. D., Hinckley, T. M., McNaughton, G., and Smit, B. A. (1992). Root cold hardiness and native distribution of subalpine conifers. *Can. J. For. Res.* 22:932–938.

Day, T. A., DeLucia, E. H., and Smith, W. K. (1990). Effect of soil temperature on stem sap flow, shoot gas exchange and water potential of *Picea engelmannii* (Parry) during snowmelt. *Oecologia* 84:474–481.

DeLucia, E. H., Day, T. A., and Vogelman, T. C. (1992). Ultraviolet-B and visible light penetration into needles of two species of subalpine conifers during foliar development. *Plant, Cell Environ.* 15:921–929.

Demmig-Adams, B., and Adams, W. W., III (1992). Photoprotection and other responses of plants to high light stress. *Annu. Rev. Plant Physiol. Plant Mol. Biol.* 43:599–626.

DeYoe, D. R., and Brown, G. N. (1979). Glycerolipid and fatty acid changes in Eastern white pine chloroplast lamellae during the onset of winter. *Plant Physiol.* 64:924–929.

Doumas, P., and Zaerr, J. B. (1988). Seasonal changes in levels of cytokinin-like compounds from Douglas-fir xylem extrudate. *Tree Physiol.* 4:1–8.

Fischer, C., and Höll, W. (1991). Food reserves of Scots pine (*Pinus sylvestris* L.). I. Seasonal changes in the carbohydrate and fat reserves of pine needles. *Trees* 5:187–195.

Fischer, C., and Höll, W. (1992). Food reserves of Scots pine (*Pinus sylvestris* L.). II. Seasonal changes and radial distribution of carbohydrate and fat reserves in pine wood. *Trees* 6:147–155.

Friedel, H. (1967). Verlauf der alpinen Waldgrenze im Rahmen anliegender Gebirgsgelände. *Mitt. Forstl. Bundesversuchsanst. Wien* 75:81–172.

Fry, D. J., and Phillips, D. J. (1977). Photosynthesis of conifers in relation to annual growth cycles and dry matter production. II. Seasonal photosynthetic capacity and mesophyll ultrastructure in *Abies grandis*, *Picea sitchensis*, *Tsuga heterophylla* and *Larix leptolepis* growing in S.W. England. *Physiol. Plant.* 40:300–306.

Fuchigami, L. H., Weiser, C. J., Kobayashi, K., Timmis, R., and Gusta, L. V. (1982). A degree growth stage (°GS) model and cold acclimation in temperate woody plants. *In* "Plant Cold Hardiness and Freezing Stress" (P. H. Li and A. Sakai, eds.), Vol. 2, pp. 93–116. Academic Press, New York and London.

George, M. F., Burke, M. J., Pellet, H. M., and Johnson, A. G. (1974). Low temperature exotherms and woody plant distribution. *HortScience* 9:519–522.

Gillies, S. L., and Vidaver, W. (1990). Resistance to photodamage in evergreen conifers. *Physiol. Plant.* 80:148–153.

Glerum, C. (1973). "The Relationship Between Frost Hardiness and Dormancy in Trees," Int. Symp. Dormancy Trees, 1973, Kornik, Poland. Polish Academy of Sciences, Inst. Dendrol. and Kornik Arboretum.

Godnev, T. N., and Hodasevic, E. V. (1965). Biosynthesis of pigments in some evergreen plants at temperatures below 0°C. *Dokl. Akad. Nauk. SSSR.* 160:1206–1208.

Gorchakovsky, P. L., and Shiyatov, S. G. (1978). The upper forest limit in the mountains of the boreal zone of the USSR. *Arct. Alp. Res.* 10:349–363.

Grill, D., and Pfeifhofer, H. W. (1985). Carotinoide in Fichtennadeln. II. Quantitative Untersuchungen. *Phyton (Horn, Austria)* 25:1–15.

Gross, K., and Koch, W. (1991). Water relations of *Picea abies. Physiol. Plant.* 83:290–303.

Gross, M. (1989). "Untersuchungen an Fichten der alpinen Waldgrenze," Diss. Bot. 139. Cramer/G Borntraeger, Berlin and Stuttgart.

Gross, M., Rainer, I., and Tranquillini, W. (1991). Über die Frostresistenz der Fichte mit besonderer Berücksichtigung der Zahl der Gefrierzyklen und der Geschwindigkeit der Temperaturänderung beim Frieren und Auftauen. *Forstwiss. Centralbl.* 110:207–217.

Guehl, J. M. (1985). Comparative study of the winter photosynthetic potential of three evergreen conifers of the temperate zone (*Pseudotsuga menziesii* Mirb., *Abies alba* Mill. and *Picea excelsa* Link.). *Ann. Sci. For.* 42:23–38.

Hadley, J. L., and Amundson, R. G. (1992). Effects of radiational heating at low air temperature on water balance, cold tolerance and visible injury of red spruce foliage. *Tree Physiol.* 11:1–17.

Hadley, J. L., and Smith, W. K. (1986). Wind effects on needles of timberline conifers: seasonal influence on mortality. *Ecology* 67:12–19.

Hadley, J. L., and Smith, W. K. (1987). Wind erosion of leaf surface wax in alpine timberline conifers. *Arctic Alpine Res.* 21:382–389.

Hadley, J. L., and Smith, W. K. (1990). Influence of leaf surface wax and leaf area to water content ratio on cuticular transpiration in western conifers, USA. *Can. J. For. Res.* 20:1306–1311.

Hansen, J., and Beck, E. (1994). Seasonal changes in the utilization and turnover of assimilation products in 8-year-old Scots pine (*Pinus sylvestris* L.) trees. *Trees* 8:172–182.

Hansen, D. H., and Klikoff, L. G. (1972). Water stress in krummholz, Wasatch Mountains, Utah. *Bot. Gaz. (Chicago)* 133:392–394.

Härtel, O., and Eisenzopf, R. (1953). Zur Physiologie und Ökologie der kutikulären Wasseraufnahme durch Koniferennadeln. *Zentralbl. gesamte Forst- Holzwirtsch.* 72:47–59.

Häsler, R. (1991). Vergleich der Gaswechselmessungen der drei Jahre (Juli 1986–Juni 1989). In "Luftschadstoffe und Wald" (M. Stark, ed.), pp. 177–184. Verlag der Fachvereine, Zürich.

Havis, J. R. (1971). Water movement in woody stems during freezing. *Cryobiology* 8:581–585.

Havranek, W. M. (1972). Über die Bedeutung der Bodentemperatur für die Photosynthese und Transpiration junger Forstpflanzen und für die Stoffproduktion an der Waldgrenze. *Angew. Bot.* 46:101–116.

Havranek, W. M. (1987). Physiologische Reaktionen auf Klimastress bei Bäumen an der Waldgrenze. *GSF-Ber.* 10:115–130.

Herrick, G. T., and Friedland, A. J. (1991). Winter desiccation and injury of subalpine red spruce. *Tree Physiol.* 8:23–36.

Höll, W. (1985). Seasonal fluctuation of reserve materials in the trunkwood of spruce [*Picea abies* (L.) Karst.]. *J. Plant Physiol.* 117:355–362.

Holtmeier, F.-K. (1971). Waldgrenzstudien im nördlichen Finnisch-Lappland und angrenzenden Nordnorwegen. *Rep. Kevo Subarctic Res. Sta.* 8:53–62.

Holzer, K. (1958). Die winterlichen Veränderungen der Assimilationszellen von Zirbe (*Pi-*

nus cembra L.) und Fichte (*Picea excelsa* Link) an der alpinen Waldgrenze. *Oesterr. Bot. Z.* 105:323–346.

Holzer, K. (1959). Winterliche Schäden an Zirben nahe der alpinen Baumgrenze. *Centralbl. Gesamte Forstwes.* 76:232–244.

Jordan, D. N., and Smith, W. K. (1994a). Energy balance analysis of nighttime leaf temperatures and frost formation in a subalpine environment. *Agric. For. Meteorol.* (*in press*).

Jordan, D. N., and Smith, W. K. (1994b). Microclimate factors influencing the frequency and duration of growth season frost for subalpine plants (*submitted*).

Jurik, T. W., Briggs, G. M., and Gates, D. M. (1988). Springtime recovery of photosynthetic activity of white pine in Michigan. *Can. J. Bot.* 66:138–141.

Kandler, O., Dover, C., and Ziegler, P. (1979). Kälteresistenz der Fichte. I. Steuerung von Kälteresistenz, Kohlehydrat- und Proteinstoffwechsel durch Photoperiode und Temperature. *Ber. Dtsch. Bot. Ges.* 92:225–241.

Katz, C., Oren, R., Schulze, E.-D., and Milburn, J. A. (1989). Uptake of water and solutes through twigs of *Picea abies* (L.) Karst. *Trees* 3:33–37.

Kimura, M. (1969). Ecological and physiological studies on the vegetation of Mt. Shimagare. VII. Analysis of production processes of young *Abies* stand based on the carbohydrate economy. *Bot. Mag.* 82:6–19.

Kincaid, D. T., and Lyons, E. E. (1981). Winter water relations of red spruce on Mount Monadnock, New Hampshire. *Ecology* 62:1155–1161.

Körner, C., and Perterer, J. (1988). Nehmen immergrüne Waldbäume im Winter Schadgase auf? *GSF-Ber.* 17(88):400–414.

Korotaev, A. A. (1994). Investigations of frost resistance of tree roots. *Forstarchiv* 65:93–95 (in german).

Kullman, L. (1989). Cold-induced dieback of montane spruce forests in the Swedish Scandes—A modern analogue of paleoenvironmental processes. *New Phytol.* 113:377–389.

Larcher, W. (1957). Frosttrocknis an der Waldgrenze und in der alpinen Zwergstrauchheide auf dem Patscherkofel bei Innsbruck. *Veröff. Museum Ferdinandeum Innsbruck*, 37:49–81.

Larcher, W. (1963). Zur spätwinterlichen Erschwerung der Wasserbilanz von Holzpflanzen an der Waldgrenze. *Ber. Naturwiss.-Med. Ver. Innsbruck* 53:125–137.

Larcher, W. (1980). "Physiological Plant Ecology," 2nd ed. Springer-Verlag, Berlin.

Larcher, W. (1981). Effects of low temperature stress and frost injury on plant productivty. *In* "Physiological Processes Limiting Plant Productivity" (C. B. Johnson, ed.), pp. 253–269. Butterworth, London.

Lassoie, J. P., Hinckley, T. M., and Grier, C. C. (1985). Coniferous forests of the pacific northwest. *In* "Physiological Ecology of North American Plant Communities" (B. F. Chabot and H. A. Mooney, eds.), pp. 127–161. Chapman and Hall, New York.

Lavender, D. P., and Silim, S. N. (1987). The role of plant growth regulators in dormancy in forest trees. *Plant Growth Regul.* 6:171–191.

Levitt, J. (1980). "Response of Plants to Environmental Stresses," 2nd ed., Vol. 1. Academic Press, London and New York.

Lindsay, J. H. (1971). Annual cycle of the leaf water potential in *Picea engelmannii* and *Abies lasiocarpa* at timberline in Wyoming. *Arct. Alp. Res.* 3:131–138.

Little, C. H. A. (1970). Seasonal changes in carbohydrate and moisture content in needles of balsam fir (*Abies balsamea*). *Can. J. Bot.* 48:2021–2028.

Loris, K. (1981). Dickenwachstum von Zirbe, Fichte und Lärche an der alpinen Waldgrenze/Patscherkofel. Ergebnisse der Dendrometermessungen 1976/79. *Mitt. Forstl. Bundesversuchsanst. Wien* 142:417–441.

Loris, K. (1991). Zonobiom VIII. Kalttemperiertes, boreales ZB in Amerika. *In* "Ökologie der Erde" (H. Walter and S. Breckle, eds.), Vol. 4, pp. 425–482. Fischer, Stuttgart.

Lundmark, T., Hällgren, J.-E., and Heden, J. (1988). Recovery from winter depression of photosynthesis in pine and spruce. *Trees* 2:110–114.

MacHattie, L. B. (1963). Winter injury of Lodgepole pine foliage. *Weather* 19:301–307.

Maronek, D. M., and Flint, H. L. (1974). Cold hardiness of needles of *Pinus strobus* L. as a function of geographic source. *For. Sci.* 20:135–141.

Mellerowicz, E. J., Coleman, W. K., Riding, R. T., and Little, C. H. A. (1992). Periodicity of cambial activity in *Abies balsamea*. I. Effects of temperature and photoperiod on cambial dormancy and frost hardiness. *Physiol. Plant.* 85:515–525.

Michael, G. (1967). Über die Beanspruchung des Wasserhaushaltes einiger immergrüner Gehölze im Mittelgebirge im Zusammenhang mit dem Frosttrocknisproblem. *Arch. Forstwes.* 16:1015–1032.

Michaelis, P. (1934). Ökologische Studien an der alpinen Baumgrenze. III. Über die winterlichen Temperaturen der pflanzlichen Organe, insbesondere der Fichte. *Beih. Bot. Zentralbl.* 52B:333–377.

Moser, M. (1958). Der Einfluss tiefer Temperaturen auf das Wachstum und die Lebenstätigkeit höherer Pilze mit spezieller Berücksichtigung von Mykorrhizapilzen. *Sydowia* [2] 12:386–399.

Nagele, M. (1989). Winterliche Veränderungen der Photosyntheseaktivität ausgewählter Holzpflanzen. Ph.D. Dissertation, Naturwiss. Fak., Univ. Innsbruck.

Neilson, R. E., Ludlow, M. M., and Jarvis, P. G. (1972). Photosynthesis in Sitka spruce [*Picea sitchensis* (Bong.) Carr.] 2. Response to temperature. *J. Appl. Ecol.* 9:721–745.

Niederer, M., Pankow, W., and Wiemken, A. (1992). Seasonal changes of soluble carbohydrates in mycorrhizas of Norway spruce and changes induced by exposure to frost and desiccation. *Eur. J. For. Pathol.* 22:291–299.

Nikolov, N., and Helmisaari, H. (1992). Silvics of the circumpolar boreal forest tree species. *In* "A Systems Analysis of the Global Boreal Forest" (H. H. Shugart *et al.*, eds.), pp. 13–84. Cambridge Univ. Press, Cambridge, UK.

Olien, C. R. (1981). Analysis of midwinter freezing stress. *In* "Analysis and Improvement of Plant Cold Hardiness" (C. R. Olien and M. N. Smith, eds.), pp. 35–59. CRC Press, Boca Raton, FL.

Öquist, G., and Strand, M. (1986). Effects of frost hardening on quantum yield, chlorophyll organization, and energy distribution between the two photosystems in Scots pine. *Can. J. Bot.* 64:748–753.

Öquist, G., Brunes, L., Hällgren, J.-E., Gezelius, K., Hallén, M., and Malmberg, G. (1980). Effects of artificial frost hardening and winter stress on net photosynthesis, photosynthetic electron transport and RuBP carboxylase activity in seedlings of *Pinus sylvestris*. *Physiol. Plant.* 48:526–531.

Ottander, C., and Öquist, G. (1991). Recovery of photosynthesis in winter-stressed Scots pine. *Plant, Cell Environ.* 14:345–349.

Pelkonen, P., and Luukkanen, O. (1974). Gas exchange in three populations of Norway spruce. *Silvae Genet.* 23:160–164.

Pisek, A., and Kemnitzer, R. (1968). Der Einfluss von Frost auf die Photosynthese der Weisstanne (*Abies alba* Mill.). *Flora (Jena), Abt. B* 157:314–326.

Pisek, A., and Larcher, W. (1954). Zusammenhang zwischen Austrocknungsresistenz und Frosthärte bei Immergrünen. *Protoplasma* 44:30–46.

Pisek, A., and Schiessl, R. (1947). Die Temperaturbeeinflussbarkeit der Frosthärte von Nadelhölzern und Zwergsträuchern an der alpinen Waldgrenze. *Ber. Naturwiss.-Med. Ver. Innsbruck* 47:33–52.

Pisek, A., and Winkler, E. (1958). Assimilationsvermögen und Respiration der Fichte (*Picea excelsa* Link) in verschiedener Höhenlage und der Zirbe (*Pinus cembra* L.) an der Waldgrenze. *Planta* 51:518–543.

Qamaruddin, M., Dormling, I., Ekberg, I., Eriksson, G., and Tillberg, E. (1993). Abscisic acid content at defined levels of bud dormancy and frost tolerance in two contrasting populations of *Picea abies* grown in a phytotron. *Physiol. Plant.* 87:203–210.

Rehfeldt, G. E. (1988). Ecological genetics of *Pinus contorta* from the Rocky Mountains (USA): A synthesis. *Silvae Genet.* 37:3–4.

Rehfeldt, G. E. (1989). Ecological adaptations in Douglas-Fir (*Pseudotsuga menziesii* var. *glauca*): A synthesis. *For. Ecol. Manage.* 28:203–215.

Repo, T. (1992). Seasonal changes of frost hardiness in *Picea abies* and *Pinus sylvestris* in Finland. *Can. J. For. Res.* 22:1949–1957.

Richards, J. H., and Bliss, L. C. (1986). Winter water relations of a deciduous timberline conifer, *Larix lyallii* Parl. *Oecologia* 69:16–24.

Ritchie, G. A., and Shula, R. G. (1984). Seasonal changes of tissue-water relations in shoots and root systems of Douglas-fir seedlings. *For. Sci.* 30:538–548.

Robson, D. J., and Petty, J. A. (1987). Freezing in conifer xylem. I. Pressure changes and growth velocity of ice. *J. Exp. Bot.* 38:1901–1908.

Robson, D. J., McHardy, W. J., and Petty, J. A. (1988). Freezing in conifer xylem. II. Pit aspiration and bubble formation. *J. Exp. Bot.* 39:1617–1621.

Rossa, M.-L., and Larsen, J. B. (1980). Die winterlichen Austrocknungsraten verschiedener Herkünfte der Douglasie (*Pseudotsuga menziesii*) und deren Abhängigkeit von der Ausbildung der Cuticula und des Spaltöffnungstiefe. *Allg. Forst- Jagdztg.* 151:137–147.

Saetersdal, L. S. (1956). Investigations on respiration and assimilation rates of various provenances of Norway spruce (*Picea excelsa*) in winter and spring. *Arbok Univ. Bergen, Naturvitensk. Rekke* 6:1–46.

Sakai, A. (1983). Comparative study on freezing resistance of conifers with special reference to cold adaptation and its evolutive aspects. *Can. J. Bot.* 9:2323–2332.

Sakai, A., and Larcher, W. (1987). "Frost Survival of Plants. Responses and Adaptation to Freezing Stress," Ecol. Stud., Vol. 62. Springer-Verlag, Berlin.

Sakai, A., and Weiser, C. J. (1973). Freezing resistance of trees in North America with reference to tree regions. *Ecology* 54:118–126.

Sarvas, R. (1974). Investigations on the annual cycle of development of forest trees. II. Autumn dormancy and winter dormancy. *Commun. Inst. For. Fenn.* 84.

Schulze, E.-D., Mooney, H. A., and Dunn, E. L. (1967). Wintertime photosynthesis of Bristlecone pine (*Pinus aristata*) in the White Mountains of California. *Ecology* 48:1044–1047.

Schwarz, W. (1968). Der Einfluss der Temperatur und Tageslänge auf die Frosthärte der Zirbe. *Tagungsber., Dtsch. Akad. Landwirtschaftswiss. Berlin* 100:55–63.

Schwarz, W. (1971). Das Photosynthesevermögen einiger Immergrüner während des Winters und seine Reaktivierungsgeschwindigkeit nach scharfen Frösten. *Ber. Dtsch. Bot. Ges.* 84:585–594.

Senser, M., and Beck, E. (1979). Kälteresistenz der Fichte. II. Einfluss von Photoperiode und Temperatur auf die Struktur und photochemischen Reaktionen von Chloroplasten. *Ber. Dtsch. Bot. Ges.* 92:243–259.

Senser, M., and Beck, E. (1982). Frost resistance in spruce (*Picea abies* (L.) Karst.). V. Influence of photoperiod and temperature on the membrane lipids of the needles. *Z. Pflanzenphysiol.* 108:71–85.

Senser, M., and Beck, E. (1984). Correlation of chloroplast ultrastructure and membrane lipid composition to the different degrees of frost resistance achieved in leaves of spinach, ivy, and spruce. *J. Plant Physiol.* 117:41–55.

Senser, M., Dittrich, P., Kandler, O., Thanbichler, A., and Kuhn, B. (1971). Isotopenstudien über den Einfluss der Jahreszeit auf den Oligosaccharidumsatz bei Coniferen. *Ber. Dtsch. Bot. Ges.* 84:445–455.

Sheppard, L. J., Smith, R. I., and Cannel, M. G. R. (1989). Frost hardiness of *Picea rubens* growing in spruce decline regions of the Appalachians. *Tree Physiol.* 5:25–37.

Siminovitch, D. (1982). Major acclimation in living bark of Sept. 16 black locust tree trunk sections after 5 weeks at 10° C in the dark—Evidence for endogenous rhythms in winter hardening. In "Plant Cold Hardiness and Freezing Stress" (P. H. Li and A. Sakai, eds.), Vol. 2, pp. 117–128. Academic Press, New York and London.

Smith, W. K. (1985). Environmental limitations on leaf conductance in Central Rocky Mountain conifers, USA. Eidg. Anst. Forstl. Versuchswes., Ber. 270:95–101.

Smith, W. K., and Knapp, A. K. (1990). Ecophysiology of high elevation forests. In "Plant Biology of the Basin and Range" (C. B. Osmond, L. F. Pitelka, and G. M. Hidy, eds.), pp. 87–142. Springer-Verlag, New York.

Smith, W. K., Young, D. R., Carter, G. A., Hadley, J. L., and McNaughton, G. M. (1984). Autumn stomatal closure in six conifer species of the Central Rocky Mountains. Oecologia 63:237–242.

Sperry, J. S., and Sullivan, J. E. M. (1992). Xylem embolism in response to freeze-thaw cycles and water stress in ring-porous, diffuse-porous, and conifer species. Plant Physiol. 100:605–613.

Sperry, J. S., and Tyree, M. T. (1990). Water-stress-induced xylem embolism in three species of conifers. Plant, Cell Environ. 13:427–436.

Stalfelt, M. G. (1944). The water consumption of the spruce. K. Lantbruksakad. Tidskr. 83:1–83.

Stone, E. C. (1963). The ecological importance of dew. Q. Rev. Biol. 38:328–341.

Strand, M., and Öquist, G. (1985). Inhibition of photosynthesis by freezing temperatures and high light levels in cold-acclimated seedlings of Scots pine (Pinus sylvestris). I. Effects on the light-limited and light-saturated rates of CO_2 assimilation. Physiol. Plant. 64:425–430.

Strand, M., and Öquist, G. (1988). Effects of frost hardening, dehardening and freezing stress on in vivo chlorophyll fluorescence of seedlings of Scots pine (Pinus sylvestris L.). Plant, Cell Environ. 11:231–238.

Strimbeck, G. R., Vann, D. R., and Johnson, A. H. (1991). In situ experimental freezing produces symptoms of winter injury in red spruce foliage. Tree Physiol. 9:359–367.

Teskey, R. O., Hinckley, T. M., and Grier, C. C. (1983). Effect of interruption of flow path on stomatal conductance of Abies amabilis. J. Exp. Bot. 34:1251–1259.

Teskey, R. O., Grier, C. C., and Hinckley, T. M. (1984a). Changes in photosynthesis and water relations with age and season in Abies amabilis. Can. J. For. Res. 14:77–84.

Teskey, R. O., Hinckley, T. M., and Grier, C. C. (1984b). Temperature-induced change in the water relations of Abies amabilis (Dougl.) Forbes. Plant Physiol. 74:77–80.

Tranquillini, W. (1957). Standortsklima, Wasserbilanz und CO_2-Gaswechsel junger Zirben (Pinus cembra L.) an der alpinen Waldgrenze. Planta 49:612–661.

Tranquillini, W. (1979). "Physiological Ecology of the Alpine Timberline," Ecol. Stud., Vol. 31. Springer-Verlag, Berlin.

Tranquillini, W. (1982). Frost drought and its ecological significance. Encyc. Plant Physiol., New Ser. 12B:379–400.

Tranquillini, W. (1987). Effects of a change in temperature on the phenology, growth, photosynthesis, frost damage and frost drought of trees growing at the forest limit in the Alps. In "European Workshop on Interrelated Bioclimatic and Land Use Changes," Vol. G, pp. 43–47. Noordwijkerhout, The Netherlands.

Tranquillini, W., and Holzer, K. (1958). Über das Gefrieren und Auftauen von Coniferennadeln. Ber. Dtsch. Bot. Ges. 71:143–156.

Tranquillini, W., and Machl-Ebner, I. (1971). Über den Einfluss von Wärme auf das Photosynthesevermögen der Zirbe (Pinus cembra L.) und der Alpenrose (Rhododendron ferrugineum L.) im Winter. Rep. Kevo Subarctic Res. Sta. 8:158–166.

Tranquillini, W., and Turner, H. (1961). Untersuchungen über die Pflanzentemperaturen in der subalpinen Stufe mit besonderer Berücksichtigung der Nadeltemperaturen der Zirbe. Mitt. Forstl. Bundesversuchsanst. Mariabrunn 59:127–151.

Troeng, E., and Linder, S. (1982). Gas exchange in a 20-year-old stand of Scots pine. I. Net photosynthesis of current and one-year-old shoots within and between seasons. *Physiol. Plant.* 54:7–14.

Tsel'niker, Y. L., and Chetverikov, A. G. (1988). Dynamics of chlorophyll content and amounts of reaction centres of photosystems 1 and 2 in *Pinus sylvestris* L. and *Picea abies* Karst. needles during a year. *Photosynthetica* 22:483–490.

Turner, H. (1988). Frostschäden und Witterungsverlauf. *Eidg. Anst. Forstl. Versuchswes., Ber.* 307:35–44.

Turner, H., and Streule, A. (1983). Wurzelwachstum und Sprossentwicklung junger Koniferen im Klimastress der alpinen Waldgrenze, mit Berücksichtigung von Mikroklima, Photosynthese und Stoffproduktion. *In* "Root Ecology and Its Practical Application," (W. Böhm, L. Kutschera, and E. Lichtenegger, eds.), pp. 617–635. Verlag Bundesanstalt Gumpenstein, Irdning, Austria.

Ungerson, J., and Scherdin, G. (1968). Jahresgang von Photosynthese und Atmung unter natürlichen Bedingungen bei *Pinus sylvestris* L. an ihrer Nordgrenze in der Subarktis. *Flora (Jena), Abt. B* 157:391–434.

Unterholzner, L. (1979). Höhenzuwachs und Knospenentwicklung bei verschiedenen österreichischen Fichtenherkünften mit besonderer Berücksichtigung der Ausreifungsvorgänge. Ph.D. Dissertation, Naturwiss. Fak., Univ. Innsbruck.

Van Der Kamp, B., and Worrall, J. (1990). An unusual case of winter bud damage in British Columbia interior conifers. *Can. J. For. Res.* 20:1640–1647.

Vanhinsberg, N. B., and Colombo, S. J. (1990). Effect of temperature on needle anatomy and transpiration of *Picea mariana* after bud initiation. *Can. J. For. Res.* 20:598–601.

Viereck, L. A. (1970). Forest succession and soil development adjacent to the Chena river in interior Alaska. *Arct. Alp. Res.* 2:1–26.

Walter, H. (1968). "Die Vegetation der Erde," Vol. 2. Fischer, Stuttgart.

Walter, H., and Breckle, S.-W., eds. (1986). "Ökologie der Erde," Vol. 3. Fischer, Stuttgart.

Weiser, C. J. (1970). Cold resistance and injury in woody plants. *Science* 169:1269–1278.

White, W. C., and Weiser, C. J. (1964). The relation of tissue desiccation, extreme cold, and rapid temperature fluctuations to winter injury of American arborvitae. *Proc. Am. Soc. Hortic. Sci.* 85:554–563.

Yoshida, S. (1984). Chemical and biophysical changes in the plasma membrane during cold acclimation of mulberry bark cells (*Morus bombycis* Koidz. cv. Goroji). *Plant Physiol.* 76:257–265.

Zelawski, W., and Kucharska, J. (1967). Winter depression of photosynthetic activity in seedlings of Scots pine (*Pinus silvestris* L.). *Photosynthetica* 1:207–213.

Zimmermann, M. H. (1983). "Xylem Structure and the Ascent of Sap." Springer-Verlag, Berlin.

6

Ecophysiology and Insect Herbivory

Karen M. Clancy, Michael R. Wagner, and Peter B. Reich

I. Acquisition and Allocation of Nutrients (Sugars, Nitrogen, Minerals)

Conifers require carbohydrates, nitrogen (N), and various mineral elements to grow. Photosynthesis in the needles produces carbohydrates, and the trees' root systems acquire N and other elements (Kramer and Kozlowski, 1979; Kozlowski *et al.*, 1991). Conifers manufacture their basic food materials, such as carbohydrates, fats, and proteins, from N, carbon dioxide, oxygen, water, and a dozen mineral elements, using light energy trapped with photosynthesis. Carbohydrates are composed of carbon, hydrogen, oxygen, and sometimes phosphorus or N. They are the chief constituents of plant dry matter and include sugars, starch, and cellulose. The most common simple sugars, or monosaccharides, in conifers are glucose and fructose; sucrose is the most important oligosaccharide because it is the most commonly translocated carbohydrate. Nitrogen-containing compounds are very important to many physiological functions in trees; N is a key component of proteins, enzymes, amides, amino acids, nucleic acids and nucleotides, and chlorophyll. Finally, good tree growth depends on adequate amounts of a dozen essential mineral elements, including the macronutrients phosphorus (P), potassium (K), calcium (Ca), magnesium (Mg), and sulfur (S), plus the micronutrients manganese (Mn), zinc (Zn), iron (Fe), copper (Cu), chlorine

(Cl), boron (B), and molybdenum (Mo). Mineral nutrients function as components of plant tissues, regulators of osmotic potential, constituents of buffer systems, activators of enzymes, and regulators of membrane permeability.

A. What Factors Determine Nutrient Levels in Plant Tissues?

1. Environmental Stress

a. Water Stress Drought stress increases levels of minerals, soluble N and soluble sugars in plant foliage, inner bark, and sapwood, because most plants lower their osmotic potential during drought by accumulating inorganic ions, amino acids, sugars, and other osmolytes (Mattson and Haack, 1987a,b). Concentrations of minerals, including Ca, Cl, K, Mg, N, and sodium (Na), can increase in the aboveground tissues of drought-stressed plants; levels of soluble N also generally increase, although total N may not. Finally, complex carbohydrates such as starch typically decrease in concentration when plants are stressed, whereas levels of soluble sugars and total foliar carbohydrates usually increase.

Observational or experimental evidence offers some support for the assumption that water stress affects levels of nutrients in conifer tissues. For example, Czapowsky (1979) reported that levels of foliar N, P, and K of balsam fir (*Abies balsamea*) decreased with a decrease in soil drainage. Similarly, Bauce and Hardy (1988) found that the raw fiber content of current-year balsam fir needles was higher on poorly drained sites, which implies that the nutrient levels would be lower because fiber dilutes nutrient concentrations (Mattson and Scriber, 1987). Mattson *et al.* (1991) also concluded that water-stressed (i.e., root-trenched) balsam fir trees had lower levels of N, Mg, Ca, Mn, and aluminum (Al) in their foliage compared to control trees, although the stressed trees tended to have higher levels of sugars. Results in Schmitt *et al.* (1983) indicated that red and white spruce (*Picea rubens* and *Picea glauca*) and balsam fir trees growing on a dry, productive site in Maine had higher levels of foliar N, glucose, and total sugars and lower levels of P, Ca, and Mg compared to trees growing on a wet, less productive site.

Douglas fir (*Pseudotsuga menziesii*) trees growing on soils with low available moisture had higher foliar concentrations of Na, P, and K and lower levels of Ca compared to trees growing on soils with more moisture (Kemp and Moody, 1984). However, Cates *et al.* (1983) reported that the N content of current-year Douglas fir needles was negatively affected by water stress—nonstressed trees had higher levels of foliar N than water-stressed trees. Conversely, experimentally induced moisture stress increased the concentration of soluble N compounds, in particular proline, in the foliage of 3-year-old Douglas fir seedlings (van den Driessche and Webber, 1975).

Niemela *et al.* (1987) found a similar trend for Scotch pine (*Pinus sylvestris*); the N content of trees increased with increasing altitude, i.e., more stressful growing conditions. However, foliage from lodgepole pine (*Pinus contorta*) trees growing in deep peat (i.e., poorly drained or water-stressed) soils had lower levels of N in both current and 1-year-old foliage compared to trees growing on better sites of iron-pan soils (Watt *et al.*, 1991). Sometimes water stress has no detectable effect on nutrients. An example is ponderosa pine (*Pinus ponderosa*) trees that were water stressed by trenching their roots; the stress treatment did not affect the levels of total or soluble N in the needles (M.R. Wagner, unpublished data).

b. Nutrient Stress It is well established that the levels of sugar, N, and minerals in plant tissues are influenced by the balances of N and minerals found in the soil in which they grow (Kramer and Kozlowski, 1979; Dale, 1988; Barbosa and Wagner, 1989; Kozlowski *et al.*, 1991). The most common limitation on plant growth, after water stress, is N deficiency. A reduced supply of N for producing proteins required for the synthesis of protoplasm results in reduced plant growth. Synthesis of enzymes and chlorophyll is reduced as well, and thus photosynthesis is reduced, which decreases the supply of carbohydrates available for growth; this may further reduce the uptake of N and minerals. Plants respond to low N availability with accelerated proteolysis and accumulation of soluble N and carbohydrates. Fertilization with N invariably leads to increases in concentrations of soluble N compounds such as amino acids and amides, although inorganic N can rise too; C:N ratios decline as a consequence.

After N, P is the most limiting element in soils. Deficiencies of P are common in alkaline soils and can lead to an inhibition of protein metabolism and auxin production. Excess P stimulates root growth more than shoot growth.

Deficiencies of K often occur in acidic soils. Potassium is involved in starch formation, translocation of sugars, development of chlorophyll, protein synthesis, cell division, and growth. Thus, K deficiency results in accumulation of soluble carbohydrates, which reduces sugars, amino acids, and protein synthesis.

Deficiencies of other elements may also occur in forest soils. For example, Fe deficiencies are common in alkaline soils; Zn, Cu, and Mn are also less soluble in alkaline than in acidic soils. In acidic soils, Ca and Mg are often deficient. In general, mineral deficiencies decrease the synthesis of carbohydrates and their translocation to growing tissues. Mineral-deficient plants growing on infertile soils also tend to have higher root:shoot ratios than plants growing on fertile soils.

Forest soils rarely contain an excess of mineral elements, although high concentrations of heavy metals such as lead, Zn, and Cu can result from particulates in air pollution. In acidic soils, the increased solubility of Al, Mn, and other ions may produce concentrations toxic to plants.

Of course, responses of plants to nutrient stress are undoubtedly connected to water stress. Kramer and Kozlowski (1979) note that water-saturated soils are subject to denitrification and toxic concentrations of such elements as Fe and Mn, whereas dry soils more commonly have deficient levels of Fe and Mn and reduced microbial activity. Phosphorus and K can become fixed in dry soils and thus unavailable to plants. A plant's water status can be positively affected by nutrient enhancement via increasing water efficiency and decreasing transpiration rate. On the other hand, water stress can limit a plant's nutrient uptake and thus affect a plant's nutrient status (Waring and Cobb, 1992).

A lot of evidence links levels of nutrients in conifer tissues to soil nutrient regimes. Kemp and Moody (1984) found differences in mineral characteristics of soil (Al, Fe, Ca, Mg, Na) and Douglas fir foliage (Na, P, Ca) which were associated with western spruce budworm (*Choristoneura occidentalis*) outbreak frequency classes.

Likewise, Stark *et al.* (1985) reported significant differences in growth rates and ion concentrations in the foliage, xylem sap, and soil for Douglas fir trees growing on nutrient-poor acid (pH 6.7) soils versus moderately fertile alkaline (pH 7.5) soils. Foliar concentrations of all 12 ions studied (Al, B, Ca, Cu, Fe, K, Mg, Mn, P, Na, Si, Zn) and N were lower for trees on the poor growth site compared to trees on the good site. Xylem sap analysis indicated the poor growth site had an excess of Mg but deficient levels of Ca, N, K, Mn, and P compared with the more rapid growth site. Soil at the good site had more Ca, whereas soil at the poor site had more water-soluble Mg.

Zasoski *et al.* (1990) also found low to perhaps deficient foliar concentrations of Fe, Cu, and Zn in Douglas fir and western hemlock (*Tsuga heterophylla*) stands growing in coastal locations in western Washington, where soils are older, more acidic, and higher in N and organic matter than many lowland interior soils. They concluded that low metal availability in the coastal sites is probably the reason tissue concentrations were low (versus an absolute deficiency), and they suggested conifers are less effective metal accumulators than hardwoods.

Bryant *et al.* (1983) noted that long-lived, slowly growing evergreen species typically occur on nutrient-deficient sites, and fertilization studies of conifers typically confirm that increases in soil N and minerals lead to increases in tree growth and foliar N and mineral levels. This is true for Douglas fir (DeBell *et al.*, 1986; Gower *et al.*, 1992; Velazquez-Martinez *et al.*, 1992), grand fir (*Abies grandis*) (Waring *et al.*, 1992; Wickman *et al.*, 1992), lodgepole pine (Watt *et al.*, 1991), jack pine (*Pinus banksiana*) (McCullough and Kulman, 1991), black spruce (*Picea mariana*)

(Mattson et al., 1983), and red spruce and balsam fir (see references in Schmitt et al., 1983). Entry et al. (1991) also found that Douglas fir trees growing in stands that were thinned and fertilized 10 years earlier had improved growth compared to trees that were only thinned. However, fertilization had little detectable effect on foliar nutrient levels 10 years after the treatment; concentrations of N, P, K, Ca, Mg, Mn, Fe, Cu, and S were not changed by fertilization, whereas levels of boron (B) and Zn were decreased, and Al was increased (Entry et al., 1991). Kozlowski et al. (1991, p. 473) note that "a voluminous literature shows that growth of established forest trees often is reduced by mineral deficiencies that frequently can be corrected by applying fertilizers," and they cite many examples involving conifers. The enhanced growth of fertilized trees reflects increases in both photosynthetic efficiency and biomass of the foliage (Kozlowski et al., 1991).

c. Air Pollution Discussions of how air pollution stress affects levels of nutrients in plant tissues are included in Hughes (1988), Riemer and Whittaker (1989), and Kozlowski et al. (1991). Pollutants can alter the nutritional quality of plants in many ways. Decreases in soil pH associated with prolonged acidic deposition can affect the activity of microbes involved in nutrient cycling, elevate levels of elements such as Fe, Mn, and Al to toxic concentrations and reduce the availability of minerals such as P, Ca, Mg, and Mo; decreased soil pH can also detrimentally affect mycorrhizal associations, which could decrease the plants' ability to absorb water and nutrients from the soil. Some pollutants, such as oxides of S or N, and some metals, can provide additional nutrients for plants under certain conditions. Acid rain can increase leaching of nutrients from foliage, e.g., K, Ca, Mg, P, and N. Finally, pollution can induce increases in levels of some primary metabolites, such as soluble amino acids, soluble sugars, and organic acids, whereas it can lead to decreases in others, such as total protein and starch.

Kozlowski et al. (1991) list several examples of the effects of air pollutants on levels of carbohydrates, amino acid pools, and protein pools in conifers. Also, Paynter et al. (1992) reported that acid rain and ozone affected carbohydrates in shortleaf pine (*Pinus echinata*), and Schonwitz et al. (1991) associated an increase in S content of needles of Norway spruce (*Picea abies*) with SO_2 pollution.

2. Plant and Tissue Age

a. Plant Age Predictable physiological changes occur as trees mature and senesce. For example, root:shoot ratios decline (Kramer and Kozlowski, 1979), the growth rate and rooting aptitude of cuttings are reduced (Haffner et al., 1991), and N and other nutrient concentrations decline at least slightly with plant age (Mattson, 1980). Annual growth increments in woody plants reach a maximum value during their juvenile, "full vigor" phase and then decline during the mature phase (Haffner et al., 1991). Mattson and Addy (1975) also noted that senescent forest systems have passed their peak efficiencies in biomass production.

Haffner et al. (1991) discuss how both the carbon and mineral metabolisms of woody plants change as they mature. Carbohydrate synthesis varies during maturation, which leads to variations in carbohydrate concentrations. Levels of many enzymes generally related to carbon metabolism also change during maturation. The concentrations and balances of Cl, K, Na, and Ca in buds of *P. abies* and *Sequoia sempervirens* are known to change with physiological age of the tree. Rejuvenation of Douglas fir needle tissue through *in vitro* subculturing decreases the K:Na ratio, and the K:Ca ratio can indicate the juvenility of Douglas fir cultivated *in vivo*. Haffner et al. (1991, p. 625) concluded that "at least the mineral elements . . . appear to be implicated in both physiological aging of the tree and ontogenetic aging of the meristem."

In Douglas fir, concentrations of sugars and several mineral elements in current-year foliage are correlated with tree age (Figs. 1–3). Levels of

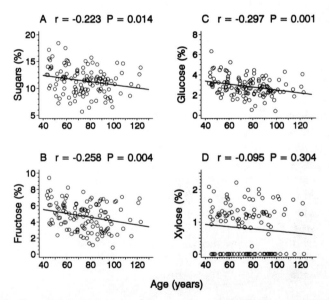

Figure 1 Correlations (r) between foliar concentrations (% dry mass) of total sugars (A), fructose (B), glucose (C), and xylose (D) versus tree age, for current-year foliage from 48 Douglas fir trees in Arizona and Colorado. Foliage was sampled when the western spruce budworm was in the late-instar feeding period in 1988 (24 trees at a Colorado site) (Clancy, 1991a), and in 1989–1990 (same 24 trees at the Colorado site, plus 24 trees at an Arizona site) (Clancy et al., 1993). P value indicates the significance level for the correlation, $n = 120$. Lines were fit with regression analysis.

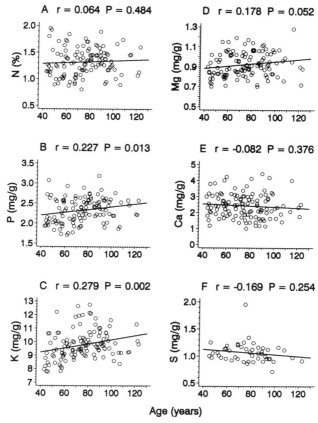

Figure 2 Correlations (*r*) between foliar concentrations (mg/g dry mass) of nitrogen (A), phosphorus (B), potassium (C), magnesium (D), calcium (E), and sulfur (F) versus tree age, for current-year foliage from 48 Douglas fir trees in Arizona and Colorado. For sample times and sizes, see Fig. 1; for sulfur (F), $n = 48$, because this mineral was monitored in 1990 only. *P* values and lines as in Fig. 1.

total sugars, plus fructose and glucose, declined with increased age; xylose concentrations were not significantly related to age (Fig. 1). The macronutrients P, K, and Mg increased with age, whereas N, Ca, and S did not vary in relation to age (Fig. 2). Conversely, the micronutrients Zn and Mn, and the ratio of Mn to Fe, all declined with age, but Cu, Fe, and Na were unrelated to age (Fig. 3).

However, Mattson *et al.* (1991) concluded that tree age has minor effects on foliar nutritional quality of balsam fir. There were no striking or consistent age effects when foliar nutrients (N, P, K, Ca, Fe, and total sugars) were compared among balsam fir trees <12 years old, >25 years old, and >40 years old.

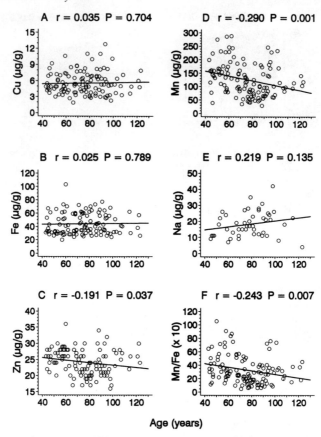

Figure 3 Correlations (*r*) between foliar concentrations (μg/g dry mass) of copper (A), iron (B), zinc (C), manganese (D), and sodium (E), or the ratio of Mn/Fe (F), versus tree age, for current-year foliage from 48 Douglas fir trees in Arizona and Colorado. For sample times and sizes, see Fig. 1; for sodium (E), *n* = 48, because this mineral was monitored in 1990 only. *P* values and lines as in Fig. 1.

b. Tissue Age Many nutritional traits of tree tissues vary with seasonal cycles of growth and dormancy, and with the phenological or ontogenetic age of leaves or needles (e.g., see Mattson and Addy, 1975; Kramer and Kozlowski, 1979; Mattson, 1980; Haack and Slansky, 1987; Horner *et al.*, 1987; Mattson and Scriber, 1987; Bernard-Dagan, 1988; Cates and Redak, 1988; Clancy *et al.*, 1988a,b; Lorio, 1988; Barbosa and Wagner, 1989; Lawrence, 1990; Fischer and Holl, 1991; Jensen, 1991; Kozlowski *et al.*, 1991; Schowalter *et al.*, 1991; Watt *et al.*, 1991; Clancy, 1992b; Kozlowski, 1992). Most of the woody tissues of trees show seasonal fluctuations in nutrient levels, particularly in the phloem and sapwood, although the fluctuations are more pronounced in deciduous hardwoods than in evergreen conifers. Concentrations of N, sugars, and many minerals in conifer needles exhibit consistent patterns of change as the current-year needles expand and mature. Furthermore, the general pattern for 1-year or older needles of conifers is typically an extension of the seasonal trends for nutrient concentration changes in current-year needles.

Concentrations of N are always highest in the young, actively growing meristematic tissues, such as expanding buds and needles, cambia, and root and stem tips. Levels of N decline substantially as new needles mature because as leaves grow older, the proportion of cell wall material increases. This causes an apparent decrease in N (and other constituents) when expressed as a percentage of dry weight (see Fig. 4). The N

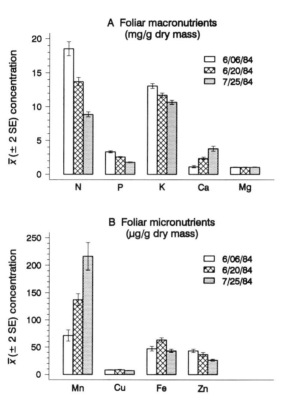

Figure 4 Differences among sample times (June 6 and 20 and July 25, 1984) in foliar concentrations ($\bar{x} \pm 2$ SE, or ≈95% confidence interval) of macronutrients (A) and micronutrients (B), for current-year foliage from Douglas fir, Engelmann spruce, and white fir (20 trees of each species). Sample times corresponded with western spruce budworm early-instar feeding, late-instar feeding, and postfeeding periods, respectively (see Clancy et al., 1988b). Seasonal trends occurring in all three species as the new needles matured were declines in N, P, K, and Zn; increases in Ca and Mn; and an increase followed by a decrease in Fe (Clancy et al., 1988b).

concentration of older tissue, such as 1-year or older needles, is typically lower than that of young tissue.

Some mobile mineral elements, such as P and K, also decline substantially as new needles mature and then grow older, whereas other relatively immobile elements, such as Ca and Mn, tend to accumulate as needles age (Fig. 4). Many micronutrients frequently show a concave concentration curve as new needles expand, but variations in the typical seasonal pattern can occur because of species and soil interactions. For example, Clancy et al. (1988b) found a decline in Zn and a convex seasonal curve for Fe levels in current-year needles of Douglas fir, white fir (*Abies concolor*), and Engelmann spruce (*Picea engelmannii*) (Fig. 4).

Seasonal trends for sugars in conifer foliage are determined by the age (current-year versus older) and maturity of the needles. Expanding new conifer needles act as carbohydrate sinks. They depend on imports of carbohydrates, primarily from older foliage. Consequently, concentrations of sugars in current-year needles are low while needle growth is in progress, and sugar levels gradually increase over time as the new needles mature and become producers and exporters of carbohydrates. Sugar concentrations in 1-year-old or older needles decline while the new needles are expanding and depend on the older needles for carbohydrates to support formation of new protoplasm.

3. Plant Genotype

a. Interspecific Variation There are consistent differences among woody plant species in their requirements for N and minerals, and in their ability to accumulate various elements; large differences in mineral content among species have been reported by many workers (Kramer and Kozlowski, 1979). Several examples document interspecific variation in nutrient levels among conifers. Clancy *et al.* (1988a) summarized data on foliar concentrations of N, P, K, Ca, Zn, and Fe for seven host species (spruces, true firs, and Douglas fir) of spruce budworms (*Choristoneura* spp.), and noted that the range in reported nutrient concentrations among species was substantial. Furthermore, there were statistically significant differences in foliar concentrations of N, P, K, Ca, Fe, and Zn among Douglas fir, Engelmann spruce, and white fir trees growing on the same site (Clancy *et al.*, 1988b) (Fig. 5). Watt *et al.* (1991) reported that N content of both current and 1-year-old foliage was higher for Scotch pine than for lodgepole pine. Zasoski *et al.* (1990) found differences in foliar levels of Zn, Fe, Mn, and Cu among western red cedar (*Thuja plicata*), Douglas fir, and western hemlock.

b. Intraspecific Variation Differences in nutrient concentrations among individual trees or populations of the same species are also

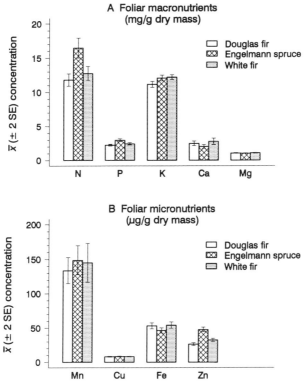

Figure 5 Differences among Douglas fir, Engelmann spruce, and white fir trees in foliar concentrations ($\bar{x} \pm 2$ SE, or ≈95% confidence interval) of macronutrients (A) and micronutrients (B), for current-year foliage from 20 trees of each species. Foliage was sampled on June 6 and 20 and July 25, 1984, when the western spruce budworm was in the early-instar feeding, late-instar feeding, and postfeeding periods, respectively (see Clancy et al., 1988b). Nutrients showing distinct differences among species included N, P, K, Ca, Fe, and Zn (Clancy et al., 1988b).

common, and at least part of this variation is probably genetically based rather than related to site factors or tree age. For example, among 48 Douglas fir genotypes from Arizona and Colorado, there are wide ranges in foliar concentrations of N, sugars, P, K, Ca, Mg, S, Cu, Mn, Zn, Fe, and Na, plus ratios of various nutrients (Table I). Coefficients of variation (CV) ranged from 11 to 63% of the mean values, indicating much variability among individual trees in foliar nutrient content, particularly for some of the micronutrients, for example, Cu, Mn, Fe, and Na (Table I).

Differences in foliar nutrients among families or populations have also been reported for conifers. Lawrence (1990) found significant differences among white spruce seed sources, or provenances, in foliar levels

Table I Variations among 48 Douglas Fir Genotypes in Concentrations and Ratios of Macro- and Micronutrients in Current-Year Foliage[a]

Nutrient (units)[b]	$\bar{x}$	SD	CV	n	Range
N (%)	1.32	0.256	19.4	120	0.80–1.95
Sugars (%)	11.39	2.489	21.8	120	5.67–18.40
Fructose (%)	4.71	1.750	37.2	120	0.82–9.42
Glucose (%)	2.86	0.952	33.3	120	0.40–6.32
Sugars:N	8.74	1.677	19.2	120	4.09–13.84
P (mg/g)	2.32	0.305	13.1	120	1.71–3.19
P:N (× 10)	1.79	0.199	11.1	120	1.34–2.40
K (mg/g)	9.77	1.058	10.8	120	7.77–12.74
K:N (× 10)	7.68	1.751	22.8	120	4.29–14.2
Ca (mg/g)	2.43	0.744	30.6	120	1.00–4.42
Ca:N (× 10)	1.92	0.722	37.6	120	0.59–4.29
Mg (mg/g)	0.92	0.120	13.0	120	0.70–1.28
Mg:N (× 100)	7.23	1.553	21.5	120	4.27–11.72
S (mg/g)	1.06	0.182	17.1	48	0.71–1.95
S:N (× 100)	7.08	1.528	21.6	48	5.05–15.35
Cu (µg/g)	5.56	2.079	37.4	120	1.9–12.8
Cu:N (× 10,000)	4.25	1.435	33.8	120	1.45–9.34
Mn (µg/g)	125.97	59.904	47.5	120	29–288
Mn:N (× 1000)	10.37	6.451	62.2	120	1.83–33.10
Zn (µg/g)	24.25	3.840	15.8	120	16–36
Zn:N (× 1000)	1.92	0.544	28.4	120	1.11–3.30
Fe (µg/g)	44.09	15.718	35.6	120	19.9–103.0
Fe:N (× 1000)	3.45	1.328	38.5	120	1.43–7.41
Na (µg/g)	18.14	7.820	43.1	48	4–42
Na:N (× 1000)	1.24	0.603	48.7	48	0.25–3.16
Mn:Fe (× 10)	32.94	20.853	63.3	120	4.86–105.36
Cu:Zn (× 100)	23.50	10.286	43.8	120	10.77–64.00
Ca:Mg (× 1)	2.62	0.685	26.2	120	1.15–5.18

[a] Foliage was sampled when the western spruce budworm was in the late-instar feeding period in 1988 (24 trees in Colorado) (see Clancy, 1991a) and in 1989–1990 (same 24 trees in Colorado, plus 24 trees in Arizona) (see Clancy et al., 1993).
[b] All concentrations are based on the dry mass of the foliage.

of P, K, Mn, and Fe, but not for total sugars, N, Ca, Mg, Cu, Na, or Zn. Cates et al. (1991) noted differences among Douglas fir populations throughout the western United States in foliar levels of N and soluble carbohydrates. Likewise, DeBell et al. (1986) discovered that foliar concentrations of N, P, K, Ca, and Mg differed significantly among 12 open-pollinated families of Douglas fir, and they suggested elemental uptake and composition may be under genetic control. Furthermore, although DeBell et al. (1986) did not find striking differences among the Douglas fir families in response to fertilizers, they cite several other studies with Douglas fir, loblolly pine (Pinus taeda), and slash pine (Pinus elliottii) that showed strong family or clone × fertilizer interactions in growth traits for seedlings or young trees. Crawford *et al.* (1991) also reported that nutrient efficiency varied among four loblolly pine families.

4. Stand Age, Density, and Structure Evidence suggests that stand age probably has important effects on levels of nutrients in woody plant tissues, although specific empirical support for conifers is lacking. Barbosa and Wagner (1989) note that variation in foliar nutrients due to tree age does appear to occur; they attribute the changes to increasing availability of nutrients in the soil as stands age. However, Stoszek (1988) proposed that older forests are subject to bottlenecks in nutrient cycling that result from undecomposed plant material and reduced soil nutrient availability. Changes in litter fall also occur as conifer stands age, with rates continually increasing in young stands, stabilizing at crown closure, and declining slightly in old stands (Trofymow *et al.*, 1991). According to Wulf and Cates (1987), climax stages of forest stands are characterized by closed mineral cycles, slow nutrient exchange rates, and a large proportion of nutrients being retained in organic matter.

As stand density increases, competition for soil nutrients increases, which can lead to decreased foliar concentrations of nutrients such as N (Dale, 1988). Thinning a forest stand typically increases foliar N in the remaining trees, unless N is not a limiting resource (Dale, 1988).

Several studies illustrate the effects that stand density can have on levels of nutrients in conifer foliage. Thinning stands of grand fir improved tree growth efficiency, but it had no detectable effect on foliar concentrations of N, free amino acids, or sugars (Waring *et al.*, 1992). Likewise, thinning immature Douglas fir stands did not affect levels of sugars in the current-year needles (Carlson and Cates, 1991). Entry *et al.* (1991) reported variable effects of thinning on Douglas-fir foliar chemistry 10 years after the treatment; levels of N, P, K, Ca, Mg, Mn, Cu, and S were not changed, whereas concentrations of Fe, B, and Al decreased and Zn increased in the thinned trees compared to the unthinned controls. Thinning improved foliar levels of N, K, and Mg in young Douglas fir plantations, and it increased the translocation of K from 1-year-old foliage to support new growth; levels of P and Ca were not affected (Velaquez-Martinez *et al.*, 1992). Horner *et al.* (1987) also reported higher foliar N levels in Douglas fir foliage from an open-canopy stand compared to a closed-canopy stand. Thinning lodgepole pine led to several changes in foliage chemistry, including an increase in foliar N (Watt *et al.*, 1991).

B. How Does Variation in Plant Nutrients Affect Insect Herbivores?

Host plants are heterogeneous substrates for insect herbivores, and there is reason to believe that this variation can significantly affect herbivore populations (Clancy *et al.*, 1988a). Mattson *et al.* (1982) reasoned that the performance of insects depends on how well their nutritional requirements, among other properties, match those of the host plant. Each individual plant has a unique set or collection of ever-changing nutritional traits that have behavioral and physiological ramifications for a potential consumer (Mattson and Scriber, 1987). For each individual or collective nutritional trait of a plant, there may be optimal ranges that maximize the herbivore's growth, survival, and reproduction (i.e., fitness). If one or more important traits are outside the insect's optimal range, a plant may be more resistant to herbivore attack than plants that have all traits within the consumer's optimal range (Mattson *et al.*, 1982). A major goal of insect nutritional ecology is to understand how plant nutrients affect the fitness of insects (Slansky, 1990). Knowledge of the physiological mechanisms involved in host plant resistance to insects, including nutritional status of the host, may lead to manipulation of these relationships to better manage insect pests (Hanover, 1975; Slansky, 1990).

However, Hanover (1975, p. 84) noted that "although resistance mechanisms involving host nutritional status may actually be quite prevalent, they are most difficult to prove because the net effect is likely to be more quantitative or subtle than that of the other resistance types." Consequently, before the role of plant nutrients as potential resistance mechanisms can be evaluated, the "nutritional niche" of insect herbivores must be defined (Clancy, 1991b). This requires detailed laboratory and field experiments to answer questions about which specific nutrients (or allelochemicals) are most important in their effects on herbivore fitness over multiple generations. In other words, the response curves of herbivores to selected compounds in plants must be determined to identify the minimum and maximum concentrations required or tolerated, plus the optimal range that may lead to population increases. Moreover, we must know how different compounds in plants (e.g., various nutrients) interact to influence the plant's quality as food for herbivores. To our knowledge, this has been attempted for only one conifer-feeding insect to date, the western spruce budworm (*Choristoneura occidentalis*) (Clancy, 1991a,b,c, 1992a,b; Clancy and King, 1993; Clancy *et al.*, 1992, 1993).

Most of the evidence for the effects of conifer nutrients on insect herbivores comes from correlations of herbivore performance with variations in the nutritional content of host plant tissues. However, these results can be misleading because correlations do not prove cause and effect; changes in the level of one nutrient in a plant, such as N, tend to be accompanied by changes in the levels of many other nutrients, plus water, fiber, and numerous allelochemicals (see references in Clancy,

1992a). The strong intercorrelations among many nutrients in plant foliage make it difficult to determine a herbivore's response to one specific nutrient. Another important limitation associated with conclusions of many published studies is that they do not evaluate herbivore fitness over multiple generations (Clancy, 1992a). Much of the inconsistent and contradictory evidence about how variations in plant nutrients affect insect herbivores probably stems from these inherent weaknesses of correlational studies that rely on incomplete measures of insect performance.

Insect herbivores that feed on conifers require N and minerals for growth and reproduction (Mattson, 1980; Mattson and Scriber, 1987; Barbosa and Wagner, 1989). Sugars, on the other hand, are important in the nutrition of many insect herbivores, but they are not known to be essential nutrients (Clancy, 1992b).

Nitrogen is a critical element in the growth of all organisms; it is an important component of the nutritional niche of many insect herbivores (White, 1976, 1978, 1984; Mattson, 1980; Clancy, 1992a, and references therein). As the N content of plant tissues increases, survival and reproduction of herbivorous insects are predicted to increase in either a linear or a convex manner. Most of the evidence supporting this hypothesis comes from positive correlations between growth and reproduction of insect herbivores and the N content of their food. Nevertheless, substantial contradictory evidence implies that the importance of N as a key nutrient is probably not strictly cause and effect, but is related to the strong link between N and many other important nutritional factors in plants. For example, Clancy (1992a) reported that the western spruce budworm's response to increased N in artificial diets was neither positively linear nor convex, and it was dependent on levels of minerals in the diets. Host plant N appears to determine the amount of food herbivores ingest, which in turn affects the amounts of other nutrients consumed; thus, a proper balance of many different nutrients is probably the most important factor in the nutritional ecology of insect herbivores (Clancy, 1992a).

More than 15 mineral elements are essential for growth and development of animals, implying that minerals are undoubtedly important in insect nutrition (Mattson and Scriber, 1987). Minerals are vital in at least three major metabolic processes in animals: enzyme activation, trigger and control mechanisms, and structure formation (Mattson and Scriber, 1987). Although minerals are important nutrients for insect herbivores, the amounts and balances required for optimal performance are largely unknown (Mattson and Scriber, 1987). A lot of experimental evidence supports important roles for minerals as nutrients for the western spruce budworm (Clancy, 1991a, 1992a; Clancy and King, 1993; Clancy *et al.*, 1993). Results from artificial diet studies indicated that several minerals have important effects on budworm population growth (convex responses for P, K, Mg, Ca; positive linear response for Zn), whereas responses to Cu (concave) and Mn (convex plus concave) were variable, and responses to Fe (concave) were not dramatic (K.M. Clancy, unpublished data). Furthermore, interactions between Mg and P imply that balances or ratios of minerals are important in budworm nutritional ecology (Clancy and King, 1993).

Sugars in conifer tissues can serve as energy-yielding substrates. Nutrition theory implies that if an insect obtains adequate energy from its food, it will direct most dietary protein into growth, rather than metabolize proteins for their energy content (Clancy, 1992b). Sugars (sucrose in particular) are also phagostimulants for the conifer-feeding eastern spruce budworm, *Choristoneura fumiferana* (Albert, 1991), and adult weights and rates of larval development increase with increasing dietary levels of some sugars (see references in Clancy, 1992b). However, Clancy (1992b) found that western spruce budworm population growth was best on artificial diets that had sugar concentrations near the lower limit observed for Douglas fir foliage, which implies that plants with higher foliar sugar may be inferior hosts for the budworm.

1. Variation Associated with Environmental Stress Destructive insect pests of coniferous forests typically reach outbreak population levels only periodically, and only in certain geographic areas and types of stands. Foresters intuitively believe the physiological condition of the tree or stand influences attack by insect pests and how much injury the tree sustains. Environmental stress imposed on trees from imbalances of water (drought, flooding) or nutrients in the soil, pollution, or climatic variation has been implicated in predisposing trees to attack by insect herbivores (White, 1969, 1976, 1978, 1984; Kemp and Moody, 1984; Mattson and Haack, 1987a,b; Niemela *et al.*, 1987; Dale, 1988; Hughes, 1988; Barbosa and Wagner, 1989; Riemer and Whittaker, 1989; Cates *et al.*, 1991; Thomas and Hodkinson, 1991; Waring and Cobb, 1992). Mattson and Addy (1975, p. 521) summarized this belief when they stated that "increases in the quality of host food and decreases in host resistance are apparently brought about by interactions of host age, stressful climatic conditions, low fertility of the site, and bottlenecks in the flow of certain vital nutrients. The combination of these events enhances insect survival or fecundity, and increases the probability of escape from natural enemies." Likewise, Waring and Cobb (1992, p. 168) concluded that "despite a lack of consensus, the view that plant stress exerts strong positive effects on herbivore populations has acquired near paradigm status in ecology." Alternatively, Price (1991, p. 245) proposed that there is "more probably a continuum, from herbivores that attack stressed plants most commonly to herbivores that attack the most vigorous plants most frequently." Nutrient levels in conifer tissues can change when trees are physiologically stressed, making them a more or less nutritious source of food for insect herbivores.

a. Water Stress A general discussion of how variation in plant water relations affects insect herbivores is presented later (Section II,B), so this section will focus on the link between water-stress-related changes in plant nutrients and herbivore responses to these nutrient changes. Water stress is commonly assumed to cause plant tissue nutrient level changes that affect insect herbivores. White (1969, 1976, 1978, 1984) generated the plant stress hypothesis, which proposes that when plants are under moisture stress (from drought or flooding), protein synthesis is decreased, and there is consequently an increase in total N in the aerial parts of the plant and changes in the relative amounts of certain amino acids. Price (1991) emphasized that the plant stress hypothesis is coupled with the climatic release hypothesis, which proposed that a series of favorable seasons for herbivores (e.g., unusually warm, dry summers in north temperate regions) may stress plants and impact their nutritional quality.

The assumption that water stress enhances the nutritional quality of foliage for insect herbivores is frequently invoked to explain insect outbreaks on conifers, or differences among stands in susceptibility and vulnerability to damage from various species of herbivores. Some examples from the literature include Mattson and Addy (1975), Schmitt *et al.* (1983), Kemp and Moody (1984), Haack and Slansky (1987), Mattson and Haack (1987a,b), Wulf and Cates (1987), Cates *et al.* (1991), Dupont *et al.* (1991), and Waring and Cobb (1992).

However, Schowalter *et al.* (1986) suggest forest stands under stress from drought (or crowding or disease) may derive some protection from herbivores by producing less nutritious foliage. Results from a study by Bauce and Hardy (1988) support this view; balsam fir trees growing on a poorly drained site had higher raw fiber content in their current-year needles, implying that lower levels of nutrients caused a decrease in spruce budworm weight, development rate, and survival on this site.

Furthermore, Waring and Cobb (1992) reported that herbivores showed no clear response to water-stress-related increases in the N fraction of plant tissues in the 13 studies that measured water-stress-induced changes in N, protein, or amino acids, and herbivore responses to these nutrient changes. The studies they reviewed included all plants, not just conifers. They noted that this contradicts the proposition that increased amino acid concentrations lead to increased herbivore performance in stressed plants.

It is noteworthy that results from several empirical studies on conifers failed to support the predictions from the plant stress hypothesis. Mattson *et al.* (1983) found that their induced stress treatments (trenching) of spruce and fir host trees had only a minor enhancing effect on spruce budworm growth and no effect on survival rates. Likewise, although water-stressed balsam fir trees had different levels of several nutrients in their foliage compared to control trees, Mattson *et al.* (1991) concluded that there was little evidence that drought stress makes fir trees more nutritionally favorable for the spruce budworm. In fact, moisture stress and root pruning treatments had negligible or negative effects on host-plant quality for spruce budworm; the negative effects were probably due to lowered levels of N and various minerals in the foliage of severely stressed trees (Mattson *et al.*, 1991). Finally, Watt *et al.* (1991) used many different approaches to test the hypothesis that outbreaks of the pine beauty moth (*Panolis flammea*) are caused by an increase in the nutritive quality of the foliage of lodgepole pine resulting from "stress" imposed by growing in deep peat (poorly drained) soils. They found no evidence to support the "stress" hypothesis. The stress imposed on lodgepole pine growing in deep peat did not positively affect pine beauty moth larvae: the N content of the foliage was lower, larval growth and survival were poorer or no better, and population survival of the pine beauty moth during the summer months was no better (Watt *et al.*, 1991).

b. Nutrient Stress Deficiencies or excesses of nutrients in the soil can have important effects on herbivorous insects because soil nutrients affect the nutrient levels in plant tissues (Kemp and Moody, 1984; Mattson and Haack, 1987b; Dale, 1988; Barbosa and Wagner, 1989; Waring and Cobb, 1992). Interestingly, coniferous trees were the only group of plants that did not elicit strong positive responses from herbivores when presumed "nutrient stress" was alleviated through fertilization, suggesting that conifers are equally likely to become more or less resistant to herbivores when fertilized (Waring and Cobb, 1992).

Dale (1988) makes several important points in his review on plant-mediated effects of soil mineral stresses on insects. Many observations relate patterns of insect incidence on plants to the nutritional status of the soils on which the host plants grow, but knowledge on the basic nature of soil mineral–plant–insect interactions is weak. It is difficult to identify the direct link among nutrient stress in the soil, changes in the plant caused by the stress, and the many effects these have on the biology of insect herbivores. Furthermore, the sensitivity of forest insects to variations in host plant quality has not been amply studied, although many reports indicate massive insect outbreaks typically start in middle-aged to old forests, which are in the waning years of their productivity. Soil nutrient levels and the incidence of forest insects are very often inversely related.

The association between nutrient stress and insect outbreaks in conifer forests is widely cited in the literature. For example, Kemp and Moody (1984) found that frequent outbreaks of the western spruce budworm occurred on soils with no volcanic ash, low available moisture, and low extractable Al; tissues of Douglas fir growing on these sites had high Na and P and low Ca. Their results strongly suggested a link between soil conditions, Douglas fir stress, budworm survival, and outbreak frequency (Kemp and Moody, 1984). Mattson and Addy (1975) cite numerous reports that link outbreaks of insect herbivores to soil litter substrates in which nutrient and moisture regimes are less than optimal, including several examples that involve conifer-feeding insects [e.g., Swaine jack-pine sawfly (*Neodiprion swainie*), lodgepole needle miner (*Coleotechnites milleri*), spruce budworm, red-headed pine sawfly, several species of pine defoliators, and root weevils in conifer plantations]. Scotch pine trees growing at upper slope in "stressful conditions" (wind, low temperature, plus nutrient-deficient soils) had higher foliar N, and *Neodiprion sertifer* larvae grew best on needles from these trees (Niemela *et al.*, 1987). Stoszek (1988) concluded that insect outbreaks, or chronically high levels of insects (including western spruce budworm), indicate the forest ecosystems are under stress from nutrient limitations.

If soil nutrient stress predisposes conifers to attack by insect herbivores, then alleviation of the stress via fertilization should make conifer trees less suitable hosts for insects. The 20 cases reported by Stark (1965) for insects feeding on pines support the prediction that in general forest fertilization should reduce insect populations, although the specific mode of action of fertilizer treatments was not clear. However, Schowalter *et al.* (1986) noted that the effects of forest fertilization on herbivory are little understood and are contradictory, perhaps because few studies have examined plant allocation of subsidized nutrients. For example, Tuomi *et al.* (1984) suggest that when plants are nutrient stressed, excess carbon is diverted to produce defenses because it cannot be used for growth.

Recent experiments typically indicate beneficial effects of fertilizers on conifer-feeding herbivores via improved foliar nutritional quality. Application of N fertilizer to Douglas fir or white fir seedlings increased foliar N levels and resulted in enhanced survivorship, development, and reproductive success of western spruce budworm at low to intermediate treatment levels (Brewer *et al.*, 1985, 1987). Nitrogen fertilization of grand fir led to increased concentrations of free amino acids and total N in the foliage (Waring *et al.*, 1992), which was associated with significant increases in biomass of insect defoliators, plus greater weights for western spruce budworm larvae and pupae for at least 4 years after treatment (Mason et al., 1992). Likewise, fertilization of western larch nearly doubled the amount of feeding by western spruce budworm larvae, with N fertilizer eliciting the greatest response (Schmidt and Fellin, 1983). Shaw et al. (1978) found increased pupal weights and fecundity for spruce budworm feeding on N fertilized balsam fir. Studies by Schmitt et al. (1983) also suggested that foliage quality-altering silvicultural practices such as fertilization may stimulate populations of the spruce budworm, and Mattson et al. (1983) reported that fertilization of small firs and spruces with urea increased foliar N levels and the growth of spruce budworm larvae. However, McCullough and Kulman (1991) reported that increased N content of jack pine foliage from fertilization did not significantly affect the survival of jack pine budworm (*Choristoneura pinus*) larvae. Pinyon pine (*Pinus edulis*) trees that received water and fertilizer had higher foliar N than control trees, and pinyon sawflies (*Neodiprion edulicolis*) performed better on these trees, refuting the notion that abiotic stress increases plant quality and benefits phytophagous insects (Mopper and Whitham, 1992). In a similar vein, Watt et al. (1991) found that thinning and fertilization treatments to alleviate nutrient stress increased the N content of lodgepole pine foliage, but pine beauty moth numbers did not decline in response to the treatments as predicted.

c. Air Pollution Much evidence supports a connection between air pollution and changes in insect attacks on plants (Hughes, 1988; Riemer and Whittaker, 1989). Much of it is observational, e.g., descriptions of outbreaks of forest insects in the vicinity of industrial facilities, or studies along pollution gradients. These studies often find a negative correlation between pollutant and insect levels in areas of very high pollutant concentrations, and a positive association in areas with moderate or low pollution levels. Hughes (1988) noted that early studies on how air pollution affects insect–plant relationships were virtually all observational and concerned forest ecosystems. Although experimental studies have been conducted since 1980, Riemer and Whittaker (1989) listed only one study with a conifer host plant. Furthermore, virtually no research has been directed toward identifying the mechanisms by which herbivore success is altered in the presence of air pollution (Hughes, 1988). Host nutritional quality is clearly one important mechanism by which pollutants might alter plants as hosts for insects, but experimental evidence from conifers is lacking.

2. Variation Associated with Plant and Tissue Age Phenological and ontogenetic changes in host plants have important effects on the quality of food available to insect herbivores (Mattson and Scriber, 1987; Barbosa and Wagner, 1989). Many species or guilds of forest insect herbivores only attack seedlings, young trees, or older trees (Barbosa and Wagner, 1989). Within a suitable host age class, phenological age of developing plant tissues can have profound effects on the herbivore's growth, survival, and reproduction (e.g., see Clancy *et al.*, 1988a,b; Lawrence, 1990). A large part of the variation in host selection and subsequent herbivore performance on plants and tissues of different ages may be associated with ontogenetic and phenological changes in the levels of N, minerals, and sugars in plant tissues, along with concomitant variations in tissue morphology (e.g., fiber or toughness), water content, and defensive chemistry.

Strong empirical evidence linking variation in conifer nutrients, caused by tree age, to effects on insect herbivores is lacking. Mattson *et al.* (1991) noted that tree age has been implicated as a factor influencing host quality for spruce budworm because outbreaks usually occur in mature spruce–fir forests. Results reported by Mattson *et al.* (1983) supported this assumption, because the smallest age/size class of balsam fir and white spruce trees produced the smallest spruce budworms. However, Mattson *et al.* (1991) concluded tree age had minor effects on foliar nutritional quality of balsam fir, and consequently, it did not affect spruce budworm survival rates. On the other hand, Cates *et al.* (1983) found a greater infestation intensity (defoliation × number of larvae) for western spruce budworm on the older trees in a Douglas fir stand; there was also a significant positive correlation between tree age and dry weight of adult females. This suggests that the age-related variation in Douglas fir foliar nutrients illustrated in Figs. 1–3 may affect western spruce budworm performance.

There is much evidence that variation in conifer nutrients associated with phenological age of plant tissues has dramatic effects on insect herbivores. For example, both the (eastern) spruce and western spruce budworms prefer to feed on early season, new foliage; as the current-year host foliage matures, it becomes a progressively worse source of food for budworm larvae (Clancy *et al.*, 1988a,b). This decrease in nutritional suitability is associated with declining foliar levels of N, P, K, and Zn and increasing concentrations of Ca and Mn (Fig. 4; Clancy *et al.*, 1988a,b). Concentrations of sugars also typically increase as current-year host needles expand and mature, which may decrease their nutritional value to budworm larvae (Clancy, 1992b). Lawrence (1990) provided strong empirical proof that there are "phenological windows of susceptibility" during which white spruce is a suitable host for spruce budworm. Reduced susceptibility to spruce budworm during spring and summer was associated with declining foliar levels of N, P, K, sugars, and water and increasing leaf toughness; optimal budworm performance was strongly correlated with high needle water content and low leaf toughness (Lawrence, 1990). [See Mattson *et al.* (1983), Haack and Slansky (1987), and Codella *et al.* (1991) for additional evidence linking tissue age-related variation in conifer nutrients to effects on insect herbivore performance.]

3. Variation Associated with Plant Genotype Mattson and Scriber (1987, p. 106) emphasized that "variations in food due to . . . genetic differences between species are usually so substantial that one . . . insect species cannot acclimate to consume just any kind of leaf." Likewise, variations in nutritional quality among individual trees or populations within a species have been commonly reported (Barbosa and Wagner, 1989), and such intraspecific variation has been linked to resistance to insect herbivores (e.g., see Clancy, 1991a; Clancy *et al.*, 1993).

Much of the published work to date that connects variation in conifer nutrients caused by plant genotype with susceptibility and vulnerability to insect herbivores is focused on spruce budworms. Mattson and Koller (1983) reported both inter- and intraspecific variation for black spruce and balsam fir trees in the levels of N and various minerals in their foliage. Furthermore, spruce budworm female body size was correlated with variation in foliar N, K, Mg, and Ca, whereas budworm survival rates were linked to levels of N, K, and Fe (Mattson and Koller, 1983). Mattson *et al.* (1983) also reported variable spruce budworm growth among balsam fir, lowland and upland black spruce, and white spruce, which was related to foliar N, Fe, and K. Likewise, Schmitt *et al.* (1983) found that budworm pupae and moths from balsam fir trees were significantly heavier than those from red spruce, and that weight was correlated overall with foliar nutritional variables. However, Mattson *et al.* (1991) concluded there was no evidence for "substantial variation" among balsam fir and white spruce trees from different populations in their inherent susceptibility to spruce budworm, although the authors noted (p. 184) that "early work suggested that host-plant quality for spruce budworm was not constant, but varied among host species."

Clancy *et al.* (1988b) summarized several lines of evidence that imply that host plant species may have important effects on western spruce budworm performance (see Beckwith, 1983; McLean *et al.*, 1983; Wagner and Blake, 1983; Wagner *et al.*, 1987; Wulf and Cates, 1987). Furthermore, they documented interspecific variation among Douglas fir, white fir, and Engelmann spruce in concentrations of N, P, K, Ca, Fe, and Zn, which may account for variation among host species in susceptibility (Clancy *et al.*, 1988b). However, they noted that the question of which host species is the most preferred or best for budworm growth and reproduction is unresolved (Clancy *et al.*, 1988b).

Intraspecific variation among individual trees or populations of Douglas fir in resistance to western spruce budworm defoliation has also been frequently reported (e.g., see Johnson and Denton, 1975; McDonald, 1983; Perry and Pitman, 1983; Wulf and Cates, 1987; Clancy, 1991a; Cates *et al.*, 1991; Clancy *et al.*, 1993). Moreover, the phenotypic variation in resistance found among individual Douglas fir trees by Clancy (1991a) and Clancy *et al.* (1993) was strongly associated with variation in foliar nutrients (i.e., N, sugars, and minerals).

Finally, there is at least one example relating interspecific variation in conifer nutrients to herbivory by a nonbudworm species. Watt *et al.* (1991) concluded that Scotch pine is a nutritionally better host for pine beauty moth than is lodgepole pine because the N content of foliage was higher on Scotch pine than on lodgepole pine.

4. Variation Associated with Stand Age, Density, and Structure The age, density, and structure of forest stands have been frequently implicated in determining both susceptibility to infestation and vulnerability to damage from forest insect herbivores (Barbosa and Wagner, 1989). For example, the conditions that lead to conifer stands that are susceptible and vulnerable to damage from the eastern (*Choristoneura fumiferana*) and western (*Choristoneura occidentalis*) spruce budworms are well known (Witter *et al.*, 1983; Wulf and Cates, 1987; Carlson and Wulf, 1989). It is generally believed that susceptible stands provide abundant, high-quality food for herbivores, although the specific relationships between stand characteristics and optimal levels of nutrients in conifer tissues have not been well defined.

Two studies were conducted to test the hypothesis that thinning would induce a change in foliar chemistry that would be detrimental to the western spruce budworm. Thinning stands of grand fir had no detectable effect on foliar concentrations of N, free amino acids, or sugars (Waring *et al.*, 1992), and little if any effect on budworm performance (Mason *et al.*, 1992). Likewise, Carlson and Cates (1991) concluded there was no support for the contention that thinning induces a change in Douglas fir foliar chemistry (including levels of sugars) that results in reduced defoliation by the budworm.

The susceptibility and vulnerability of spruce–fir stands to western spruce budworm damage depend on stand density, crown-class structure, and tree and stand maturity (Wulf and Cates, 1987; Carlson and Wulf, 1989). Overstocked, dense conifer host forests are high-quality western spruce budworm habitat; dense stands favor increasing budworm populations, open stands do not. Possible relationships of these factors to variation in conifer nutrients are implicated by Wulf and Cates (1987) via a general role for "foliage quality." Carlson and Wulf (1989) also hypothesized that "stressed trees" in overstocked stands may be nutritionally better for budworm larvae. They also reason that susceptibility to budworm may increase as trees and stands mature because older trees have more foliar biomass. However, older trees may also be better matches to the budworm's nutritional requirements (see Figs. 1–3).

Douglas fir tussock moth (*Orgyia pseudotsugata*) defoliation levels increased with site elevation, stand age, site occupancy, proportion of grand fir in the stand, and decreasing depth of the volcanic ash mantle, leading Stoszek (1988) to postulate that defoliation levels may increase with tree age because of reduced availability of nutrients. The scenario presented is that as stands develop to the pole-sized stage, their nutrient demands increase and lead to imbalances in leaf concentrations of marginally available nutrients, such as B and K. Stoszek (1988) contends that this hypothesized deficiency of nutrients could alter the nutritional quality of tissues being consumed by larvae.

C. How Does Herbivory Affect the Acquisition and Allocation of Nutrients by Plants?

The most obvious immediate effect of defoliation by insect herbivores is the reduction of a plant's ability to acquire carbon. This is done through removing the photosynthetic leaf area, which reduces the amount of available carbohydrates (Johnson and Denton, 1975). Severe defoliation of conifers can also deplete the plant's carbohydrate reserves because conifers store carbon primarily in leaves rather than in roots (Tuomi *et al.*, 1988). For example, Webb and Karchesy (1977) found that the starch content of Douglas fir was reduced proportional to the intensity of defoliation by the Douglas fir tussock moth. When carbohydrate content is reduced, tree growth declines. Although deciduous trees often refoliate following complete defoliation by insect herbivores, evergreen conifers often lose substantial canopy volume or die as a result of intense herbivory (Schowalter *et al.*, 1986).

Johnson and Denton (1975) describe the process by which western spruce budworm defoliation can affect deterioration of host trees: (1) Stems produce less wood; (2) fewer vegetative buds, flowers, and cones are produced; (3) parts of the aerial and root structures die; (4) abnormal budding develops, leading to deformed crowns and stems; (5) root rots or bark beetles may infest weakened trees; and (6) whole trees may die. However, Van Sickle (1987) concluded that there is little known about changes in host foliage chemistry brought about by an extended period of defoliation by the budworm. Likewise, Piene and Little (1990, p. 902) noted that "the physiological explanation for the growth reduction induced by the feeding of spruce budworm and other phytophagous insects is complex and poorly understood . . . it has been established with forest tree species that defoliation affects the availability of essential growth factors such as carbohydrates, water, minerals, and the plant hormone indole-3-acetic acid by variously altering their synthesis, transport, allocation, and conversion."

Substantial root mortality of woody plants can result from heavy defoliation (Kozlowski, 1971; Bassman and Dickman, 1985), and may reduce nutrient absorption (Bryant *et al.*, 1988). For example, Piene (1989) cited evidence that a 70–100% defoliation of the current-year foliage resulted in a 30–75% mortality of rootlets in balsam fir.

Furthermore, herbivory can negatively affect the mutualism between ectomycorrhizal fungi associated with the roots of woody perennials that enhance nutrient uptake in exchange for a portion of the photosynthate produced by the host plant (Kozlowski, 1992). Removal of photosynthetic tissue by herbivores can reduce the amount of photosynthate available for maintaining this mutualism, as demonstrated by Gehring and Whitham (1991) for pinyon pine trees that were susceptible to chronic insect attack.

Because of the negative effects of defoliation on root growth and mycorrhizal associations, one would predict that herbivory would decrease levels of N and minerals in plant tissues, in addition to reducing the amounts of carbohydrates in the host tree. Results from Bauce and Hardy (1988) support this assumption; two consecutive years of defoliation of balsam fir led to higher raw fiber content of the current year's foliage, implying that foliar nutrient concentrations decreased. However, Piene (1980) reported that intact needles from defoliated balsam fir trees had increased concentrations of N, P, K, Ca, and Mg compared to needles from protected trees. Piene and Percy (1984) compared the foliar chemistry of balsam fir trees that were severely defoliated by spruce budworm and then protected for 5 years to that of trees that were undefoliated. The defoliated trees had significantly higher foliar N content and a tendency toward higher P and K; Ca and Mg did not differ (Piene and Percy, 1984). Previous work cited by Piene and Percy (1984) had also found that heavy defoliation of white spruce by sawflies resulted in elevated foliar concentrations of N, P, K, and Mg. Tuomi *et al.* (1988) cite several additional examples suggesting that foliar nutrients can increase in conifers following defoliation. Likewise, Wagner and Evans (1985) found that protein levels increased in ponderosa pine seedlings following mechanical defoliation, although protein levels decreased in pole-sized trees (Wagner, 1988).

A common response of plants to defoliation is compensatory growth, which typically involves stimulation of photosynthesis, increased leaf N, and greater allocation of carbohydrate reserves to growth of new foliage rather than roots (see Section III, C). Most plants compensate for herbivory to some extent, but the degree of compensation for arthropod damage varies widely from partial to overcompensation (Trumble et al., 1993). Kozlowski (1992) noted that the mechanisms that lead to compensatory growth are complex and may be intrinsic (changes in physiology and development of plants) or extrinsic (changes in environment). Compensatory responses of plants to herbivory are discussed in detail in Section III, C.

Mattson and Addy (1975) were among the first ecologists to recognize the potential role that insect herbivores can play in recycling nutrients in forest ecosystems through effects on plant growth, foliage loss, and litter decomposition. They used the spruce budworm and its spruce and fir host forests as an example of how insects can act as regulators of primary productivity and nutrient cycling in forest ecosystems. Mattson and Addy (1975) concluded that wood production can actually be increased 10 to 15 years following a spruce budworm outbreak due to enhanced circulation of important growth elements (N, P, and K) in the litter fall. The litter fall from defoliation was postulated to be both greater in volume and richer than normal because of the exceptionally high concentrations of nutrients in dead insect bodies, insect excrement, and wasted food parts. Furthermore, Mattson and Addy (1975) noted that insect grazing could increase the rate of nutrient leaching from foliage and stimulate the activity of decomposer organisms.

The role of insects as potential regulators of primary productivity in forest ecosystems has subsequently been echoed or investigated by many others (e.g., see Larsson and Tenow, 1980; Schowalter, 1981, 1988; Swank et al. 1981; Schowalter and Crossley, 1983; Lamb, 1985; Schowalter et al., 1986, 1991; Crossley et al., 1988; Stoszek, 1988; Risley, 1990; Wheeler et al., 1992). Larsson and Tenow (1980) provide empirical evidence that needle-eating insects on *P. sylvestris* contributed to the transfer of N, P, and K from the canopy to the soil via their frass and green litter (needle litter cut off by the larvae). They estimated that insect frass and green litter transferred about 1% of the C, Ca, and Na, 2% of the N, P, Mg, and S, and 4% of the K carried annually to the forest floor by total pine litter; the effect of this transfer of N and other bioelements on soil processes was unknown (Larsson and Tenow, 1980). Results from a study by Schowalter et al. (1991) also indicate that low-to-moderate levels of herbivory by forest insects have significant effects on nutrient recycling, principally through influences on throughfall leaching and litter fall. Schowalter et al. (1991) tested the effects of herbivore density and functional group (a sap sucker versus a defoliator) on primary production, nutrient turnover, and litter decomposition of young Douglas fir. They concluded that nutrient turnover from foliage to litter was related to phytophage abundance, but herbivores did not appear to affect tree growth or exogenous litter decomposition.

II. Water Relations

The importance of water relations to all basic physiological processes in plants can hardly be overstated. Water is the essential solvent for many biochemical reactions in plants that serve as the basis of all plant life. The amount of available water and plant adaptations to extremes in water availability determine the global distribution of plants. Likewise, water relations can be profoundly important in explaining a variety of patterns of interaction between coniferous trees and insect herbivores.

A. What Factors Affect Plant–Water Relations?

1. Environmental Stress Water stress is perhaps the best studied of the various environmental stresses that affect plant physiological processes. Environmental stress, or simply plant stress, has numerous and diverse definitions (Ivanovici and Wiebe, 1981). The most common denominator in these diverse views is that stress is any condition that limits assimilation and requires plants to utilize previously assimilated carbon (Levitt, 1972). This shift to utilizing stored carbon results in reduced plant productivity expressed as a growth reduction in various plant parts (Hale and Orcutt, 1987; Kozlowski, 1979). Although we do not wish to challenge the basic premise that growth reduction is directly related to stress, we feel there are multiple dimensions to how reduced productivity comes about from stress and that these need further comment. A more thoughtful assessment of how stress affects plants may explain some of the considerable ambiguity that exists in the effect of plant stress on insect herbivory.

The direct link between stress and growth reduction has led many ecologists to use growth rate as a measure of stress (Coley, 1983; Coley *et al.*, 1985; Price, 1991). A recent review of measuring plant water stress (Bennett, 1990) does not even mention growth as a measure of stress. Growth is a cumulative effect of many environmental factors, and the importance of any given factor cannot be assessed. The use of growth rate as an index of vigor is especially problematic when comparing plant species.

For example, it is known that the impact of water deficits on shoot growth varies considerably depending on whether the tree species exhibits fixed growth or free growth (Kozlowski *et al.*, 1991). A coniferous species with fixed growth may be growth limited by stress that occurred during bud formation and consequently the rate of shoot growth the following season may not correlate with other measures of water stress. Shoot growth of a free-growth conifer such as *Pinus taeda* is largely dependent on water availability during the summer growth period. Measuring shoot growth as an index of water stress could yield a quite erroneous interpretation of the water status of the majority of conifers that exhibit the fixed-growth pattern. Clearly, comparisons among conifer species with different growth patterns using the growth rate index of stress are totally inappropriate.

Even within species our interpretation of the effect of water stress on conifer susceptibility to herbivory can be confounded by precondition of the tree prior to stress. Plants grown under favorable conditions, such as in gardens and greenhouses, tend to have thinner cuticles and lower root:shoot ratios than do hardened plants. When subjected to stress, such as outplanting, these plants may be greatly affected by stress to the point of mortality. Hardened plants, on the other hand, can withstand stress because of their preconditioning stress. Identical stress in these two conditions would have very different impacts on trees and could change herbivory dramatically. A slow-growing plant that is well adapted to its current environment could likely withstand considerably greater stress without increased susceptibility to herbivory, in contrast to a poorly adapted plant undergoing much less stress.

The above examples serve to illustrate the importance of recognizing the many dimensions of plant stress that could have very different effects on how plant susceptibility to herbivory changes with stress. Perhaps adopting the view that stress is any factor that affects the carbon balance of the plant, either positive or negative (Amthor and McCree, 1990), would clarify when plants are stressed and whether herbivores will respond to stress. Likewise, the view of Coleman *et al.* (1992), that stress is an external perturbation that results in an alteration of plant function, is probably far more appropriate. These latter two views of stress incorporate the ability of a plant to adjust physiologically to stress and maintain growth. For example, a plant that has a genetic predisposition to place high priority on shoot growth may respond to stress by reducing allocation to defensive compounds that might reduce resistance to a herbivore without reducing growth. In this case, the plant is not stressed by classical definition, but stress has altered the resistance level to herbivory. The use of aboveground growth as a measure of stress is clearly problematic and should be rejected.

2. Plant and Tissue Age The response and tolerance of conifers to water stress depends on tree age. Very young seedlings of most species, even species that are drought tolerant when mature, have very little tolerance to dry conditions.

A variety of conifer traits are adaptations to water stress that also likely vary considerably with age. For example, deep roots and high root:shoot ratios are generally considered adaptations to water stress. Very young seedlings often begin growth with a free-growth phase. During this period the seedlings must balance allocations to aboveground growth to increase assimilation potential with allocation to roots to gain water and mineral absorption. This necessary trade-off reduces stress tolerance of seedlings. Likewise, cuticle thickening and suberization of roots represent differentiation processes that compete with growth for carbon. Consequently, young seedlings and younger age tissues that are actively growing generally have lower tolerance to water stress. Hobbs and Wearstler (1983) observed that Douglas fir seedlings in a common planting area were more stressed when 1 year old than when 2 years old. The larger mass of older conifers allows for the storage of water and greater isolation of stored water from the dry outside atmosphere than is possible for younger plants.

Much of our understanding of basic water relations relates to understanding the movement of liquid across a resistance gradient. Water flow determines the water potential differences in various parts of the plant. Needles on the tops of conifers are subjected to more water stress than lower needles because there is a decrease in water potential of 0.01 MPa per meter of height (Kozlowski *et al.*, 1991). Root conductivity of *Pinus resinosa* is higher than that of stems (Stone and Stone, 1975). Water flow resistance varies among primary branches, secondary branches, and the main stem of *Picea sitchensis* (Hellkvist *et al.*, 1974). Because size and allocation to roots, branches, etc., vary with tree age, and these factors all influence water relations, it is reasonable to assume tree age has considerable influence on water relations.

3. Plant Genotype The plant genotype most assuredly impacts many aspects of plant–water relations. Most work in this area has focused on intraspecific genetic variation in drought tolerance (see Newton *et al.*, 1991, and references therein). Because drought tolerance is affected by root:shoot ratios, rooting pattern, cuticle thickness, root suberization, etc., it is likely that these factors are also under genetic control.

Examples of western conifers for which drought tolerance is under genetic control include ponderosa pine (Feret, 1982; Baldwin and Barney, 1976), Douglas fir (Pharis and Ferrell, 1966; Ferrell and Woodward, 1966), and lodgepole pine (Dykstra, 1974; Perry *et al.*, 1978). In the case of Douglas fir, transpiration rates were lower on seed sources from xeric habitats when compared with seed sources from more mesic habitats, which could explain the differences in drought tolerance (Zavitkovski and Ferrell, 1970). Seasonal growth patterns of ponderosa pine roots are also under genetic control (Jenkinson, 1980). Seasonal growth patterns can greatly influence the drought hardiness of ponderosa pine.

Genetic regulation of drought tolerance has been the focus of most research on the effect of genotype on conifer water relations. There is no reason to assume, however, that many other basic plant water physiological processes are not also under genetic control. Perhaps the current interest in genetic engineering of conifer species will encourage more physiologists to focus on the genetic regulation of specific physiological processes that confer drought tolerance. The link between genetic regulation of plant–water relations and herbivory has not been examined.

4. Stand Age, Density, and Structure The role of stand factors in plant–water relations is the subject of other chapters in this volume. Stand factors are known to be major determinants of population dynamics of forest insects; indeed, this knowledge is used as the basis for the widely applied silvicultural approach to forest insect management (Berryman, 1986; Coulson and Witter, 1984; Barbosa and Wagner, 1989). Perhaps the best understood relationship is that of the effect of stand density on plant–water relations and susceptibility to pine bark beetles. It is believed that the primary mechanism of conifer defense against bark beetles is physical resinosis. The production of resin is highly dependent on the water status of the tree; trees under high water stress produce less resin than trees under low to moderate water stress. Consequently, reducing stand density and water stress decreases the susceptibility of the stand to bark beetles. This relationship is generally accepted because there is a strong link between water stress and the primary defensive mechanism of the tree. However, considerable ambiguity exists with regard to how stress influences susceptibility to insects in other cases, which is the topic of the next section.

B. How Does Variation in Plant–Water Relations Affect Insect Herbivores?

There is long-standing conventional wisdom that water stress in plants as a result of atypical weather increases plant susceptibility to insects and disease. Kozlowski *et al.* (1991, p. 282) project this conventional wisdom by stating "water stress usually increases the susceptibility of plants to attack by insects and fungi." Several authors have recently reviewed the literature on this topic (Jones and Coleman, 1991; Mattson and Haack, 1987a; Holtzer *et al.*, 1988; Waring and Cobb, 1992). Although the conventional view is still widely held by scientists, there is increasing evidence, largely from experimental studies, that stress does not inevitably increase the susceptibility of a plant to herbivores (Mattson and Haack, 1987b; Myers, 1988; Larsson, 1989; McCullough and Wagner, 1987; Craig *et al.*, 1991; Wagner and Frantz, 1990). Some elaboration and explanations of the conflicting evidence are the subject of this section.

Stress is only one of many factors that may be responsible for dramatic increases in insect populations, commonly referred to as outbreaks. In Barbosa and Schultz (1987), at least six theories were outlined to explain insect outbreaks. There is evidence to support each of these ideas under certain circumstances. Because only a subset of all outbreaks is potentially caused by stress (assuming by random chance that other factors are equally responsible), then it would be surprising to find a good correlation between outbreaks and stress. The key point is that other factors are frequently responsible for outbreaks independent of stress; consequently, any correlation between forest stress and insect outbreaks is necessarily confounded.

Plant water stress is not a unidimensional phenomenon. The duration and intensity of stress and plant preconditioning and osmoregulation are all factors that can modify how the plant responds to stress and the subsequent impact on herbivores. The basic phrase "stress increases susceptibility to insects" implies that this relationship is either linear or a threshold relationship. In other words, the greater the stress the greater the susceptibility, or "once stressed then susceptible." In reality, the relationship is probably more of a normal or bell-shaped curve. There is likely a point along the stress continuum that is optimal for the herbivore. Stress levels below or above that point are suboptimal (Wagner, 1991). For example, Lorio and Sommers (1986) concluded that moderate stress was unfavorable for the southern pine beetle, because at moderate stress pine growth declined more than resin production, which is the basis for tree defense. Ferrell (1978) observed that a minimum threshold water stress of -1.5 MPa was necessary to make white fir susceptible to engraver beetles. Few studies have ever attempted to quantify water stress using basic physiological methods such as those reviewed by Bennett (1990). Entomologists will need to employ considerably more quantitative measures of stress and work more closely with physiologists to address this problem.

Plant growth form appears to play a significant role in the relationship between water stress and herbivore response (Fig. 6). The general pattern appears to be that annual plants that are water stressed do not become more susceptible to herbivores and may in fact become more resistant. Perennial plants, such as woody species, as a group become more susceptible to herbivores when stressed. However, there is a strong contrast between conifers and hardwoods (Fig. 6). The conclusions in Waring and Cobb (1992) are based on a broad survey of the literature (total of 66 studies on woody plants) and must be followed by rigorous comparative analyses. Despite the potential limitations of an analysis of the literature, there is compelling evidence to warrant further study of this issue.

The feeding guild of the insect herbivore has also been identified as an important trait that influences the effect of water stress on herbivores (Larsson, 1989; Waring and Cobb, 1992). Wood-boring insects and bark beetles have consistently been viewed as responding positively to in-

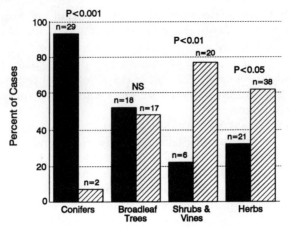

Figure 6 Percentages of positive (solid bars) and negative (striped bars) responses of herbivores to water stress in conifers, broadleaf trees, shrubs and vines, and herbs. NS, No significant difference. (Redrawn from Waring and Cobb, 1992.)

creased water stress. This was confirmed by Waring and Cobb (1992) when they found 100% of all cases showed a positive relationship between stress and herbivore response. The relationship for other guilds is much less clear (Fig. 7). Less than 50% of all mite and galling insect studies showed a positive relationship on either hardwoods or conifers. For other feeding guilds there is a high positive relationship to stress on conifers and little or no relationship for hardwoods. This analysis reveals

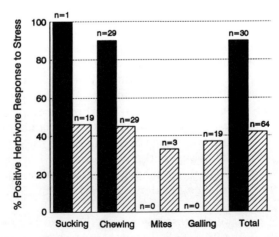

Figure 7 Percentages of conifer (solid bars) and hardwood (striped bars) studies that indicate that herbivore performance (by feeding guilds) improves because of stress. (Modified from Waring and Cobb, 1992.)

that different feeding guilds do not respond uniformly to water stress in trees.

Finally, an issue that has contributed to the ambiguity surrounding stress/herbivore studies is that responses tend to vary depending on whether the studies are natural drought events or experimentally created stress situations (Waring and Cobb, 1992). In general, herbivores respond positively to natural drought events but negatively to experimental water stress treatments (Fig. 8). There are two broad interpretations of these patterns. First, it may be that there are other factors associated with drought besides the direct effects of water stress that are responsible for increased herbivore survival and growth. Many of these alternative ideas are discussed in Barbosa and Schultz (1987). A second interpretation may be that the experimental treatments are not adequately mimicking the water stress effects caused by a natural drought. In the case of research on conifers, only three studies (McCullough and Wagner, 1987; Craig *et al.*, 1991; Wagner and Frantz, 1990) have attempted to use experimentally manipulated water stress monitored by xylem water potential to test for the effect of stress on herbivores; herbivores did not respond or responded negatively in these cases. In all cases of natural drought on conifers, herbivores responded positively (Waring and Cobb, 1992). Obviously considerably more research, especially collaborative research between entomologists and physiologists, is required to further understand these relationships.

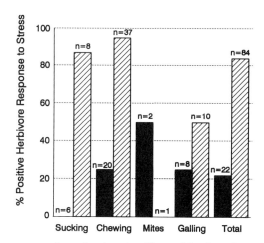

Figure 8 Percentages of woody plant (conifer and hardwood) research studies that indicate that herbivore performance (by feeding guilds) improves because of natural drought (striped bars) or experimental water stress (solid bars). (Modified from Waring and Cobb, 1992.)

C. How Does Herbivory Affect Plant–Water Relations?

The focus of this section thus far is on how plant–water relations affect insect herbivory. Although herbivory is both a cause and consequence of stress, it can also have positive and negative effects on plant–water relations.

Conventional wisdom is that defoliation stresses plants by removing carbohydrate reserves. The impact of defoliation in terms of growth reduction varies for conifers and hardwoods because of general differences in carbohydrate storage. Hardwoods store more carbohydrate in roots and stems and consequently can sustain more defoliation more frequently than conifers. Conifers store carbohydrates more in foliage, and severe defoliation can greatly reduce growth (Kulman, 1971). Any negative effects of herbivory on water relations are probably directly related to the effect on nutrient stress. We could find no evidence to support a direct negative effect of herbivory on water relations that is not confounded by the effect on nutrient stress.

There is considerably more evidence to suggest that herbivory has a positive effect on plant–water relations. Leaf shedding has long been recognized as an adaptation that plants use to regulate water use (Kozlowski et al., 1991; Jones et al., 1981). Shedding of leaves reduces leaf area and transpiration, which will decrease water stress. For example, there are many tropical tree species in West Africa, such as *Milicia excelsa*, that shed their leaves during the dry season to minimize water loss (Hall and Swaine, 1981). Insect defoliation has been shown to decrease transpiration and water stress (Stephens et al., 1972).

The potential beneficial effect of defoliation on plant–water relations may be an important and underappreciated phenomenon. Increased defoliation following a drought might be an important biofeedback process that allows some plant species to occupy dry sites. When a plant that is marginally adapted to a dry site is subjected to drought, it is probably unable to reallocate carbon to roots to compensate for higher evaporative demand. Defoliation could be very beneficial in reducing transpiration and evaporative demand, thereby improving the water relations of the tree at a critical period. Such a relationship might explain why some species of insects, for example, sawflies (Hymenoptera: Diprionidae), tend to occur on trees growing at low density on marginal sites for that tree species (McMillin and Wagner, 1993). For example, populations of *Neodiprion autumnalis* in Arizona are relatively stable on ponderosa pine growing in transition zones between pine forest and grasslands. Even though these are poor sites, the trees sustain significant defoliation of up to 10 years without tree death. The defoliation occurs at the peak of the typical summer drought. In this case the reduced water stress during the summer drought caused by defoliation might be very beneficial to the trees because of lower water stress and increased nutrient cycling. This and other beneficial effects of herbivory on plant–water relations remain largely unstudied.

III. Carbon Acquisition and Allocation

A. What Factors Affect the Acquisition and Allocation of Carbon in Relation to Conifer–Herbivore Interactions?

Within any given plant, certain factors affect the allocation of carbon to different tissue types and to defensive versus growth functions. Variations in resources (e.g., light), stresses (e.g., air pollution), and genotypes (e.g., among species) can influence carbon acquisition and allocation. In this section we will examine environmental, developmental, and genotypic influences on carbon acquisition and allocation within the context of interactions between coniferous trees and their herbivores. After reviewing the causes and consequences of variation in carbon uptake and allocation, we will discuss how variation in these processes affects insect herbivores and how herbivory affects the subsequent acquisition and allocation of carbon by coniferous trees.

Differences in carbon uptake and allocation among species have been hypothesized to be related to species differences in resistance to herbivory (e.g., Coley *et al.*, 1985) as well as in responses to herbivory (not necessarily the same thing). In addition, differences in carbon uptake and acquisition resulting from environmental or developmental factors can also affect resistance and response to herbivory within a given species. Carbon is critical to plant defense both because of the importance of carbon-based allelochemicals and because of the role of carbon in structural defenses. Balances of carbon and nutrients in plants are thought to determine differences among species in the type (i.e., carbon or nitrogen based), costs, and efficacies of their defenses, and whether defenses are induced in response to damage from herbivores. This linkage makes it difficult to separate carbon and nutrient relations. However, we will attempt to address carbon separately in this section, and subsequently integrate carbon and nutrient dynamics in relation to herbivory and conifers.

Carbon-based allelochemicals such as lignin, tannins, terpenes, and resins are related to plant–herbivore interactions and are widely distributed in trees (Bryant *et al.*, 1991). The relative proportion of carbon allocated to growth (primary production) versus to carbon-based defenses (secondary metabolite production) is hypothesized to be a result of the balance between carbon sink strength for growth and carbon source supply from photosynthesis (Lorio, 1988; Tuomi *et al.*, 1988). In other words, carbon not used for growth is used for defensive chemistry. One issue that is not often considered, but perhaps should be, is the extent to which allocation toward physical integrity and durability and allocation toward biological defense differ. Although they are identical in some scenarios, they can be completely distinct in others. It may be useful to keep this context in mind when thinking about allocation, morphology, and plant chemistry.

The hypothesized trade-off between allocation to growth versus defense would lead to faster growing plants having lesser carbon defenses and slower growing plants with greater defenses. This is consistent with empirical data for tropical angiosperm trees (Coley, 1988), but such data are presently not available for conifers. Perhaps carbon that is not devoted to growth processes is used for defensive and/or structural purposes. This would imply that species adapted to low-resource environments are likely to have biotically and abiotically resistant and persistent tissues in order to amortize their construction costs (Kikuzawa, 1991). In such a context, certain species might intrinsically tend to allocate carbon within a leaf to defense (Coley, 1988), regardless of overall supply, whereas in other species allocation to growth would be the first "priority." Empirical data support these hypotheses (Coley, 1988; Reich et al., 1991). A trade-off between productivity and persistence of foliage appears to be an unavoidable and highly regular aspect of nature (Reich et al., 1991, 1992). Differential patterns of whole-plant carbon allocation are apparent among species (and within species when grown in contrasting environments), which explains observed differences in relative growth rates (e.g., Reich et al., 1992; Walters et al., 1993). It has also been hypothesized that maximizing allocation toward growth rate can only occur at the cost of defenses. To provide perspective relative to plant–herbivore interaction, we will briefly review how carbon acquisition and allocation are affected by environment, development, and genotype, and consider in more detail the consequences that might ensue for carbon and plant defense.

In general terms, plants phenotypically respond to variation in environmental resource availability in a fashion that balances the supply of water, nutrients, and carbon—usually observed as a preferential allocation of new biomass to the tissue type(s) involved in acquisition of the "most limiting" resource. Typical examples in trees include increased allocation to roots in response to a shortfall of edaphic resources (low water and/or nutrient availability) and increased allocation to shoots in response to stresses that reduce carbon acquisition (e.g., low light availability, ozone pollution, loss of leaf area due to defoliation).

Although such generalities usually hold when one factor is limiting and all else is held equal, two caveats should be considered. First, plants in nature often face less than optimal conditions for many factors simultaneously. Is there any way that we can generalize how allocation might differ for a plant grown under optimal conditions of all factors versus under elevated CO_2 on an infertile, droughty site in the shade with substantial ozone pollution and periodic leaf herbivory? The first three factors might tend to "push" the plant toward greater root growth, whereas the latter three factors would tend toward the reverse. Second, there is increasing evidence that shifts in allocation may be sometimes limited in scope in comparison with (and/or accompanied by) plasticity in morphology that can alter uptake of resource, and probably also alter the defensive properties of such tissues. Examples might include a greater specific leaf area (SLA; area:mass ratio) under shaded conditions, which increases light interception per unit biomass allocation but results in physically weaker and perhaps chemically less well-defended foliage.

Given that plants may allocate more carbon per leaf under certain sets of circumstances, it is important to ask whether or how such plasticity might alter allocation of carbon to growth versus defense. We might begin with the assumption that if the character of a given tissue type does not change during a shift in allocation, then there is no reason why allocation of carbon within that tissue type should necessarily shift. In essence, a slight difference in a given factor (e.g., light or water availability) might result in a slightly greater allocation of new biomass to foliage, without changing the nutrient status, morphology (SLA), or allelochemistry of the foliage. It is also plausible that tissue quality might change somewhat even without any change in carbon allocation within that tissue type. An example might be a plant subject to slightly greater shade than its neighbor, such that it produced somewhat thinner leaves (lower SLA) that were able to produce an equal amount of carbon per unit leaf tissue as its neighbor (that has denser "sun leaf" foliage) in higher light. In this scenario, the relative availability of carbon and nutrient within the leaf might not be any different, and yet the physical strength and durability of the tissue have been altered, which may lower its physical defenses.

However, if a given factor varies considerably (e.g., deep shade versus high light), not only will allocation be different, but the character of the tissue type will differ substantially, and so will the relative availability of carbon. In deep shade, a given plant will produce leaves that are thinner (higher SLA), but light is in such short supply that this plasticity and shifted allocation cannot totally compensate for reduced carbon gain. Thus, total plant carbon gain is reduced, potentially reducing carbon/nutrient availability, and reducing the allocation toward carbon-based allelochemicals (Lorio, 1988). Given a physically less durable foliage, and with lesser levels of allelochemicals, one might argue that defensive properties have been reduced (with the trade-off of attempting to keep growth rate as high as possible). In essence, given low resources, growth is maximized rather than defense.

B. How Does Variation in Carbon Acquisition and Allocation Affect Insect Herbivores?

Carbon is the dominant component in plant tissue and represents the medium of energy storage and transport. Trees show strong seasonality in carbon allocation to root, shoot, stem, and reproductive growth (Waring, 1987). Trees must balance carbon costs of new growth with assimilation and maintain a positive carbon balance. There is evidence that various plant functions compete for carbon, and carbon allocation patterns can dramatically shift as a result of environmental factors such as water deficit (Tschaplinski and Blake, 1989; Nguyen and Lamant, 1989; Comeau and Kimmins, 1989), insect defoliation (Mihaliak and Lincoln, 1989), and soil nutrient status (Lyr and Hoffman, 1967; Bassman, 1988). There is also evidence that carbon allocation patterns are under genetic control (Joly *et al.*, 1989; Tschaplinski and Blake, 1989; Merritt, 1968; Pregitzer *et al.*, 1990; Kuuluvainen and Kahninen, 1992).

Carbon assimilation and allocation create the seasonal and ontogenetic patterns of secondary compounds, fibers, and nutrients that are important determinants of herbivory. Extrinsic perturbations such as stress may divert organic solutes from growth to osmotic adjustment—which is a plant adaptation to stress—and increase the availability of these products to certain herbivores. This well-recognized phenomenon in plants is the basis for the White hypothesis of drought inducing outbreaks of insects (White, 1984).

A more direct effect of carbon allocation on herbivory is related to the stage of the carbon sink/carbon source transition. Leaves that are at full expansion and shifting from a carbon sink to a carbon source have been shown to be highly preferred by chewing and sucking insects over leaves that are either strong carbon sinks or carbon sources (Coleman, 1986; Jones and Coleman, 1991). The assumption is that these leaves are nutritionally adequate but chemically or structurally undefended (Coleman, 1986). The extent to which this phenomenon occurs outside of *Populus* has yet to be investigated.

C. How Does Herbivory Affect the Acquisition and Allocation of Carbon by Coniferous Trees?

1. Photosynthesis and Carbon Gain Defoliation by herbivores reduces the photosynthetic area per plant and causes a variety of physiological responses generic in most plants. Many of these responses have been interpreted as compensatory mechanisms, largely because they enhance the rate of carbon capture and presumably the rate of growth relative to a defoliated plant that does not exhibit these responses (Caldwell et al., 1981; McNaughton, 1983; Belsky, 1986). These mechanisms include changes in carbohydrate and biomass allocation patterns favoring growth (McNaughton and Chapin, 1985; Oesterheld and McNaughton, 1988) and increases in leaf N or related protein concentrations, as observed often (e.g., Piene, 1980; Caldwell et al., 1981; Heichel and Turner, 1983; Wagner and Evans, 1985), but not always (e.g., Valentine et al., 1983; Tuomi et al., 1984; Reich et al., 1993). Increased photosynthesis of residual and/or regrowth foliage has also been reported in defoliated trees (Heichel and Turner, 1983; Reich et al., 1993; Lovett and Tobiessen, 1993) and other species (e.g., Detling et al., 1979; Painter and Detling, 1981; Caldwell et al., 1981; Wallace et al., 1984). Because reduced leaf area will typically reduce plant carbon gain, the net result on whole-plant growth may be negative, neutral, or positive, depending on the intensity of defoliation and other factors (Reich et al., 1993).

Plant response to defoliation is influenced by the timing of defoliation (Ericsson et al., 1980; Olson et al., 1989; Reich et al., 1993) as well as by the nutritional status of the plant (Mattson, 1980). The time after defoliation (recovery time) should also be considered in interpreting plant response (Oesterheld and McNaughton, 1988). However, the specific influence of such factors is not firmly established in general, and few studies have examined these different issues for a single species. Despite this lack, perhaps the best way to illustrate the impact of defoliation on carbon acquisition and allocation is to depict the step-by-step scenario, using information from different studies as available.

The immediate impact of defoliation is to reduce whole-plant leaf area. Given that

$$\bar{x} \text{ leaf net photosynthetic rate} \times \text{total leaf area} = \text{whole-plant photosynthesis},$$

reduced leaf area should immediately result in diminished carbon gain and growth. Obviously, the position and age of needles in a conifer tree or forest canopy influence the impact of their losses. If the foliage being consumed were shaded needles in the lower canopy (or the oldest cohorts on a branch), their loss might have minor impact on whole-tree net carbon gain, because the shaded two-thirds of a canopy typically contribute roughly one-third toward whole-tree carbon gain (Reich et al., 1990; Ellsworth and Reich, 1993). Conversely, if a tree is young and open-grown and has almost all foliage in high light environments, loss of any foliage may diminish whole-tree carbon gain. The trickier question involves trees or forests with dense layers of foliage that intercept most radiant energy. Even if foliage from the upper sunlit portion of such a canopy were removed, it is possible that lower layers, normally partially to deeply shaded, would now intercept more light and fix more carbon. If only a small portion of the upper canopy were lost, it is feasible that lower layers could make up for their lost productivity without any net loss to the tree or stand. Unfortunately, good data evaluating this scenario are rare or nonexistent, and the exact outcome in any given situation will depend on how much upper canopy foliage is lost, how much lower canopy foliage is in place to "take up the slack," and how effective that foliage is in doing that job. Perhaps the most important point of this scenario is that for complex, multilayered canopies, our simplified carbon balance models (as given specifically for defoliation below) may not be effective descriptors of reality.

Assuming a hypothetical tree, sapling, or seedling without many shaded layers of foliage, defoliation will result in reduced net carbon gain very roughly proportional to the loss in leaf area. However, acclimation (or accidental) responses of trees can compensate for these losses by increasing net carbon gain per unit of remaining foliage. For instance, reduced leaf area following defoliation will often reduce water deficits and stress, alleviating a limitation to carbon assimilation. Probably more important in terms of carbon gain, photosynthetic rates per unit foliage mass or area usually increase following defoliation in unstressed trees (Heichel and Turner, 1983; Lovett and Tobiessen, 1993), including conifers (red pine) (Reich *et al.*, 1993). However, recent data (D. W. Vanderklein and P. B. Reich, unpublished data) suggest no such stimulation of photosynthetic rate in *Larix*.

Net photosynthesis may increase following defoliation due to one of several reasons, including improved root-to-shoot balance and increased leaf N concentrations. Greater root-to-shoot balance might result in greater root-to-leaf hydraulic conductivity, which has been associated with greater stomatal conductance and net photosynthesis in woody plants (but primarily hardwoods) (e.g., Kruger and Reich, 1993; Reich and Hinckley, 1989). However, in a direct test of this hypothesis in red pine, root-to-leaf hydraulic conductance was only slightly and briefly increased following defoliation; at most, defoliation weakly and temporarily stimulated gas exchange rates in needles (Vanderklein and Reich, 1993).

An increase in leaf N following defoliation by herbivores would logically lead to higher net photosynthesis, because photosynthetic capacity and leaf N are functionally related (Field and Mooney, 1986; Reich *et al.*, 1992) in most species. For example, regrowth foliage of trees (Heichel and Turner, 1983) and grasses (Caldwell *et al.*, 1981) had increases in both leaf N and photosynthesis, and increases in N have been observed in conifers (Piene, 1980) and other species following defoliation

(Oesterheld and McNaughton, 1988). However, conifers as a group appear to follow the typical net photosynthesis-to-leaf N relationship less consistently than hardwoods or herbaceous plants (Reich and Schoettle, 1988; P. B. Reich, unpublished data; R. O. Teskey, personal communication; S. T. Gower, personal communication). Moreover, in red pine, net photosynthesis was stimulated by defoliation despite no increase in needle N (Reich *et al.*, 1993), and in red oak, a large increase in net photosynthesis following defoliation was unrelated to leaf N concentration among experimental N availability and defoliation treatments (Lovett and Tobiessen, 1993). Clearly, more data on the gas exchange responses of defoliated trees are needed before we can generalize about the patterns and causes of their responses in this respect (e.g., see Welter, 1989).

Given that stimulation of net photosynthesis is common soon after defoliation, how might this affect whole-plant net photosynthesis? Greater net photosynthesis may partially or fully offset lowered leaf area or mass (e.g., if leaf area is reduced by 20% and net photosynthesis is increased by 10 or 30%, total net carbon gain will be lesser or greater, respectively). This static situation, however, may change rapidly following defoliation. In an experiment with red pine, Reich *et al.* (1993) found that the stimulation of net photosynthesis by defoliation was relatively short-lived (Fig. 9). At 6 weeks following defoliation, 2-year-old seedlings in three different nutrient regimes all had net photosynthesis rates 25–50%

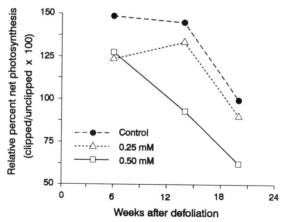

Figure 9 Proportional impact of defoliation on net photosynthetic rate of 2-year-old red pine seedlings in three nutrient availability treatments at 6, 14, and 20 weeks following a 50% defoliation. Photosynthetic rates of clipped plants are shown as a percent of rates for unclipped plants within the same nutrient regime. Nutrient availability treatments: control/very low (●), low (△), and intermediate (□) obtained by no supplement or addition of a ratio of 100:14:50 N:P:K, using 0.25 or 0.50 mM N, to a 4:1 mixture of silica sand and forest soil. (Data from Reich *et al.*, 1993.)

Table II Maximum Change in Net Photosynthetic Rate and Diffusive Leaf Conductance following Defoliation in Young Red Pine Trees[a]

Tree age (years)	Defoliation treatment	Change following defoliation (%)	
		Net photosynthesis	Leaf conductance
2	50%	+25 to 50%	+25 to 50%
4	25, 50, 75%	+10 to 30%	+15 to 50%
11	33, 66%	−5 to +10%	0 to 10%

[a] Data from Reich et al. (1993).

greater than unclipped plants, but by 20 weeks after defoliation, net photosynthesis was equal to or lower than unclipped plants in all three nutrient regimes. In larger, 11-year-old trees, differences in net photosynthesis and leaf diffusive conductance were observed 1 year following defoliation (Reich et al., 1993), suggesting that enhancement may be longer lived in larger plants. However, the degree of enhancement of photosynthesis also seems to be greater in younger, smaller plants (Table II).

2. Growth Shifts in carbon allocation, particularly in distribution of new biomass, limit our "static" model for describing how changes in net photosynthesis and leaf area affect growth following defoliation. During the subsequent period of shoot and needle growth, defoliated pines preferentially shift allocation to new needles rather than to roots (Fig. 10, Reich et al., 1993), consistent with studies in other species. This proportional increase in needle area or mass increases photosynthesis on a whole-plant basis, but at the same time the stimulation of photosynthetic rate gradually disappears because the imbalances that cause it in the first place are slowly resolved.

It has been generally established that for plants in relatively high light, relative growth rate (RGR) is a function of the proportion of a plant in foliage (leaf area or leaf weight ratio, LAR or LWR) multiplied by the photosynthetic rate A on the same basis (i.e., LAR × A_{area} or LWR × A_{mass}) (Poorter and Remkes, 1990; Reich et al., 1992; Walters et al., 1993). Thus, if shifts in allocation of new biomass to foliage, as seen in red pine, can offset losses in leaf area, it is possible that a defoliated plant can emerge with a greater leaf weight ratio (foliage dry mass: whole plant dry mass) than a nondefoliated one after the next period of shoot growth.

Regardless of defoliation intensity, red pine seedlings all preferentially

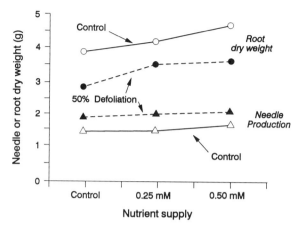

Figure 10 Needle production (dry mass at harvest plus needles removed during defoliation, triangles) and root dry mass (circles) of 2-year-old red pine seedlings 6 months following defoliation. Plants were grown in three nutrient availability treatments: control/very low, low, and intermediate obtained by no supplement or addition of a ratio of 100:14:50 N:P:K, using 0.25 or 0.50 mM N, to a 4:1 mixture of silica sand and forest soil. Data shown for plants either clipped (50% defoliation) (filled symbols) or unclipped (open symbols) just prior to the growing season. (Data from Reich *et al.*, 1993.)

allocated new growth to foliage rather than to roots (Reich *et al.*, 1993). In 2- or 4-year-old seedlings defoliated by 50%, this shift in allocation offset the loss in needle area, and plants recovered root:shoot and LWRs similar to unclipped plants. In severely defoliated plants (75%), the large loss in foliage was only partially overcome by shifted allocation. However, in 4-year-old seedlings defoliated by 25%, this shift in allocation was more than sufficient to offset the relatively slight needle area loss, and the plants had lesser root:shoot ratios and greater LWR compared to the unclipped plants (Fig. 11). The ability to recover needle-to-root balance is important and, given the strong relation between RGR and LWR (Reich *et al.*, 1992; Walters *et al.*, 1993), it may determine the outcome of growth responses to defoliation.

For example, in red pine unclipped or defoliated by 25, 50, or 75%, those with greater LWRs had greatest growth and those with lowest LWRs grew least (Fig. 11). The groups with the greatest growth and least growth were both defoliated plants. Thus, shifts in biomass allocation occur and can be extremely effective in maintaining high RGR, and substantial needle losses can be partially or fully compensated. For example, in red pine, defoliation of less than 30% did not reduce growth, and defoliation had to be at least 50% to induce a 20% growth decrease (Fig. 12) (Kulman, 1965; Reich *et al.*, 1993). But what are the risks to plants of such compensatory responses? The responses described above

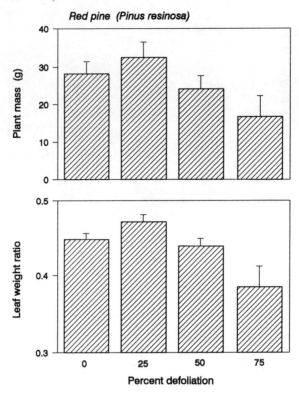

Figure 11 (Top) Total plant dry mass (grams, ± SE) and (bottom) leaf weight ratio (grams of needles: grams of total plant, ± SE) of 4-year-old red pines 1 year following four defoliation treatments. (Data from Reich *et al.*, 1993.)

have a net energy cost, and carbohydrate reserves (TNC, total nonstructural carbohydrates) are lower in plants that have been recently defoliated (O'Neill, 1962; Ericsson *et al.*, 1980; Reich *et al.*, 1993). However, the relative importance of stored carbon to regrowth is still not well understood (Richards and Caldwell, 1985; Busso *et al.*, 1990). Also, once defoliated, red pine stores a greater proportion of its TNC in its roots, which would lessen stored energy losses in subsequent defoliation events (Reich *et al.*, 1993).

In summary, observed carbon assimilation and growth responses of conifers to defoliation appear to follow a consistent pattern. First, photosynthetic rate and leaf conductance are increased due to an altered root-to-needle balance. Second, during the next period of shoot and needle growth, defoliated plants preferentially shift allocation to new needles rather than to roots. This proportional increase in needle area increases photosynthesis on a whole-plant basis, but the stimulation of

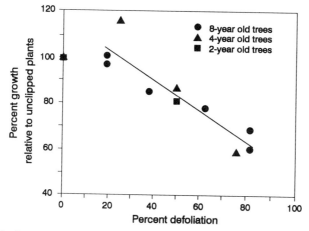

Figure 12 Percent growth relative to unclipped plants for 2-, 4-, and 8-year-old red pine trees as influenced by percent defoliation. This value is linearly related ($r^2 = 0.84$, $P < 0.001$) to defoliation intensity using data from all defoliation treatments (i.e., omitting control plants, which were set equal to 100). (Data from Kulman, 1965, and Reich et al., 1993.)

photosynthetic rate gradually disappears. The enhancement of photosynthesis seems to be greater in younger, smaller plants, but the enhancement also disappears sooner.

IV. Conclusions: Effects of Herbivory on Conifer Forest Ecology

We have examined some of the effects that insect herbivores have on conifer ecophysiology (see Sections I,C, II,C, and III,C). Herbivory can have negative or positive effects on important physiological processes in conifers, including acquisition and allocation of nutrients and carbon, and plant–water relations. Defoliation by insect herbivores reduces a plant's ability to acquire carbon through removing the photosynthetic leaf area. This can lead to substantial root mortality, which impairs the tree's ability to obtain N and other mineral elements from the soil. Thus, herbivory is predicted to decrease levels of carbohydrates, N, and minerals in conifer tissues. However, foliar nutrients can also increase in conifers following defoliation. Furthermore, insect herbivores can help recycle nutrients in conifer forests through effects on plant growth, foliage loss, and litter decomposition. A common response of conifers to defoliation is compensatory growth via stimulation of photosynthesis, increased leaf N, and greater allocation of stored carbohydrates to growth of new foliage versus roots. Defoliation can increase plant moisture stress through increased root mortality, but it can also decrease water stress by reducing leaf area and transpiration.

Given the variable effects of herbivory on ecophysiological processes in conifers, it is difficult to predict how global change might alter insect–plant relationships. For example, increased atmospheric CO_2 tends to increase carbon:nitrogen ratios in plant tissues (Kozlowski *et al.*, 1991). If plants allocate the extra carbon to increased production of carbon-based allelochemicals, host trees should become more resistant to damage from herbivores. On the other hand, herbivores frequently compensate for lower levels of N in host tissues by increasing the amount of food they consume (Clancy, 1992a), implying more damage will occur under conditions of elevated CO_2 (Kozlowski *et al.*, 1991).

Haack and Byler (1993, p. 36) concluded that "the activities of forest insects and pathogens are complex and varied, and can have positive or negative effects depending on management objectives. In an ecosystem context, they can contribute to forest diversity, soil fertility, and long-term forest health and sustainability." Native forest insect herbivores have coevolved with their host trees over thousands of years, and they are undoubtedly important components of all forest ecosystems, functioning as recyclers of nutrients, agents of disturbance, members of food chains, and regulators of the productivity, diversity, and density of plants. There is a growing recognition of the need to better understand the roles that insects (and diseases) play in forest ecosystems in order to use an ecological approach to forest management (e.g., see Schowalter *et al.*, 1986; Schowalter, 1988; Clancy, 1994; Haack and Byler, 1993, and references therein). Improved knowledge of the important ecophysiological processes that determine interactions between insect herbivores and their host plants is central to developing this understanding.

References

Albert, P. J. (1991). A review of some host-plant chemicals affecting the feeding and oviposition behaviors of the eastern spruce budworm, *Choristoneura fumiferana* Clem. (Lepidoptera: Tortricidae). *Mem. Entomol. Soc. Can.* 159:13–18.

Amthor, J. S., and McCree, K. J. (1990). Carbon balance of stressed plants: A conceptual model for integrating research results. *In* "Stress Responses in Plants: Adaptation and Acclimation Mechanisms" (R. G. Alscher and J. R. Cumming, eds.), pp. 1–15. Wiley-Liss, New York.

Baldwin, V. C., and Barney, C. W. (1976). Leaf water potential in planted ponderosa and lodgepole pines. *For. Sci.* 22:344–350.

Barbosa, P., and Schultz, J. C., eds. (1987). "Insect Outbreaks." Academic Press, San Diego.

Barbosa, P., and Wagner, M. R. (1989). "Introduction to Forest and Shade Tree Insects." Academic Press, San Diego.

Bassman, J. H. (1988). Photosynthesis and water relations of ponderosa pine. *In* "Pon-

derosa Pine—The Species and Its Management," Coop. Ext. Symp. Proc., pp. 45–58. Washington State University, Pullman.
Bassman, J. H., and Dickman, D. I. (1985). Effects of defoliation in the developing leaf zone on young *Populus* × *euramericana* plants. II. Distribution of ^{14}C-photosynthate after defoliation. *For. Sci.* 31:358–366.
Bauce, E., and Hardy, Y. (1988). Effects of drainage and severe defoliation on the rawfiber content of balsam fir needles and growth of the spruce budworm (Lepidoptera: Tortricidae). *Environ. Entomol.* 17:671–674.
Beckwith, R. C. (1983). Western larch as a host of the western spruce budworm: A comparison of caged larvae on susceptible conifers. *USDA For. Serv. Gen. Tech. Rep. NE* 85:21–23.
Belsky, A. J. (1986). Does herbivory benefit plants? A review of the evidence. *Am. Nat.* 127: 870–892.
Bennett, J. M. (1990). Problems associated with measuring plant water status. *HortScience* 25:1551–1554.
Bernard-Dagan, C. (1988). Seasonal variations in energy sources and biosynthesis of terpenes in maritime pine. *In* "Mechanisms of Woody Plant Defenses Against Insects: Search for Pattern" (W. J. Mattson, J. Levieux, and C. Bernard-Dagan, eds.), pp. 93–116. Springer-Verlag, New York.
Berryman, A. A. (1986). "Forest Insects: Principles and Practice of Population Management." Plenum, New York.
Brewer, J. W., Capinera, J. L., Deshon, R. E., Jr., and Walmsley, M. L. (1985). Influence of foliar nitrogen levels on survival, development, and reproduction of western spruce budworm, *Choristoneura occidentalis* (Lepidoptera: Tortricidae). *Can. Entomol.* 117: 23–32.
Brewer, J. W., O'Neill, K. M., and Deshon, R. E., Jr. (1987). Effects of artificially altered foliar nitrogen levels on development and survival of young instars of western spruce budworm, *Choristoneura occidentalis* Freeman. *J. Appl. Entomol.* 104:121–130.
Bryant, J. P., Chapin, F. S., III, and Klein, D. R. (1983). Carbon/nutrient balance of boreal plants in relation to vertebrate herbivory. *Oikos* 40:357–368.
Bryant, J. P., Tuomi, J., and Niemela, P. (1988). Environmental constraint of constitutive and long-term inducible defenses in woody plants. *In* "Chemical Mediation of Coevolution" (K. Spencer, ed.), pp. 376–389. Academic Press, San Diego.
Bryant, J. P., Provenza, F. D., Pastor, J., Reichardt, P. R., Clausen, T. P., and du Toit, J. T. (1991). Interactions between woody plants and browsing mammals mediated by secondary metabolites. *Annu. Rev. Ecol. Syst.* 22:431–446.
Busso, C. A., Richards, J. H., and Chatterton, N. J. (1990). Nonstructural carbohydrates and spring regrowth of two cool-season grasses: Interaction of drought and clipping. *J. Range Manage.* 43:336–343.
Caldwell, M. M., Richards, J. H., Johnson, D. A., Nowak, R. S., and Dzuree, R. S. (1981). Coping with herbivory: Photosynthetic capacity and resource allocation in two semiarid *Agropyron* bunchgrasses. *Oecologia* 50:14–24.
Carlson, C. E., and Cates, R. G. (1991). Thinning, foliage chemistry, and defoliation by budworm in immature northern Rocky Mountain Douglas-fir stands: A preliminary assessment. *In* "Interior Douglas-Fir: The Species and Its Management" (D. M. Baumgartner and J. E. Lotan, eds.), pp. 135–140. Washington State University, Spokane.
Carlson, C. E., and Wulf, N. W. (1989). Silvicultural strategies to reduce stand and forest susceptibility to the western spruce budworm. *U.S. Dep. Agric., Agric. Handb.* 676.
Cates, R. G., and Redak, R. A. (1988). Variation in the terpene chemistry of Douglas-fir and its relationship to western spruce budworm success. *In* "Chemical Mediation of Coevolution" (K. C. Spencer, ed.), pp. 317–344. Academic Press, San Diego.

Cates, R. G., Redak, R. A., and Henderson, C. B. (1983). Patterns in defensive natural product chemistry: Douglas-fir and western spruce budworm interactions. In "Plant Resistance to Insects" (P. A. Hedin, ed.), pp. 3–20. Am. Chem. Soc., Washington, DC.

Cates, R. G., Zou, J., and Carlson, C. (1991). The role of variation in Douglas-fir foliage quality in the silvicultural management of the western spruce budworm. In "Interior Douglas-Fir: The Species and Its Management" (D. M. Baumgartner and J. E. Lotan, eds.), pp. 115–127. Washington State University, Spokane.

Clancy, K. M. (1991a). Douglas-fir nutrients and terpenes as potential factors influencing western spruce budworm defoliation. USDA For. Serv. Gen. Tech. Rep. NE 153: 123–133.

Clancy, K. M. (1991b). Multiple-generation bioassay for investigating western spruce budworm nutritional ecology. Environ. Entomol. 20:1363–1374.

Clancy, K. M. (1991c). Western spruce budworm response to different moisture levels in artificial diets. For. Ecol. Manage. 39:223–235.

Clancy, K. M. (1992a). Response of western spruce budworm (Lepidoptera: Tortricidae) to increased nitrogen in artificial diets. Environ. Entomol. 21:331–344.

Clancy, K. M. (1992b). The role of sugars in western spruce budworm nutritional ecology. Ecol. Entomol. 17:189–197.

Clancy, K. M. (1994). Research approaches to understanding the roles of insect defoliators in forest ecosystems. USDA For. Serv. Gen. Tech. Rep. RM-247:211–217.

Clancy, K. M., and King, R. M. (1993). Defining the western spruce budworm's nutritional niche with response surface methodology. Ecology 74:442–454.

Clancy, K. M., Wagner, M. R., and Tinus, R. W. (1988a). Variations in nutrient levels as a defense: Identifying key nutritional traits of host plants of the western spruce budworm. In "Mechanisms of Woody Plant Defenses Against Insects: Search for Pattern" (W. J. Mattson, J. Levieux, and C. Bernard-Dagan, eds.), pp. 201–213. Springer-Verlag, New York.

Clancy, K. M., Wagner, M. R., and Tinus, R. W. (1988b). Variation in host foliage nutrient concentrations in relation to western spruce budworm herbivory. Can. J. For. Res. 18: 530–539.

Clancy, K. M., Foust, R. D., Huntsberger, T. G., Whitaker, J. G., and Whitaker, D. M. (1992). Technique for using microencapsulated terpenes in lepidopteran artificial diets. J. Chem. Ecol. 18:543–560.

Clancy, K. M., Itami, J. K., and Huebner, D. P. (1993). Douglas-fir nutrients and terpenes: Potential resistance factors to western spruce budworm defoliation. For. Sci. 39:78–94.

Codella, S. G., Jr., Fogal, W. H., and Raffa, K. F. (1991). The effect of host variability on growth and performance of the introduced pine sawfly, *Diprion similis*. Can. J. For. Res. 21:1668–1674.

Coleman, J. S. (1986). Leaf development and leaf stress: Increased susceptibility associated with sink-source transition. Tree Physiol. 2:289–299.

Coleman, J. S., Jones, C. G., and Krischik, V. A. (1992). Phytocentric and exploiter perspectives of phytopathology. Adv. Plant Pathol. 8:149–194.

Coley, P. D. (1983). Herbivore and defensive characteristics of tree species in lowland tropical forest. Ecol. Monogr. 53:209–233.

Coley, P. D. (1988). Effects of plant growth rate and leaf lifetime on the amount and type of anti-herbivore defense. Oecologia 74:531–536.

Coley, P. D., Bryant, J. P., and Chapin, F. S., III (1985). Resource availability and plant antiherbivore defense. Science 230:895–899.

Comeau, P. G., and Kimmins, J. P. (1989). Above- and below-ground biomass and production of lodgepole pine on sites with differing soil moisture regimes. Can. J. For. Res. 19: 447–454.

Coulson, R. N., and Witter, J. A. (1984). "Forest Entomology: Ecology and Management." Wiley (Interscience), New York.
Craig, T. P., Wagner, M. R., McCullough, D. G., and Frantz, D. P. (1991). Effects of experimentally altered plant moisture stress on the performance of *Neodiprion* sawflies. *For. Ecol. Manage.* 39:247–261.
Crawford, D. T., Lockaby, B. G., and Somers, G. L. (1991). Genotype–nutrition interactions in field-planted loblolly pine. *Can. J. For. Res.* 21:1523–1532.
Crossley, D. A., Jr., Gist, C. S., Hargrove, W. W., Risley, L. S., Schowalter, T. D., and Seastedt, T. R. (1988). Foliage consumption and nutrient dynamics in canopy insects. *Ecol. Stud.* 66:193–205.
Czapowsky, M. M. (1979). Foliar nutrient concentration in balsam fir as affected by soil drainage and methods of slash disposal. *USDA For. Serv. Res. Pap. NE* 278.
Dale, D. (1988). Plant-mediated effects of soil mineral stresses on insects. *In* "Plant Stress–Insect Interactions" (E. A. Heinrichs, ed.), pp. 35–110. Wiley, New York.
DeBell, D. S., Silen, R. R., Radwan, M. A., and Mandel, N. L. (1986). Effect of family and nitrogen fertilizer on growth and foliar nutrients of Douglas-fir saplings. *For. Sci.* 32:643–652.
Detling, J. K., Dyer, M. I., and Winn, D. T. (1979). Net photosynthesis, root respiration, and regrowth of *Bouteloua gracilis* following simulated grazing. *Oecologia* 41:127–134.
Dupont, A., Bélanger, L., and Bousquet, J. (1991). Relationships between balsam fir vulnerability to spruce budworm and ecological site conditions of fir stands in central Quebec. *Can. J. For. Res.* 21:1752–1759.
Dykstra, G. F. (1974). Photosynthesis and carbon dioxide transfer resistance of lodgepole pine seedlings in relation to irradiance, temperature, and water potential. *Can. J. For. Res.* 4:201–206.
Ellsworth, D. S., and Reich, P. B. (1993). Canopy structure and vertical patterns of photosynthesis and related leaf traits in a deciduous forest. *Oecologia* 96:169–178.
Entry, J. A., Cromack, K., Jr., Kelsey, R. G., and Martin, N. E. (1991). Effect of thinning and thinning plus fertilization on Douglas-fir: Response to *Armillaria ostoyae* infection. *In* "Interior Douglas-Fir: The Species and Its Management" (D. M. Baumgartner and J. E. Lotan, eds.), pp. 147–153. Washington State University, Spokane.
Ericsson, A., Hellkvist, J., Hellerdal-Hagstromer, K., Larsson, S., Mattson-Djos, E., and Tenow, O. (1980). Consumption and pine growth—Hypothesis on effects on growth processes by needle-eating insects. *Ecol. Bull.* 32:537–545.
Feret, P. P. (1982). Effect of moisture stress on growth of *Pinus ponderosa* Dougl. ex. Laws. seedlings in relation to their field performance. *Plant Soil* 69:177–186.
Ferrell, G. T. (1978). Moisture stress threshold of susceptibility to fir engraver beetles in pole-sized white fir. *For. Sci.* 24:85–94.
Ferrell, W. K., and Woodward, E. S. (1966). Effects of seed origin on drought resistance of Douglas-fir (*Pseudotsuga menziesii* (Mirb.) Franco). *Ecology* 47:499–503.
Field, C., and Mooney, H. A. (1986). The photosynthesis-nitrogen relationship in wild plants. *In* "On the Economy of Plant Form and Function" (T. J. Givnish, ed.), pp. 25–55. Cambridge Univ. Press, New York.
Fischer, C., and Holl, W. (1991). Food reserves of Scots pine (*Pinus sylvestris* L.). I. Seasonal changes in the carbohydrate and fat reserves of pine needles. *Trees* 5:187–195.
Gehring, C. A., and Whitham, T. G. (1991). Herbivore-driven mycorrhizal mutualism in insect-susceptible pinyon pine. *Nature (London)* 353:556–557.
Gower, S. T., Vogt, K. A., and Grier, C. C. (1992). Carbon dynamics of Rocky Mountain Douglas-fir: Influence of water and nutrient availability. *Ecol. Monogr.* 62:43–65.
Haack, R. A., and Byler, J. W. (1993). Insects and pathogens: Regulators of forest ecosystems. *J. For.* 91:32–37.

Haack, R. A., and Slansky, F., Jr. (1987). Nutritional ecology of wood-feeding Coleoptera, Lepidoptera, and Hymenoptera. *In* "Nutritional Ecology of Insects, Mites, Spiders, and Related Invertebrates" (F. Slansky, Jr. and J. G. Rodriguez, eds.), pp. 449–486. Wiley, New York.

Haffner, V., Enjalric, F., Lardet, L., and Carron, M. P. (1991). Maturation of woody plants: A review of metabolic and genomic aspects. *Ann. Sci. For.* 48:615–630.

Hale, M. G., and Orcutt, D. M. (1987). "The Physiology of Plants Under Stress." Wiley (Interscience), New York.

Hall, J. B., and Swaine, M. D. (1981). "Distribution and Ecology of Vascular Plants in a Tropical Forest: Forest Vegetation of Ghana." Junk Publishers, London.

Hanover, J. W. (1975). Physiology of tree resistance to insects. *Annu. Rev. Entomol.* 20: 75–95.

Heichel, G. H., and Turner, N. (1983). CO_2 assimilation of primary and regrowth foliage of red maple and red oak: Response to defoliation. *Oecologia* 57:14–19.

Hellkvist, J., Richards, G. P., and Jarvis, P. G. (1974). Vertical gradients of water relations in Sitka spruce trees measured with the pressure chamber. *J. Appl. Ecol.* 11:637–667.

Hobbs, S. D., and Wearstler, K. A., Jr. (1983). Performance of three Douglas-fir stocktypes on skeletal soil. *Tree Planters Notes* 34:11–14.

Holtzer, T. O., Archer, T. L., and Norman, J. M. (1988). Host suitability in relation to water stress *In* "Plant Stress-Insect Interactions" (E. A. Heinrich, ed.), pp. 111–138. Wiley, New York.

Horner, J. D., Cates, R. G., and Gosz, J. R. (1987). Tannin, nitrogen, and cell wall composition of green vs. senescent Douglas-fir foliage. *Oecologia* 72:515–519.

Hughes, P. R. (1988). Insect populations on host plants subjected to air pollution. *In* "Plant Stress–Insect Interactions" (E. A. Heinrichs, ed.), pp. 249–319. Wiley, New York.

Ivanovici, A. M., and Wiebe, W. J. (1981). Towards a working definition of stress: A review and critique. *In* "Stress Effects on Natural Ecosystems" (G. W. Barrett and R. Rosenberg, eds.), pp. 13–27. Wiley, New York.

Jenkinson, J. L. (1980). Improving plantation establishment by optimizing growth capacity and planting time of western yellow pines. *USDA For. Serv. Res. Pap. PSW* 154.

Jensen, T. S. (1991). Patterns of nutrient utilization in the needle-feeding guild. *USDA For. Serv. Gen. Tech. Rep. NE* 153:134–144.

Johnson, P. C., and Denton, R. E. (1975). Outbreaks of the western spruce budworm in the American northern Rocky Mountain area from 1922 through 1971. *USDA For. Serv. Gen. Tech. Rep. INT* 20.

Joly, R. J., Adams, W. T., and Stafford, S. G. (1989). Phenological and morphological responses of mesic and dry site sources of coastal Douglas-fir to water deficit. *For. Sci.* 35:987–1005.

Jones, C. G., and Coleman, J. S. (1991). Plant stress and insect herbivory; toward an integrated perspective. *In* "Response of Plants to Multiple Stresses" (H. Mooney *et al.*, eds.), pp. 249–280. Academic Press, San Diego.

Jones, M. M., Turner, N. C., and Osmond, C. B. (1981). Mechanisms of drought resistance. *In* "The Physiology and Biochemistry of Drought Resistance in Plants" (L. C. Paleg and D. Aspinall, eds.), pp. 15–35. Academic Press, New York.

Kemp, W. P., and Moody, U. L. (1984). Relationships between regional soils and foliage characteristics and western spruce budworm (Lepidoptera: Tortricidae) outbreak frequency. *Environ. Entomol.* 13:1291–1297.

Kikuzawa, K. (1991). A cost–benefit analysis of leaf habit and leaf longevity of trees and their geographical pattern. *Am. Nat.* 138:1250–1263.

Kozlowski, T. T. (1971). "Growth and Development of Trees," Vol. 1. Academic Press, New York.

Kozlowski, T. T. (1979). "Tree Growth and Environmental Stresses." Univ. of Washington Press, Seattle.

Kozlowski, T. T. (1992). Carbohydrate sources and sinks in woody plants. *Bot. Rev.* 58: 107–222.

Kozlowski, T. T., Kramer, P. J., and Pallardy, S. G. (1991). "The Physiological Ecology of Woody Plants." Academic Press, San Diego.

Kramer, P. J., and Kozlowski, T. T. (1979). "Physiology of Woody Plants." Academic Press, New York.

Kruger, E. L., and Reich, P. B. (1993). Coppicing affects growth, root–shoot relations, and ecophysiology of potted *Quercus rubra* seedlings. *Physiol. Plant.* 89:751–760.

Kulman, H. M. (1965). Effects of artificial defoliation of pine on subsequent shoot and needle growth. *For. Sci.* 11:90–98.

Kulman, H. M. (1971). Effects of insect defoliation on growth and mortality of trees. *Annu. Rev. Entomol.* 16:289–324.

Kuuluvainen, T., and Kahninen, M. (1992). Patterns in aboveground carbon allocation and tree architecture that favor stem growth in young Scots pine from high latitudes. *Tree Physiol.* 10:69–80.

Lamb, D. (1985). The influence of insects on nutrient cycling in eucalypt forests: A beneficial role? *Aust. J. Ecol.* 10:1–5.

Larsson, S. (1989). Stressful times for the plant stress—Insect performance hypothesis. *Oikos* 56:277–283.

Larsson, S., and Tenow, O. (1980). Needle-eating insects and grazing dynamics in a mature Scots pine forest in central Sweden. *Ecol. Bull.* 32:269–306.

Lawrence, R. K. (1990). Phenological variation in the susceptibility of white spruce to the spruce budworm. Ph.D. Dissertation, Michigan State University, East Lansing.

Levitt, J. (1972). "Responses of Plants to Environmental Stress." Academic Press, New York.

Lorio, P. L., Jr. (1988). Growth differentiation-balance relationships in pines affect their resistance to bark beetles (Coleoptera: Scolytidae). *In* "Mechanisms of Woody Plant Defenses Against Insects: Search for Pattern" (W. J. Mattson, J. Levieux, and C. Bernard-Dagan, eds.), pp. 73–92. Springer-Verlag, New York.

Lorio, P. L., Jr., and Sommers, R. A. (1986). Evidence of competition for photosynthesis between growth processes and oleoresin synthesis in *Pinus taeda* L. *Tree Physiol.* 2:301–306.

Lovett, G. M., and Tobiessen, P. (1993). Carbon and nitrogen assimilation in red oaks (*Quercus rubra* L.) subject to defoliation and nitrogen stress. *Tree Physiol.* 12:259–270.

Lyr, H., and Hoffman, G. (1967). Growth rates and growth periodicity of tree roots. *Int. Rev. For. Res.* 2: 181–236.

Mason, R. R., Wickman, B. E., Beckwith, R. C., and Paul, H. G. (1992). Thinning and nitrogen fertilization in a grand fir stand infested with western spruce budworm. Part I: Insect response. *For. Sci.* 38:235–251.

Mattson, W. J. (1980). Herbivory in relation to plant nitrogen content. *Annu. Rev. Ecol. Syst.* 11:119–161.

Mattson, W. J., and Addy, N. D. (1975). Phytophagous insects as regulators of forest primary production. *Science* 190:515–522.

Mattson, W. J., and Haack, R. A. (1987a). The role of drought in outbreaks of plant-eating insects. *BioScience* 37:110–118.

Mattson, W. J., and Haack, R. A. (1987b). The role of drought stress in provoking outbreaks of phytophagous insects. *In* "Insect Outbreaks" (P. Barbosa and J. C. Schultz, eds.), pp. 365–407. Academic Press, San Diego.

Mattson, W. J., and Koller, C. N. (1983). Spruce budworm performance in relation to matching selected chemical traits of its hosts. *In* "The Role of Insect–Plant Relationships in the Population Dynamics of Forest Pests" (A. S. Isaev, ed.), pp. 138–148. International Union of Forestry Research Organizations/USSR Academy of Sciences, Krasnoyarsk, USSR.

Mattson, W. J., and Scriber, J. M. (1987). Nutritional ecology of insect folivores of woody plants: Nitrogen, water, fiber, and mineral considerations. *In* "Nutritional Ecology of Insects, Mites, Spiders, and Related Invertebrates" (F. Slansky, Jr. and J. G. Rodriguez, eds.), pp. 105–146. Wiley, New York.

Mattson, W. J., Lorimer, N., and Leary, R. A. (1982). Role of plant variability (trait vector dynamics and diversity) in plant/herbivore interactions. *In* "Resistance to Diseases and Pests in Forest Trees" (H. M. Heybroek, B. R. Stephan, and K. von Weissenberg, eds.), pp. 295–303. Pudoc, Wageningen, The Netherlands.

Mattson, W. J., Slocum, S. S., and Koller, C. N. (1983). Spruce budworm (*Choristoneura fumiferana*) performance in relation to foliar chemistry of its host plant. *USDA For. Serv. Gen. Tech. Rep. NE* 85:55–65.

Mattson, W. J., Haack, R. A., Lawrence, R. K., and Slocum, S. S. (1991). Considering the nutritional ecology of the spruce budworm in its management. *For. Ecol. Manage.* 39: 183–210.

McCullough, D. G., and Kulman, H. M. (1991). Effects of nitrogen fertilization on young jack pine (*Pinus banksiana*) and on its suitability as a host for jack pine budworm (*Choristoneura pinus pinus*) (Lepidoptera: Tortricidae). *Can. J. For. Res.* 21:1447–1458.

McCullough, D. G., and Wagner, M. R. (1987). Influence of watering and trenching ponderosa pine on a pine sawfly. *Oecologia* 71:382–387.

McDonald, G. I. (1983). Douglas-fir progeny testing for resistance to western spruce budworm. *USDA For. Serv. Gen. Tech. Rep. NE* 85:15–16.

McLean, J. A., Laks, P., and Shore, T. L. (1983). Comparisons of elemental profiles of the western spruce budworm reared on three host foliages and artificial medium. *USDA For. Serv. Gen. Tech. Rep. NE* 85:33–40.

McMillin, J. D., and Wagner, M. R. (1993). Influence of stand characteristics and site quality on sawfly population dynamics. *In* "Sawfly Life History Adaptations to Woody Plants" (M. R. Wagner and K. F. Raffa, eds.), pp. 333–361. Academic Press, San Diego.

McNaughton, S. J. (1983). Compensatory plant growth as a response to herbivory. *Oikos* 40:329–336.

McNaughton, S. J., and Chapin, F. S. (1985). Effects of phosphorus nutrition and defoliation on C_4 graminoids from the Serengeti plains. *Ecology* 66:1617–1629.

Merritt, C. (1968). Effect of environment and heredity on the root-growth pattern of red pine. *Ecology* 49:34–40.

Mihaliak, C. A., and Lincoln, D. E. (1989). Plant biomass partitioning and chemical defense: Response to defoliation and nitrate limitation. *Oecologia* 80:120–126.

Mopper, S., and Whitham, T. G. (1992). The plant stress paradox: Effects on pinyon sawfly sex ratios and fecundity. *Ecology* 73:515–525.

Myers, J. H. (1988). Can a general hypothesis explain population cycles in a forest Lepidotera? *Adv. Ecol. Res.* 18:179–242.

Newton, R. J., Funkhouser, E. A., Fong, F., and Tauer, C. G. (1991). Molecular and physiological genetics of drought tolerance in forest species. *For. Ecol. Manage.* 43:225–250.

Nguyen, A. and Lamant, A. (1989). Variation in growth and osmotic regulation of roots of water stressed maritime pine (*Pinus pinaster* A.T.) provenances. *Tree Physiol.* 5: 123–133.

Niemela, P., Rousi, M., and Saarenmaa, H. (1987). Topographical delimitation of *Neodiprion sertifer* (Hym., Diprionidae) outbreaks on Scots pine in relation to needle quality. *J. Appl. Entomol.* 103:84–91.

Oesterheld, M., and McNaughton, S. J. (1988). Intraspecific variation in the response of *Themeda triandra* to defoliation: The effect of time of recovery and growth rates on compensatory growth. *Oecologia* 77:181–186.

Olson, B. E., Senft, R. L., and Richards, J. H. (1989). A test of grazing compensation and optimization of crested wheatgrass using a simulation model. *J. Range Manage.* 42: 458–467.

O'Neill, L. C. (1962). Some effects of artificial defoliation on the growth of jack pine (*Pinus banksiana* Lamb.). *Can. J. Bot.* 40:273–280.

Painter, E. L., and Detling, J. K. (1981). Effects of defoliation on net photosynthesis and regrowth of western wheatgrass. *J. Range Manage.* 34:68–71.

Paynter, V. A., Reardon, J. C., and Shelburne, V. B. (1992). Changing carbohydrate profiles in shortleaf pine (*Pinus echinata*) after prolonged exposure to acid rain and ozone. *Can. J. For. Res.* 22:1556–1561.

Perry, D. A., and Pitman, G. B. (1983). Genetic and environmental influences in host resistance to herbivory: Douglas-fir and the western spruce budworm. *Z. Angew. Entomol.* 96:217–228.

Perry, D. A., Lotan, J. E., Hinz, P., and Hamilton, M. A. (1978). Variation in lodgepole pine family response to stressed induced by polyethylene glycol 6000. *For. Sci.* 24: 523–536.

Pharis, R. P., and Ferrell, W. K. (1966). Differences in drought resistance between coastal and inland sources of Douglas-fir. *Can. J. Bot.* 44:1651–1659.

Piene, H. (1980). Effects of insect defoliation on growth and foliar nutrients of young balsam fir. *For. Sci.* 26:665–673.

Piene, H. (1989). Spruce budworm defoliation and growth loss in young balsam fir: Recovery of growth in spaced stands. *Can. J. For. Res.* 19:1616–1624.

Piene, H., and Little, C. H. A. (1990). Spruce budworm defoliation and growth loss in young balsam fir: Artificial defoliation of potted trees. *Can. J. For. Res.* 20:902–909.

Piene, H., and Percy, K. E. (1984). Changes in needle morphology, anatomy, and mineral content during the recovery of protected balsam fir trees initially defoliated by the spruce budworm. *Can. J. For. Res.* 14:238–245.

Poorter, H., and Remkes, C. (1990). Leaf area ratio and net assimilation rate of 24 wild species differing in relative growth rate. *Oecologia* 83:553–559.

Pregitzer, K. S., Dickman, D. I., Hendrick, R., and Nguyen, P. V. (1990). Whole tree carbon and nitrogen partitioning in young hybrid poplars. *Tree Physiol.* 7:79–93.

Price, P. W. (1991). The plant vigor hypothesis and herbivore attack. *Oikos* 62:244–251.

Reich, P. B., and Hinckley, T. M. (1989). Relationships between leaf diffusive conductance, leaf water potential, and soil-to-leaf hydraulic conductance in oak. *Funct. Ecol.* 3: 719–726.

Reich, P. B., and Schoettle, A. W. (1988). Role of phosphorus and nitrogen in photosynthetic and whole plant carbon gain and nutrient-use efficiency in eastern white pine. *Oecologia* 77:25–33.

Reich, P. B., Ellsworth, D. S., Kloeppel, B. D., Fownes, J. H., and Gower, S. T. (1990). Vertical variation in canopy structure and CO_2 exchange of oak–maple forests: Influence of ozone, nitrogen and other factors on simulated canopy carbon gain. *Tree Physiol.* 7:329–345.

Reich, P. B., Uhl, C., Walters, M. B., and Ellsworth, D. S. (1991). Leaf lifespan as a determinant of leaf structure and function among 23 tree species in Amazonian forest communities. *Oecologia* 86:16–24.

Reich, P. B., Walters, M. B., and Ellsworth, D. S. (1992). Leaf lifespan in relation to leaf, plant and stand characteristics among diverse ecosystems. *Ecol. Monogr.* 62:365–392.

Reich, P. B., Walters, M. B., Krause, S. C., Vanderklein, D., and Raffa, K. F. (1993).

Growth, nutrition and gas exchange of *Pinus resinosa* following artificial defoliation. *Trees* 7:67–77.

Richards, J. H., and Caldwell, M. M. (1985). Soluble carbohydrates, concurrent photosynthesis and efficiency in regrowth following defoliation: A field study with *Agropyron* species. *J. Appl. Ecol.* 22:907–920.

Riemer, J., and Whittaker, J. B. (1989). Air pollution and insect herbivores: Observed interactions and possible mechanisms. *In* "Insect–Plant Interactions" (E. A. Bernays, ed.), Vol. 1, pp. 73–105. CRC Press, Boca Raton, Fla.

Risley, L. S. (1990). Relationships among potassium, calcium and trace elements in tree leaves and associated canopy arthropods. *J. Entomol. Sci.* 25:439–449.

Schmidt, W. C., and Fellin, G. (1983). Effect of fertilization on western spruce budworm feeding in young western larch stands. *USDA For. Serv. Gen. Tech. Rep. NE* 85: 87–95.

Schmitt, M. D. C., Czapowskyj, M. M., Allen, D. C., White, E. H., and Montgomery, M. E. (1983). Spruce budworm fecundity and foliar chemistry: Influence of site. *USDA For. Serv. Gen. Tech. Rep. NE* 85:97–103.

Schonwitz, R., Merk, L., Kloos, M., and Ziegler, H. (1991). Influence of needle loss, yellowing and mineral content on monoterpenes in the needles of *Picea abies* (L.) Karst. *Trees* 5:208–214.

Schowalter, T.D. (1981). Insect herbivore relationship to the state of the host plant: Biotic regulation of ecosystem nutrient cycling through ecological succession. *Oikos* 37: 126–130.

Schowalter, T. D. (1988). Forest pest management: A synopsis. *Northwest Environ. J.* 4: 313–318.

Schowalter, T. D., and Crossley, D. A., Jr. (1983). Forest canopy arthropods as sodium, potassium, magnesium and calcium pools in forests. *For. Ecol. Manage.* 7:143–148.

Schowalter, T. D., Hargrove, W. W., and Crossley, D. A., Jr. (1986). Herbivory in forested ecosystems. *Annu. Rev. Entomol.* 31:177–196.

Schowalter, T. D., Sabin, T. E., Stafford, S. G., and Sexton, J. M. (1991). Phytophage effects on primary production, nutrient turnover, and litter decomposition of young Douglas-fir in western Oregon. *For. Ecol. Manage.* 42:229–243.

Shaw, G. G., Little, C. H. A., and Durzan, D. J. (1978). Effect of fertilization of balsam fir trees on spruce budworm nutrition and development. *Can. J. For. Res.* 8:364–374.

Slansky, F., Jr. (1990). Insect nutritional ecology as a basis for studying host plant resistance. *Fla. Entomol.* 73:359–378.

Stark, N., Spitzner, C., and Essig, D. (1985). Xylem sap analysis for determining nutritional status of trees: *Pseudotsuga menziesii*. *Can. J. For. Res.* 15:429–437.

Stark, R. W. (1965). Recent trends in forest entomology. *Annu. Rev. Entomol.* 10:303–324.

Stephens, G. R., Turner, N. C., and DeRoo, H. C. (1972). Some effects of defoliation by gypsy moth (*Porthetria dispar* L.) and elm spanworm (*Ennomos subsignarius* Hbn.) on water balance and growth of deciduous trees. *For. Sci.* 18:326–330.

Stone, J. E., and Stone, E. L. (1975). Water conduction in lateral roots of red pine. *For. Sci.* 21:53–60.

Stoszek, K. J. (1988). Forests under stress and insect outbreaks. *Northwest Environ. J.* 4: 247–261.

Swank, W. T., Waide, J. B., Crossley, D. A., Jr., and Todd, R. L. (1981). Insect defoliation enhances nitrate export from forest ecosystems. *Oecologia* 51:297–299.

Thomas, A. T., and Hodkinson, I. D. (1991). Nitrogen, water stress and the feeding efficiency of lepidopteran herbivores. *J. Appl. Ecol.* 28:703–720.

Trofymow, J. A., Barclay, H. J., and McCullough, K. M. (1991). Annual rates and elemental concentrations of litter fall in thinned and fertilized Douglas-fir. *Can. J. For. Res.* 21: 1601–1615.

Trumble, J. T., Kolodny-Hirsch, D. M., and Ting, I. P. (1993). Plant compensation for arthropod herbivory. *Annu. Rev. Entomol.* 38:93–119.
Tschaplinski, T. J., and Blake, T. J. (1989). Water relations, photosynthetic as determinants of productivity in hybrid poplars. *Can. J. Bot.* 67:1689–1697.
Tuomi, J., Niemela, P., Haukioja, E., Siren, S., and Neuvonen, S. (1984). Nutrient stress: An explanation for plant anti-herbivore responses to defoliation. *Oecologia* 61: 208–210.
Tuomi, J., Niemela, P., Chapin, F. S., III, Bryant, J. P., and Siren, S. (1988). Defensive responses of trees in relation to their carbon/nutrient balance. *In* "Mechanisms of Woody Plant Defenses Against Insects: Search for Pattern" (W. J. Mattson, J. Levieux, and C. Bernard-Dagan, eds.), pp. 57–72. Springer-Verlag, New York.
Valentine, H. T., Wallner, W. E., and Wargo, P. W. (1983). Nutritional changes in host foliage during and after defoliation, and their relation to the weight of gypsy moth pupae. *Oecologia* 57:298–302.
van den Driessche, R., and Webber, J. E. (1975). Total and soluble nitrogen in Douglas fir in relation to plant nitrogen status. *Can. J. For. Res.* 5:580–585.
Vanderklein, D. W., and Reich, P. B. (1993). Hydraulic conductance and water relations of defoliated red pine seedlings. *Bull. Ecol. Soc. Am., Suppl.* 74:468.
Van Sickle, G. A. (1987). Host responses. *U.S. Dep. Agric., Tech. Bull.* 1694:57–70.
Velazquez-Martinez, A., Perry, D. A., and Bell, T. E. (1992). Response of aboveground biomass increment, growth efficiency, and foliar nutrients to thinning, fertilization, and pruning in young Douglas-fir plantations in the central Oregon Cascades. *Can. J. For. Res.* 22:1278–1289.
Wagner, M. R. (1988). Induced defenses in ponderosa pine against defoliating insects. *In* "Mechanisms of Woody Plant Defenses Against Insects: Search for Pattern" (W. J. Mattson, J. Levieux, and C. Bernard-Dagan, eds.), pp. 141–155. Springer-Verlag, New York.
Wagner, M. R. (1991). Sawflies and ponderosa pine: Hypothetical response surfaces for pine genotype, ontogenic stage, and stress level. *USDA For. Serv. Gen. Tech. Rep. NE* 153:21–34.
Wagner, M. R., and Blake, E. A. (1983). Western spruce budworm consumption—Effects of host species and foliage chemistry. *USDA For. Serv. Gen. Tech. Rep. NE* 85:49–54.
Wagner, M. R., and Evans, P. D. (1985). Defoliation increases nutritional quality and allelochemics of pine seedlings. *Oecologia* 67:235–237.
Wagner, M. R., and Frantz, D. P. (1990). Influence of induced water stress in ponderosa pine on pine sawflies. *Oecologia* 83:452–457.
Wagner, M. R., Clancy, K. M., and Kirkbride, D. M. (1987). Predicting number of oocytes in adult western spruce budworm, *Choristoneura occidentalis* (Lepidoptera: Tortricidae). *Environ. Entomol.* 16:551–555.
Wallace, L. L., McNaughton, S. J., and Coughenour, M. B. (1984). Compensatory photosynthetic responses of three African graminoids to different fertilization, watering, and clipping regimes. *Bot. Gaz. (Chicago)* 145:151–156.
Walters, M. B., Kruger, E. L., and Reich, P. B. (1993). Growth, biomass distribution and CO_2 exchange of northern hardwood seedlings in high and low light: Relationships with successional status and shade tolerance. *Oecologia* 94:7–16.
Waring, G. L., and Cobb, N. S. (1992). The impact of plant stress on herbivore population dynamics. *In* "Insect–Plant Interactions" (E. A. Bernays, ed.), Vol. 4., pp. 167–226. CRC Press, Boca Raton, FL.
Waring, R. H. (1987). Characteristics of trees predisposed to die. *BioScience* 37:569–574.
Waring, R. H., Savage, T., Cromack, K., Jr., and Rose, C. (1992). Thinning and nitrogen fertilization in a grand fir stand infested with western spruce budworm. Part IV: An ecosystem management perspective. *For. Sci.* 38:275–286.
Watt, A. D., Leather, S. R., and Evans, H. F. (1991). Outbreaks of the pine beauty moth

on pine in Scotland: Influence of host plant species and site factors. *For. Ecol. Manage.* 39:211–221.

Webb, W. L., and Karchesy, J. J. (1977). Starch content of Douglas-fir defoliated by the tussock moth. *Can. J. For. Res.* 7:186–188.

Welter, S. C. (1989). Arthropod impact on plant gas exchange. *In* "Insect–Plant Interactions" (E. A. Bernays, ed.), Vol. 1, pp. 135–150. CRC Press, Boca Raton, FL.

Wheeler, G. L., Williams, K. S., and Smith, K. G. (1992). Role of periodical cidadas (Homoptera: Cicadidae: *Magicicada*) in forest nutrient cycles. *For. Ecol. Manage.* 51: 339–346.

White, T. C. R. (1969). An index to measure weather-induced stress of trees associated with outbreaks of psyllids in Australia. *Ecology* 50:905–909.

White, T. C. R. (1976). Weather, food and plagues of locusts. *Oecologia* 22:119–134.

White, T. C. R. (1978). The importance of a relative shortage of food in animal ecology. *Oecologia* 33:71–86.

White, T. C. R. (1984). The abundance of invertebrate herbivores in relation to the availability of nitrogen in stressed food plants. *Oecologia* 63:90–105.

Wickman, B. E., Mason, R. R., and Paul, H. G. (1992). Thinning and nitrogen fertilization in a grand fir stand infested with western spruce budworm. Part II: Tree growth response. *For. Sci.* 38:252–264.

Witter, J. A., Lynch, A. M., and Montgomery, B. A. (1983). Management implications of interactions between the spruce budworm and spruce–fir stands. *USDA For. Serv. Gen. Tech. Rep. NE* 85:127–132.

Wulf, N. W., and Cates, R. G. (1987). Site and stand characteristics. *U.S., Dep. Agric., Tech. Bull.* 1694:89–115.

Zasoski, R. J., Porada, H. J., Ryan, P. J., Greenleaf-Jenkins, J., and Gessel, S. P. (1990). Observations of copper, zinc, iron and manganese status in western Washington forests. *For. Ecol. Manage.* 37:7–25.

Zavitkovski, J., and Ferrell, W. K. (1970). Effect of drought upon rates of photosynthesis, respiration, and transpiration of seedlings of two ecotypes of Douglas-fir. *Photosynthetica* 4:58–67.

7

Leaf Area Dynamics of Conifer Forests

Hank Margolis, Ram Oren, David Whitehead, and Merrill R. Kaufmann

I. Introduction

Estimating the surface area of foliage supported by a coniferous forest canopy is critical for modeling its biological properties. Leaf area represents the surface area available for the interception of energy, the absorption of carbon dioxide, and the diffusion of water from the leaf to the atmosphere. The concept of leaf area is pertinent to the physiological and ecological dynamics of conifers at a wide range of spatial scales, from individual leaves to entire biomes. In fact, the leaf area of vegetation at a global level can be thought of as a carbon-absorbing, water-emitting membrane of variable thickness, which can have an important influence on the dynamics and chemistry of the Earth's atmosphere over both the short and the long term.

Unless otherwise specified, references to leaf area herein refer to projected leaf area, i.e., the vertical projection of needles placed on a flat plane. Total leaf surface area is generally from 2.0 to 3.14 times that of projected leaf area for conifers. It has recently been suggested that hemisurface leaf area, i.e., one-half of the total surface area of a leaf, is a more useful basis for expressing leaf area than is projected area. This is because it is thought to be more appropriate for the scaling of leaf level measurements to the level of the canopy and avoids certain ambiguities inherent in the measurement of projected area of irregularly shaped leaves. However, since most of the existing literature uses projected leaf area, we will continue its use throughout this chapter. The surface area of foliage per unit land surface is referred to as leaf area index (LAI) and is expressed in the dimensionless unit of m^2/m^2. Values of LAI vary enormously with species, climate, stand development, and silvicultural treatment. Relationships between dry matter production and LAI have been studied widely and leaf area is frequently used as an initializing variable in models of forest productivity.

This chapter is concerned with the dynamics of coniferous forest leaf area at different spatial and temporal scales. In the first part, we consider various hypotheses related to the control of leaf area development, ranging from simple allometric relations with tree size to more complex mechanistic models that consider the movement of water and nutrients to tree canopies. In the second part, we consider various aspects of leaf area dynamics at varying spatial and temporal scales, including responses to perturbation, seasonal dynamics, genetic variation in crown architecture, the responses to silvicultural treatments, the causes and consequences of senescence, and the direct measurement of coniferous leaf area at large spatial scales using remote sensing.

II. Leaf Area: Structural and Functional Relationships

A. Simple Allometry

Biologists involved in modeling growth processes of conifer stands need to estimate the leaf area present at a given point in time as well as its rate of change over the course of a season or as the stand grows. Although there are numerous examples wherein LAI has been measured directly in stands, the work is laborious and time consuming. Methods that allow the prediction of foliage area from simpler measurements of stand characteristics are clearly desirable. A further reason for establishing relationships that can be used to predict LAI from easily measurable characteristics is to allow measurements made on a limited number of individual trees in small plots to be scaled on an areal basis up to stand and landscape levels.

Simple allometric relationships have been used widely to relate foliage biomass or area to stem dimensions. The first use of a power function to relate foliage weight, W_f, to the more convenient measurement of tree diameter at breast height (DBH; 1.3 m above ground level) is attributed to Kittredge (1944) working with data from *Pinus ponderosa*. Use of a logarithmic transformation,

$$\log_e(W) = \log_e(a) + b \log_e(\text{DBH}), \qquad (1)$$

where a and b are empirical coefficients, ensures that statistical criteria concerning the distribution of variance are met (Causton, 1985). Although DBH is widely used in the linear allometric relationship, the inclusion of alternative stem characteristics into the equation has improved its predictability for some species (Whitehead and Jarvis, 1981). More recently, improvements to Eq. (1) have been proposed that allow the coefficient b to vary linearly with DBH (Geron and Ruark, 1988).

Coefficients have been developed for old-growth forest communities, e.g., 450-year-old stands dominated by *Pseudotsuga menziesii* in the western United States (Grier and Logan, 1977). The linear allometric equation has also been used successfully in stands with a large number of small trees, e.g., with 29-year-old *Chamaecyparis obtusa* (Mori and Hagihara, 1991). The same approach has been used with branches to estimate foliage distribution within *Pinus taeda* crowns (Hepp and Brister, 1982). However, it is clear that while being very useful at the sites where they are derived, the coefficients cannot be assumed to be constant for the same species growing at different sites even within the same region. Differences in the coefficients have been determined by Madgwick (1983b) working with *Pinus radiata* in Australia, New Zealand, and South Africa as well as by Baker *et al.* (1984) working with the same species in Victoria, Australia. Difficulties also arise when the equations are used for trees outside the diameter range for which the coefficients were determined (Gower *et al.*, 1987).

B. Pipe Model Theory

The pipe model theory is based on the concept that a given unit of transpiring foliage is supplied with water by a corresponding unit of conducting sapwood. The existence of a relationship between the quantity of foliage supported by a tree and its sapwood area first appeared in the literature in the early part of the twentieth century (Huber, 1928; Büsgen and Münch, 1929, cited in Maguire and Hann, 1987), but Shinozaki *et al.* (1964a,b) were the first to propose the pipe model theory. Through much of the 1970s and 1980s, the theory has served as the basis for the development of linear equations for the prediction of leaf biomass or leaf area (A_f) from cross-sectional sapwood area (A_s) for a wide range of coniferous species (e.g., Grier and Waring, 1974; Waring *et al.*, 1977; Whitehead, 1978; Kendall Snell and Brown, 1978; Kaufmann and Troendle, 1981; Marchand, 1984; Oren *et al.*, 1986b).

A_f:A_s ratios reported for various conifers range between 0.08 m^2/cm^2 for *Juniperus monosperma* (Schuler and Smith, 1988) to 0.72 m^2/cm^2 for *Abies balsamea* (Coyea and Margolis, 1992) and 0.75 for *Abies lasiocarpa* (Kaufmann and Troendle, 1981, as adjusted by Waring *et al.*, 1982). Table I is an updated list, first published in Waring *et al.* (1982), of linear

Table I Examples of Published Leaf Area to Sapwood Area Ratios for Conifers[a]

Species	Common name	$A_f:A_s$ (m²:cm²)	Sources
Abies balsamea	Balsam fir	0.67–0.71	Coyea and Margolis (1992); Marchand (1984)
Abies amabilis	Pacific silver fir	0.63	Waring *et al.* (1982)
Abies grandis	Grand fir	0.51	Waring *et al.* (1982)
Abies lasiocarpa	Subalpine fir	0.75	Kaufmann and Troendle (1981)[a]
Abies procera	Noble fir	0.27	Cited in Waring (1980); Grier and Waring (1974)
Juniperus monosperma	One-seeded juniper	0.08	Schuler and Smith (1988)
Juniperus occidentalis	Western juniper	0.18	Gholz (1980)
Larix occidentalis	Western larch	0.50	Bidlake and Black (1989)
Picea abies	Norway spruce	0.46	Oren *et al.* (1986b)
Picea engelmanni	Engelmann spruce	0.29–0.34	Waring *et al.* (1982); Kaufmann and Troendle (1981)[b]
Picea sitchensis	Sitka spruce	0.45	As cited in Waring *et al.* (1982)
Pinus contorta	Lodgepole pine	0.11–0.30[c]	Dean *et al.* (1988); Hungerford (1987); Keane and Weetman (1987); Dean and Long (1986); Kaufmann and Troendle (1981)[b]
Pinus edulis	Pinyon pine	0.25	Schuler and Smith (1988)
Pinus nigra var. *maritima*	Austrian pine	0.15	As cited in Waring *et al.* (1982)
Pinus ponderosa	Ponderosa pine	0.19	Waring *et al.* (1982)
Pinus sylvestris	Scotch pine	0.137	Whitehead (1978)
Pinus taeda	Loblolly pine	0.13–0.30	Blanche *et al.* (1985); Jacobs (1988)
Pseudotsuga menziesii	Douglas fir	0.38–0.70	Borghetti *et al.* (1986); Espinosa-Bancalari *et al.* (1987); Brix and Mitchell (1983); Waring *et al.* (1982)
Tsuga heterophylla	Western hemlock	0.46	Waring *et al.* (1982)
Tsuga mertensiana	Mountain hemlock	0.16	Waring *et al.* (1982)

[a] Updated from Waring (1983). Note that the validity of applying published $A_f:A_s$ ratios to specific situations should always be verified. Some studies (e.g., Dean *et al.*, 1988; Long and Smith, 1988) have shown that the application of linear $A_f:A_s$ ratios can be inappropriate for certain species and situations.
[b] Projected leaf area calculated as total leaf area divided by 2.5 as cited in Waring *et al.* (1982).
[c] Not including senescent trees.

$A_f:A_s$ ratios reported for a number of different conifers. Ratios of $A_f:A_s$ tend to be lower for species found in more xeric environments and greater for those in more humid environments (Table I).

The $A_f:A_s$ equation takes the general form:

$$A_f = (bA_s) - a, \qquad (2)$$

where A_f is foliage area and a and b are coefficients, although a is often assumed to be zero. Bormann (1990) found this approach particularly useful for estimating foliage area for contrasting *Picea sitchensis* stands in Alaska, where the trees were equal in diameter but differed in the proportion of the diameter comprised of sapwood. Previous estimates of foliage area in stands dominated by 450-year-old *Pseudotsuga menziesii* based on allometric relationships with tree diameter overestimated leaf area in comparison to estimates made using Eq. (2) (Marshall and Waring, 1986).

Strict application of the pipe model theory demands that if A_s is measured at the base of the live crown, then the $A_f:A_s$ ratio must be constant for a given species. Furthermore, the theory implies that A_s should decrease in proportion to the A_f remaining above any given point on the main stem from the base of the live crown to the top of individual trees (e.g., Long *et al.*, 1981).

Waring *et al.* (1982) proposed that the taper of sapwood cross-sectional area between breast height and the base of the live crown needs to be taken into account for trees with a significant amount of branch-free bole. They also suggested that linear sapwood taper equations between breast height and the base of the live crown could be incorporated to allow estimates of leaf area from sapwood area at breast height. Subsequently, Maguire and Hann (1987) demonstrated that quadratic–quadratic segmented polynomials were the most appropriate sapwood taper equations for *P. menziesii*.

As general interest increased in obtaining rapid measurements of leaf area of forest trees, the pipe model theory was tested over a wider range of site conditions. It soon became clear that the $A_f:A_s$ ratio could be fairly site specific and that factors such as stand density and site quality can exert a significant effect (e.g., Binkley and Reid, 1984; Pearson *et al.*, 1984; Keane and Weetman, 1987; Espinosa-Bancalari *et al.*, 1987). Other studies, however, showed little effect of these factors (e.g., Whitehead, 1978; Blanche *et al.*, 1985; Hungerford, 1987; Coyea and Margolis, 1992).

Brix and Mitchell (1983) found that the addition of nitrogen fertilizer, particularly when combined with thinning, significantly increased $A_f:A_s$ ratios of *P. menziesii* when A_s was measured at breast height (Fig. 1). An influence of thinning for *P. menziesii* was also reported by Granier

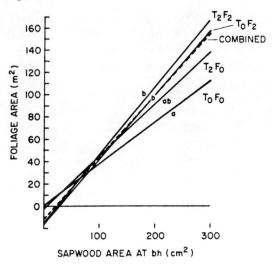

Figure 1 Relationships of projected foliage area and sapwood area at breast height (bh) for *Pseudotsuga menziesii* under different thinning and nitrogen fertilization treatments (from Brix and Mitchell, 1983). F_0, No fertilizer applied; F_2, 448 kg N/ha applied as urea; T_0, no spacing conducted; T_2, spacing reduced basal area from 23.1 to 8.3 m²/ha. Lines with different letters have significantly different slopes ($P < 0.05$).

(1981). Similarly, Albrektsson (1984) improved the precision of predicting A_f from A_s for 16 different *Pinus sylvestris* stands by including the mean annual ring width of sapwood as a covariate. Espinosa-Bancalari *et al.* (1987) measured differences among $A_f:A_s$ ratios for three 22-year-old *P. menziesii* plantations that had exhibited slow, intermediate, and fast rates of growth in the early years after plantation establishment. The variations in $A_f:A_s$ ratios in these stands were strongly correlated with mean annual ring width of the sapwood (Fig. 2). Furthermore, Thompson (1989) showed that the $A_f:A_s$ ratios of dominant and subdominant *Pinus contorta* trees differed within a single stand and appeared to be nonlinear (Fig. 3). Dean and Long (1986) noted similar differences in $A_f:A_s$ ratios between suppressed trees and other crown classes, as did Oren *et al.* (1986b) for *Picea abies* saplings growing in an understory versus a nearby clearcut.

Long and Smith (1988) proposed that the $A_f:A_s$ relationship for *P. contorta* is nonlinear when a wide range of tree sizes and ecological conditions is considered. They demonstrate that apparent effects of tree spacing and site quality are a consequence of the nonlinearity of the $A_f:A_s$ relation with increasing A_s. On the other hand, Coyea and Margolis (1992) found that $A_f:A_s$ ratios in a wide range of *A. balsamea* stands was

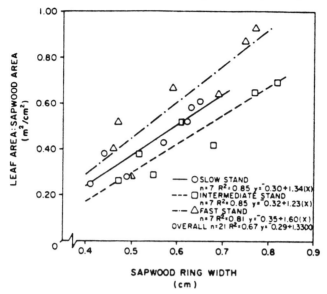

Figure 2 Relationship of the leaf area to sapwood area ratio and sapwood ring width for seven stem sections of *Pseudotsuga menziesii* growing in plantations exhibiting slow, intermediate, and fast growth in the early years following planting. (Adapted from Espinosa-Bancalari *et al.*, 1987.)

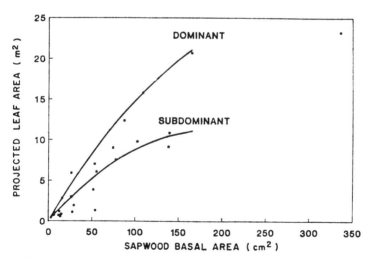

Figure 3 Relationship between projected leaf area and sapwood basal area for dominant and subdominant *Pinus contorta*. (From Thompson, 1989.)

linear in relation to increasing A_s. It is possible that differences in sapwood taper between breast height and the base of the live crown or differences in the maximum size attained by the two species could explain the apparent contradiction between the two studies.

In summary, caution is recommended when applying constant values of $A_f:A_s$, particularly across a large geographical area. The risk of systematic bias is high and this can result in misleading conclusions. For example, Dean et al. (1988) showed that when $A_f:A_s$ is obtained from a sample of *P. contorta* trees in a stand, and the average value is applied to all the trees in the stand, stand growth rate was strongly dependent on LAI. However, when LAI was calculated using the variation in $A_f:A_s$ between individual trees, then stand growth rate was only weakly dependent on LAI. Therefore, the procedure chosen must ensure an unbiased estimate of LAI. Despite these caveats, the application of the pipe model theory to predict stand leaf area index is useful, and its application to silvicultural problems has led to substantial progress in developing the capacity to manage forest stands on a physiological basis (e.g., Waring, 1983; Mäkelä, 1986; Vose and Swank, 1990) (see also Section III,D).

C. The Hydraulic Model

1. Theory Differences in $A_f:A_s$ between trees of the same species growing on different sites and with height within the crowns of individual trees have led to attempts to create more general models of leaf area development in forest trees. The hydraulic model proposed by Whitehead et al. (1984) is based on Darcy's law and, in contrast to the pipe model theory, it explicitly takes into account many of the physical characteristics of the hydraulic pathway between the stem and the foliage. It builds on previous work by Zimmermann (1978) and Ewers and Zimmermann (1984). For an individual tree, the volume flow rate of water (q) through a tree stem of length (l) is related to the cross-sectional sapwood area (A_s), the water potential difference ($\Delta\Psi$), the saturated permeability of the sapwood (k), and the viscosity (ζ) of the water such that

$$q = \frac{kA_s \Delta\Psi}{\zeta l}. \tag{3}$$

The transpiration rate (E_t) from an aerodynamically rough conifer canopy can be modeled adequately by Eq. (4):

$$E_t = \frac{c_p \rho_a}{\lambda \gamma} D \bar{g}_s A_f \tag{4}$$

(McNaughton and Black, 1973; Jarvis and Stewart, 1979), where D is the water vapor pressure deficit of the air, $\bar{g}_s$ is the mean stomatal conductance weighted in relation to the physiological activity of each cohort of foliage (Leverenz et al., 1982), and A_f is the foliage area of the average tree (LAI/n). The coefficients, c_p, ρ_a, λ, and γ are the specific heat of air, the density of air, the latent heat of vaporization of water, and the psychrometric constant, respectively, and are all weakly dependent on temperature.

By assuming that the transpiration rate for an average tree in a stand is equal to the flow rate of water passing through the stem,

$$q = \frac{E_t}{\rho_w n} = \frac{c_p \rho_a}{\lambda \gamma \rho_w} D\bar{g}_s A_f, \tag{5}$$

where n is the number of trees per hectare of average foliage area, and ρ_w is the density of water at the appropriate temperature.

When q in Eq. (5) is substituted into Eq. (3),

$$\frac{A_f}{A_s} = \frac{k \Delta \Psi}{c\zeta l D\bar{g}_s}, \tag{6}$$

where $c = (c_p\rho_a)/(\lambda\gamma\rho_w)$ and l refers to tree height. Thus, Whitehead et al. (1984) provide a theoretical basis for a relationship between leaf area and sapwood area that is directly proportional to sapwood permeability and to the water potential gradient in the stem and inversely proportional to the driving variables for transpiration represented by vapor pressure deficit and mean stomatal conductance.

2. Empirical Evidence in Support of the Theory When A_f:A_s ratios of 10 control *P. sitchensis*, 10 *P. sitchensis* with fertilizer added, and 10 *P. contorta* trees were examined, they were different for the two species (Fig. 4a) (Whitehead et al., 1984). However, when the relationship between A_f and the product $(A_s k)$ was examined, differences between the two species disappeared and all data could be described by a single line (Fig. 4b). The vertical distribution of foliage within one *Picea* with fertilizer added, one control *Picea*, and one *Pinus* tree was also examined in relation to the sapwood area at each respective internode. As suggested by Eq. (6), considerable improvement in the relationship between the two variables was obtained when both k and the internode length were included in the model.

Similarly, Coyea and Margolis (1992) found that A_f:A_s ratios in balsam fir were positively influenced by k. As suggested by the hydraulic model, there was also a negative correlation between the ratio of A_f to $A_s k$ with both tree height and crown length.

3. Influence of Site Quality and Age on Sapwood Permeability (k) The inclusion of k to account for differences in A_f:A_s ratios in forest stands

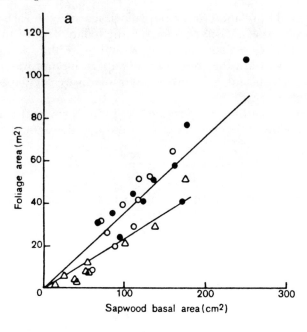
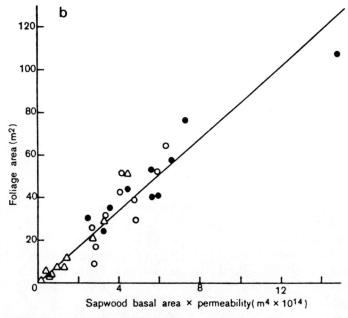

Figure 4 Relationships between (a) foliage area and sapwood basal area (1.3 m above ground level); (b) foliage area and the product of sapwood basal area and permeability. Symbols denote (○) 10 control *Picea sitchensis*, (●) 10 *Picea sitchensis* with fertilizer added, and (△) 10 control *Pinus contorta*. (From Whitehead *et al.*, 1984.)

suggests that it is particularly pertinent to understand the factors influencing k. Several studies have suggested that k increases proportionally to tree growth rate (Booker and Kininmonth, 1978; Edwards and Jarvis, 1982; Whitehead *et al.*, 1984). On the other hand, Comstock (1970) found that k in *Tsuga canadensis* was not influenced by growth rate. These contrasting results might be related to differences in the distribution of tree age within the stands measured.

After examining the combined effects of age and growth rate on k for even-aged jack pine stands, Pothier *et al.* (1989a) proposed a general model to show how k changes with stand development on sites of different quality (Fig. 5). There was a strong increase in k with increasing age and height but the rate of increase for k was greater on sites of better quality. Changes in sapwood permeability over time are similar in this regard to other aspects of stand development such as total height and leaf area development (Oren *et al.*, 1987), i.e., rates of increase are greater on better quality sites. Thus, it is possible for a slow-growing, 120-year-old tree to have a greater k than a fast-growing 15-year-old tree. Some of the anatomical reasons for this phenomenon are discussed in the following section.

When the low k values typical of young stands are combined with the

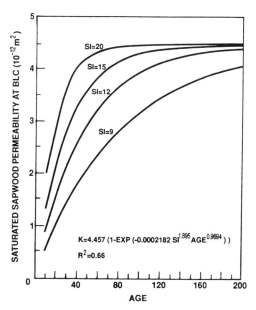

Figure 5 Model of sapwood permeability as a function of age and site quality for *Pinus banksiana* stands; $n = 120$ trees from 11 stands. Site index (SI) units are expressed in meters at age 50. BLC, base of the live crown. (From Pothier *et al.*, 1989a.)

slow rates of diameter growth that occur on poor-quality sites, the result is a k value much lower than that generally found for other trees of the same species. This is likely to be related to the stagnated height growth which sometimes occurs when trees are grown in close spacing and/or under low fertility. For example, Keane and Weetman (1987) measured $A_f:A_s$ ratios for stagnated *P. contorta* that were only 50% that of more normally developed stands of the same species.

The strong increase in k with age is consistent with the observation that midday leaf water potential is approximately the same at different stages of development whereas similar maximum transpiration rates per unit leaf area are maintained. For example, values of k measured by Pothier *et al.* (1989a) were four times greater for a 35-year-old stand growing on a good-quality site than for a stand of the same age growing on a poor-quality site. Thus, if we ignore the differences in the gravitational gradient, the trees on the good-quality site could theoretically be four times taller and still move the same amount of water per unit area of sapwood for the same water potential difference along the stem.

4. Relationship of Sapwood Permeability to Wood Anatomical Characteristics Changes in k that occur due to differences in both growth rate and tree age result from changes in the anatomical characteristics of sapwood cells. Pothier *et al.* (1989b) found that the same asymptotic negative exponential model could be fitted to either k or to the length of sapwood tracheids (Figs. 5 and 6) in *P. banksiana* stands of different age and site quality. Coyea and Margolis (1992), also found that k was proportional to tracheid length in *A. balsamea*. This is probably because tracheid length determines the number of times water must pass through bordered pits. Tracheid length may also be related to k because of its positive correlation with the number of pits per tracheid. Tracheid length, however, was only slightly influenced by site quality (Fig. 6), suggesting the involvement of other anatomical factors. Thus, differences in k that occur with site quality appear to be due to factors other than tracheid length.

The two variables that best explain the remaining variation in k in *P. banksiana* were the relative water content of sapwood and the diameter of the tracheid lumens. At lower sapwood relative water contents, k is reduced because of tracheid embolism (Waring and Running, 1978, Zimmermann, 1983; Tyree and Dixon, 1986). However, the diurnal variation in relative water content is appreciable and consequently the system of cavitation and repair is quite dynamic (Waring *et al.*, 1979). Tracheid lumen diameter, on the other hand, was important in influencing k values only in younger trees. The influence of tracheid diameter on k may be related to the positive correlation of lumen diameter with the number or area of pit membranes. It may also be related to the

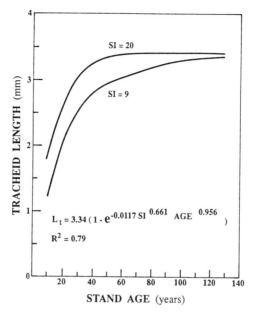

Figure 6 Model of sapwood tracheid length as a function of age and site quality for *Pinus banksiana* stands, $n = 120$ trees from 11 stands. Site index (SI) units as in Fig. 5. (From Pothier *et al.*, 1989b.)

importance of tracheid diameter on the ability of a tree to reverse embolisms (Sobrado *et al.*, 1992).

Thus, tracheid length, sapwood relative water content, and tracheid lumen diameter (particularly in younger stands) seem to be the factors that influence k in conifers. It is reasonable to hypothesize that these same anatomical characteristics play a role in determining differences in $A_f:A_s$ ratios and consequently provide an internal control on the development of leaf area in trees and forests.

D. Functional Relationship between Foliage and Roots

Little is known about the mechanisms that determine root activity in relation to water and nutrient uptake. Landsberg (1986) argued that the dynamic nature of fine roots is particularly important for allowing a flexible response to changing conditions of water and nutrient availability. Similarly, the amount and distribution of foliage area are also dynamic in response to changing environmental conditions. Thus, the concept of a functional relationship between the development of foliage and roots appears to be plausible, although there are likely to be seasonal differences in phenology and relative growth rate between the components.

Evidence from work with seedlings suggests that there is a genetic control on the development of the root system in relation to the foliage.

Allometric ratios of the absolute growth rates for foliage and roots were the same for different families of *P. taeda* during the first 2 years of development (Drew and Ledig, 1980) and during the first 3 years for different provenances of *P. sitchensis* and *P. contorta* seedlings (Cannell and Willett, 1976).

In a number of studies, good relationships between tree stem or foliage dimensions have been used to predict the amount of roots on individual trees. The volume of fine roots was linearly related to stem cross-sectional area and foliage area in 1- to 3-year-old *P. taeda* (Johnson *et al.*, 1985) and similar relationships have been found for older trees. There was a linear relationship between total cross-sectional area of coarse roots, stem cross-sectional area below the lowest living whorl, and total cross-sectional area of branches in slow-growing 6- to 44-year-old *P. sylvestris* trees (Kaipiainen and Hari, 1985). In 3- to 8-year-old *P. radiata*, the weight of fine roots could be predicted linearly from foliage weight and the weight of larger roots could be predicted from a power function using tree diameter (Jackson and Chittenden, 1981). A simpler approach is the useful correlation between total carbon allocated to roots and the quantity of litterfall developed by Raich and Nadelhoffer (1989). Hendrick and Pregitzer (1993) reported a method of studying fine root dynamics from direct observations using a root periscope.

Although the evidence suggests tight genetic regulation of the ratio of the amount of root to foliage biomass, there is also evidence from both seedlings and older trees that this ratio can be controlled by environmental factors. The ratio of root to foliage biomass has been shown to be lower in seedlings, e.g., *P. taeda* (Johnson *et al.*, 1985) and *P. contorta* (Lieffers and Titus, 1989) grown at high fertility and at wide spacing. The percentage of carbon used for fine root turnover was less on sites with higher fertility in 40-year-old *P. menziesii* (Keyes and Grier, 1981). The ratio of roots to foliage has also been shown to increase with increasing tree number per unit area in *P. contorta* stands greater than 75 years old (Pearson *et al.*, 1984), and removal of 60% of the basal area by thinning a 12-year-old *P. radiata* plantation reduced the quantity of fine roots present by half, and slightly decreased the annual production (Santantonio and Santantonio, 1987).

Increased root production has been shown to occur with stand age in mature *Abies amabilis* (Grier *et al.*, 1981) and *P. menziesii* (Santantonio, 1989) stands. Many studies have shown an increase in the ratio of root biomass to foliage biomass when conifers are grown under conditions of elevated carbon dioxide in nutrient-limiting conditions (e.g., Eamus and Jarvis, 1989; Campagna and Margolis, 1989). Worrall *et al.* (1985) showed that "stagnation" (the trees fail to develop normal height growth) in 20-year-old *P. contorta* could be reversed 3 years after grafting scions on to vigorous root stocks, supporting the suggestion that the root system exerts control on shoot activity. Thus, although the genetic capacity of a tree determines the relationship between roots and foliage under a given environment, this same relationship has been shown to have a very strong capacity for acclimating to different or changing environmental conditions.

III. Leaf Area Dynamics

A. Response of the $A_f:A_s$ Relation to Stand Perturbation

In Section II,B, it was shown that the cross-sectional area of sapwood exerts a strong control on leaf area development. It follows, therefore, that a reduction in A_f should result in a corresponding reduction in A_s. In order to test the applicability of the pipe model theory to such manipulations, Margolis et al. (1988) reduced the live crown ratios (LCR; the length of the crown divided by the height of the tree) of A. balsamea trees in the same stand from 0.8 LCR to 0.6, 0.4, and 0.2 LCR, corresponding to the removal of approximately 15, 75, and 95% of the leaf biomass, respectively. Although the two severe pruning treatments resulted in changes in A_s, the nature of the adjustment differed with the two treatments (Fig. 7). In the LCR = 0.4 pruning treatment, the reduction in A_s could be accounted for by the reduction in basal area growth after pruning, and the area of heartwood was not affected. However, the greater

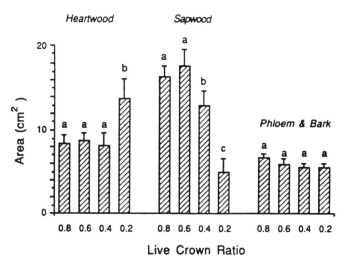

Figure 7 Cross-sectional area of heartwood, sapwood, and bark/phloem at breast height of *Abies balsamea* trees in October, 1986, after they were pruned from 0.8 (unpruned control) to 0.6, 0.4, and 0.2 live crown ratios in June, 1985. Different letters indicate significant differences at $P < 0.08$. (Adapted from Margolis et al., 1988.)

reduction in sapwood area in the LCR = 0.2 treatment resulted in both a reduction in basal area growth and an increase in the cross-sectional area of heartwood (Fig. 7).

The homeostatic adjustment of A_s in response to a reduction in A_f by decreasing the growth rate of new sapwood without increasing the area of heartwood could have an adaptive advantage, e.g., when defoliation is caused by a transient stress such as insects. This would allow more of the available carbon to be used for replacing lost leaf area. If the heartwood cross-sectional area does not increase following defoliation, the sapwood is available to support new foliage without the need to reconstruct new water-conducting tissue. Thus the nature of the sapwood adjustment in the LCR = 0.4 treatment allowed a relatively quick return to the predisturbance equilibrium between A_f and A_s. With the more severe LCR = 0.2 treatment, there was an irreversible increase in the heartwood of all surviving trees in addition to a 56% rate of tree mortality.

Further information can be gained by examining simultaneously the response of both A_f and A_s following a sudden change in environmental conditions. For example, Pothier and Margolis (1991) measured the increase in A_f and A_s for *A. balsamea* and *Betula papyrifera* following precommercial thinning (Fig. 8). Thinning in *A. balsamea* increased sapwood area growth more than leaf area growth. This is because of the restricted growth of new leaf area in the first year after treatment due to the determinant growth habit of conifers. Sapwood area growth, on the other hand, is not restricted in this manner (Lanner, 1985). Thus,

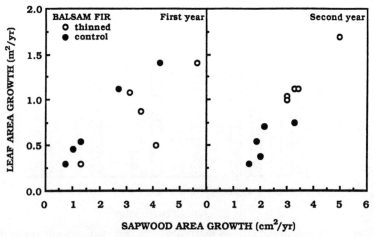

Figure 8 The leaf area growth and sapwood area growth of thinned and control *Abies balsamea* differed in the first growing season after precommercial thinning, but reestablished itself in the second year following treatment. (From Pothier and Margolis, 1991.)

the thinned *A. balsamea* produced similar amounts of new leaf area and greater amounts of new sapwood area in the first year following thinning compared with the unthinned controls. The lower leaf area to sapwood area balance in the thinned trees during the first growing season after treatment had resulted in a reduction in the resistance to water flow from roots to leaves. As hypothesized by Jarvis (1975), this would enable more rapid acclimatation and thus help the thinned trees adjust to the higher transpirational demand following thinning.

In the second year following thinning, there was no longer any difference in the ratio of leaf area growth to sapwood area growth between the thinned and control *A. balsamea* stands, indicating that the homeostasis between leaf area and sapwood area had been restored (Fig. 8). In contrast, for *B. papyrifera*, the ratio of leaf area growth to sapwood area growth of both the thinned and control stands was constant for both years. The indeterminate growth habit of *B. papyrifera* placed no restriction on the capacity to grow new leaf area in the first year following thinning.

Thus, it appears that the basic principles of the pipe model theory apply, with some modifications, even following sudden disturbances in environmental conditions brought about by silvicultural treatment. Further, the species with the indeterminate growth habit, *B. papyrifera*, reestablished the equilibrium between sapwood area growth and leaf area growth more rapidly than did the determinate growth species, *A. balsamea*.

B. Seasonal Dynamics in Leaf Area

The leaf area index in forest stands is frequently estimated from biomass measurements made at a single time of the year. However, LAI at any time is the balance between the growth of new foliage and the loss of older foliage and, due to the differences in the timing of these two phenomena, LAI does not remain constant during the year (e.g., Gholz et al., 1991). Annual carbon gain and transpiration from a tree canopy is strongly dependent on the seasonal dynamics of LAI in relation to climatic variables and other limiting factors. For example, Whitehead et al. (1993) attributed 24% of the seasonal change in transpiration (E_t) from a 7-year-old, widely spaced *P. radiata* canopy to the increase in LAI. Only 4% of E_t could be attributed to seasonal changes in stomatal conductance characteristics of the foliage, with the remainder due to changes in climatic variables.

The phenology of foliage growth and loss has been primarily studied in *Pinus* canopies, where the seasonal dynamics are more pronounced than is the case for most other conifers, i.e., seasonal changes deviate from the maximum LAI value by between 20 and 60%. For example, Rutter (1966) showed that the seasonal difference in foliage in mature

P. sylvestris from the summer maximum to the winter minimum was 40%. In a much slower growing, 120-year-old stand of the same species in Sweden, the difference in LAI was 23% between the maximum in midsummer and autumn (Lindroth, 1985). A similar value of 30% was shown by measurements of biomass at nine times during the year by Madgwick (1968) in 17-year-old *Pinus virginiana*.

Maximum foliage biomass occurred in autumn in *P. taeda*, with a difference between the maximum and minimum of 32% (Kinerson *et al.*, 1974). The seasonal variation for *P. sylvestris* was 34% (Beadle *et al.*, 1982) and was 30 and 58% for *Pinus elliottii* during the third year in control stand and in stands with fertilizer applied, respectively (Gholz *et al.*, 1991). The seasonal change in LAI was between 21 and 60% for *P. radiata* stands with different degrees of water and nutrient availability (Raison *et al.*, 1992b) and was 43% for the same species in a widely spaced 7-year-old stand that was still increasing in leaf area (Whitehead *et al.*, 1993). In contrast to *Pinus* spp., which generally retain their foliage for 2 years, *Abies*, *Picea*, and *Pseudotsuga* retain their foliage for much longer. For example, in an 89-year-old *P. abies* stand, the needles were held for up to 12 years and 40% of the needles were at least 4 years old (Schulze *et al.*, 1977).

The dynamics of LAI in conifer stands have been modeled by combining empirical expressions for foliage production and loss based on time of year. Kinerson *et al. (1974)* sampled 128 14-year-old *P. taeda* trees and fitted inverse exponential relationships to foliage growth and litterfall. Loss of foliage occurred all year, with the maximum rate in autumn after foliage growth had ceased. In 5-year-old *P. radiata*, Madgwick (1983a) measured biomass at different times during a year and showed that the weight of needles followed a sigmoidal relationship during the 4-month growing period in late spring and early summer. Beadle *et al.* (1982) used linear relationships with day number to model the dynamics of each cohort of foliage by age in a closed-canopy, 46-year-old *P. sylvestris* stand. The assumptions were that LAI was constant between years and that the same amount of leaf area was added each year. Leaf area in old needles declined slowly during the first 5 months of the year (winter and spring) but total LAI more than doubled in early summer as the current year's needles began elongating. High LAI was maintained for 5 months until early autumn then declined rapidly. During the winter and early spring period, there was only a small (10%) decline.

During a 3-year period, Gholz *et al.* (1991) modeled the seasonal changes in LAI in plots of 21-year old *P. elliottii* stands. Half of the plots had had fertilizer added and the other half served as a control. Bud burst began in early spring and needle elongation continued until late autumn (Fig. 9a). Litterfall peaked in late autumn (October and early December), when more than half of the total loss occurred (Fig. 9b).

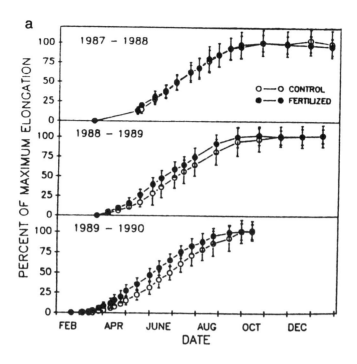
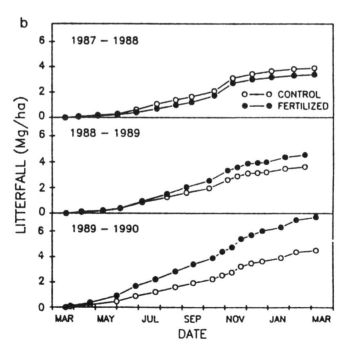

Figure 9 (a) Relative needle elongation rates (mean ± SD) and (b) cumulative needle litterfall for fertilizer-added and control plots of *Pinus elliottii*. For each treatment, $n = 8$ plots. (Adapted from Gholz *et al.*, 1991.)

Logistic functions were used to model cumulative litterfall as a function of leaf area, cumulative growing degree days, and average monthly solar radiation for the summer.

In Australia, Raison *et al.* (1992a) measured foliage dynamics throughout a 4-year period in 10- to 14-year-old *P. radiata* with different degrees of water and nitrogen stress. Needle length was a good indicator of area because weight per unit length did not change as the foliage developed. During the 4 months from spring to the end of summer, 90% of the foliage elongation occurred, and this pattern was not affected by nitrogen availability or water potential, although the rates of needle production were negatively and linearly related to the water availability defined by the water stress integral (Myers, 1988). The water stress integral is the product of water potential and accumulated time throughout the year. Foliage biomass in the winter was a good predictor of annual foliage loss the following year (Raison *et al.*, 1992b), and the timing of needlefall could be predicted from the water stress integral on a monthly basis. Water and nitrogen availability interacted positively to affect both the rate of increase in foliage biomass and its maximum value. Furthermore, the amount of foliage could be predicted at any time during a year from the nitrogen status, the water stress integral, and the foliage biomass during the previous winter.

It is important to note that a seasonal variation of 30–40% in LAI may have little effect on the amount of photosynthetically active radiation (PAR) intercepted by a forest stand. Waring (1991) notes that 95% of all incoming PAR is intercepted at a LAI of three or greater. Thus, the potential photosynthesis of a forest having a LAI of greater than five is likely to be little affected by seasonal variations in LAI even as large as 40%.

C. Genetic Variation in Crown Architecture and Leaf Area

Large differences in the growth rate of trees on similar sites and under similar stand conditions have long been observed (Zobel and Talbert, 1984). Because differences in canopy architecture and in the vertical distribution of tree foliage influence the radiation profile inside forest canopies (Oker-Blom and Kellomäki, 1982; Jarvis and Leverenz, 1983; Grace *et al.*, 1987; Hashimoto, 1991), a significant amount of the variation in growth rates within a species might be explained by genotypic variation in crown architecture (Kärki and Tigerstedt, 1985; Hinckley, *et al.*, 1992).

There have been only a few studies of the genetic differences in crown architecture between provenances, and these have rarely attempted to relate the differences to variations in growth rates. Magnussen *et al.* (1986) did not find genetic differences in crown characteristics among provenances of *P. banksiana*, but such differences were found for a number of pines from the southeastern United States (Trousdell *et al.*, 1963; Bailey *et al.*, 1974; Zobel and Talbert, 1984). Rogers *et al.* (1989) found a correlation between clonal differences in crown architecture and growth of a deciduous hardwood, *Populus trichocarpa*.

Jacobs (1988) thoroughly studied the crown characteristics of 10-year-old *P. taeda* stands on the central coastal plain of North Carolina. Of the 60 families in this trial, two of the fastest (Fast-1 and Fast-2) and one of the slowest (Slow-1) growing families were selected and investigated for both tree and stand characteristics. Trees from the faster growing families were taller and had greater diameter and biomass at age 10 than did trees in the slower growing family (Table II). Crown and canopy characteristics also varied significantly among the three families. The crowns of Fast-1 and Fast-2 trees were longer and contained greater leaf area than those of Slow-1 trees (Table II). However, the differences in leaf area were greater than the differences in crown length, implying that leaf area density (leaf area per canopy volume) was greater in the faster growing families.

The two faster growing families, however, appeared to partially compensate for their higher leaf area density, and the associated potential reduction in light penetration, by clustering their foliage. This is reflected in a higher leaf area per unit of branch length (Fig. 10). Clustered distribution of foliage (i.e., nonrandom distribution of leaves in crowns and canopies) affects light interception in the canopy by increasing the importance of sunflecks and mutual shading (Norman and Jarvis, 1975; Oker-Bloom and Kellomäki, 1983; Baldocchi and Hutchinson, 1986; Whitehead *et al.*, 1990). Kira *et al.* (1969) showed that gaps in the canopy, created by the clustered distribution of leaves, allow greater light penetration to the lower canopy strata. Kinerson *et al.* (1974) noted that the light extinction coefficient decreases when foliage is clustered. Such canopies maintain greater LAI in comparison to canopies in which foliage distribution is less clustered. Thus, either a lower leaf area density or a more clustered distribution of foliage can enhance light penetration into the canopy.

Trees from the slowest growing family (Slow-1) had the lowest leaf area. This was in part due to a lower foliage biomass, but was also a result of a greater specific leaf weight (SLW; g dry weight · cm^{-2} foliage) (Table II). Within a single species, SLW generally increases with irradiance and enhanced photosynthetic capacity (Oren *et al.*, 1986a). Trees from the slow-growing family did not exhibit the commonly found pattern of increasing SLW as irradiance increased with height in the canopy (Oren *et al.*, 1986a,b). This may indicate a genetic makeup dictating a high and relatively inelastic SLW. In support of this contention, the

Table II Characteristics of the Average Tree and of Stands of Three Open-Pollinated *Pinus taeda* Families[a]

	Average tree									Stand	
Family	Height (m)	DBH (mm)	Crown length (m)	Stemwood biomass (kg)	Stemwood growth (kg/yr)	A_l (m^2)	Specific leaf weight (g/m^2)	$A_l:A_s$ (m^2/cm^2)	A_l/above ground biomass (m^2/kg)	LAI (m^2/m^2)	Growth (kg/ha/yr)
Fast-1	12.14a	170a	6.69a	37.19a	7.03a	16.3a	351b	0.18a	0.179b	2.6a	26.85a
Fast-2	11.91a	168a	6.31a	36.61a	6.36a	14.8a	313b	0.18a	0.189a	2.3a	24.15a
Slow-1	10.78b	154b	5.81b	27.92b	4.71b	9.6b	431a	0.13b	0.148c	1.6b	20.45b

[a] Data from Jacobs (1988). Data in the same column not followed by the same letter are significantly different ($P = 0.05$).

slow-growing family had a higher SLW in its upper crown than did the other two families, even though they were exposed to the same light environment (Fig. 10). Among species, differences in SLW of foliage developing under the same irradiance reflect differences in both photosynthetic capacity and adaptation to various water and temperature regimes. Jacobs (1988) demonstrated that differences between *P. taeda* families may be as great as those that occur between some species.

Trees of the slow-growing family had a lower $A_f:A_s$ ratio (Table II), a characteristic appropriate for plants living in physiologically dry environments (Waring, 1980, 1982). The slower growing family also had the least amount of branch biomass per unit leaf area (Fig. 10). The lower relative investment in branches, in combination with greater irradiance within the canopy due to lower leaf area density, and a higher photosynthetic rate (as indicated by the higher SLW), could explain the greater growth efficiency (i.e., wood production per unit of leaf area) of trees in the slow-growing family relative to the other two families (~500 g/m^2 versus 400 g/m^2, respectively). Yet the higher LAI displayed by families Fast-1 and Fast-2 (Table II) more than compensated for their lower average growth efficiency, resulting in greater growth on a stand level relative to Slow-1.

In conclusion, several crown and canopy characteristics may be genetically controlled and, in turn, exert a certain control over growth rate.

However, total tree or canopy leaf area, which reflects both the quality

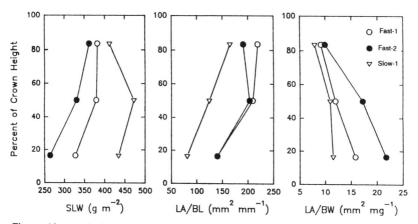

Figure 10 Vertical distribution of specific leaf weight (SLW), leaf area/branch length (LA/BL), and leaf area/branch biomass (LA/BW) of two fast-growing (Fast-1 and Fast-2) and one slow-growing (Slow-1) open-pollinated families of *Pinus taeda*. The slow-growing family did not adjust its specific leaf weight to the irradiance within the canopy, and held less leaf area per unit of branch length and more per unit of branch weight than the fast-growing families.

of sites and the extent to which trees or stands capture available resources, accounts for much of the variation in growth among families. The distribution of foliage appears to affect growth primarily by altering the quantity of foliage that can be maintained by a stand under a given site condition and tree density as well as the consequent interactions of the foliage with light.

D. Dynamics of Leaf Area Development in Natural and Managed Stands

Climate is a major determinant of both the maximum LAI a forest can attain and its rate of development (Waring et al., 1978; Waring, 1980). In a transect across Oregon, the LAI of mature conifer forests was related to the site water balance in regions limited by water during the growing season (Grier and Running, 1977). Extremes of temperature may limit the growing season and LAI of both temperate forests (Gholz, 1979) and arctic tundra (Van Cleve et al., 1983). Site fertility, particularly nitrogen availability, was also shown to exert control over LAI development in several coniferous forests (Miller and Miller, 1976; Albrektsson et al., 1977; Brix, 1981; Binkley and Reid, 1984; Vose and Allen, 1988).

Tree spacing (Brix, 1981; Oren et al., 1987; Vose and Allen, 1988) and successional changes in species composition (Schroeder, 1983) may interact with climatic and site factors to affect both maximum LAI and the rate at which LAI develops. In 24-year-old *P. menziesii* stands, reducing basal area from 23.1 to 8.3 m^2/ha lowered LAI such that 7 years after treatment the LAI of the spaced stands was still 3 m^2/m^2 lower than the control, which displayed a LAI of 6 m^2/m^2 (Brix, 1981). However, at the same point in time, the LAI of thinned plots fertilized with 448 kg N/ha increased to equal that of the unmanipulated stands. Thus, the addition of large quantities of nitrogen may compensate for thinning and significantly increase LAI at a given tree spacing. Long-term studies are necessary to quantify these responses adequately.

Such a long-term study was established in 1958 by Barrett (1982) in eastern Oregon, and measurements continued through 1983 (Oren et al., 1987). The study was installed using 40- to 70-year-old advanced regeneration of *P. ponderosa* growing in the understory of an old-growth *P. ponderosa* stand. The old-growth trees were removed and the advanced growth was subsequently thinned to five spacing levels. Half of these stands were maintained free of understory brush species during the following 25 years. Measurements of diameter, height to crown base, and total height of all trees every 4 years in each stand permitted estimation of sapwood area, and based on the pipe model theory, LAI during the study period was calculated (Oren et al., 1987).

Maximum LAI can be estimated based on the relationship between

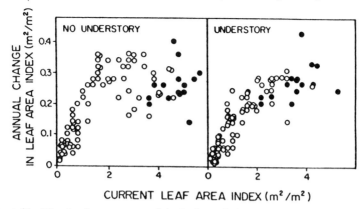

Figure 11 The development rate of the leaf area index during the growing season in *Pinus ponderosa* stands was related to the leaf area index present at the beginning of the season. Controlling the understory vegetation increased the rate of development in leaf area index and the maximum rate attainable. (Adapted from Oren *et al.* (1987), from the *Forest Science* (vol. 33, no. 2) published by the Society of American Foresters, 5400 Grosvenor Lane, Bethesda, MD 20814-2198.)

the annual increase in LAI and the LAI at the start of the growing season. Based on the relationship seen in stands with understories present and absent in Fig. 11, the annual increase in LAI should near zero as the LAI approaches 7 to 7.5 m^2/m^2. Thus, we obtain an estimate of maximum LAI of 7.5 m^2/m^2. LAI of all thinning treatments increased throughout the 25-year study (Fig. 12), without reaching the estimated maximum. However, a more detailed analysis of the data indicated that the maximum LAI attainable in the most severely thinned plots is lower than that of the more lightly thinned plots. In very open stands, where trees are unavailable to support a horizontally uniform canopy, maximum LAI is reduced because the amount of foliage that a tree can hold is limited in part by the width of its crown. Thus at some point, decreases in stand density result in a situation whereby maximum stand LAI cannot be realized. Although the leaf area of individual trees in severely thinned stands is higher, their leaf area density is also high, reducing the maximum LAI as well as the growth per unit leaf area of foliage due to increased mutual shading (Ford, 1975; Oren *et al.*, 1986a).

Understory vegetation reduced the rate of stand LAI development below that of stands in which understory was removed. It is not clear, however, that understory vegetation has an effect on the final value of LAI. It seems reasonable that in stands that are relatively open due to water or nutrient limitations, or in stands composed of shade-intolerant species, the understory may reduce the final LAI because it will persist and compete for water and nutrients with the overstory trees. However, if conditions promote the development of a horizontally uniform, dense

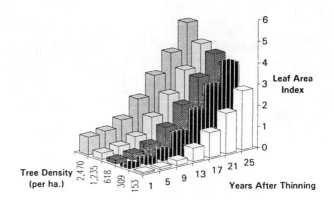

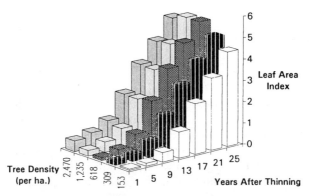

Figure 12 Leaf area index in *Pinus ponderosa* stands developed over time in relation to both the density of trees and the presence or absence of the understory vegetation.

canopy, the understory vegetation is likely to be shaded out at high LAI. In such situations, the final value of LAI is not likely to change, but the stand will require a longer time to reach it.

In their study, Oren *et al.* (1987) found that in stands in which the understory vegetation was absent, the proportion of available water used during the growing season was initially directly related to LAI (Fig. 13), i.e., about 90% of the water was used at the end of the linear portion of the curve, which occurred at an LAI of 1.5 m^2/m^2. This indicates that when the stands reached about one-third to one-half of their maximum LAI, all of their available water was exhausted before the end of the

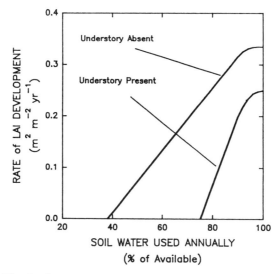

Figure 13 The development rate of leaf area index in *Pinus ponderosa* stands in Oregon (1960–1966) at a given level of water use was higher when the understory vegetation was absent. With the increasing proportion of available water used by the end of the growing season, the difference in the development rate decreased between stands with and without understory. Even when all available water was used, the development rate continued to increase (not shown), indicating improved water use efficiency, and redistribution of water from the understory to the pine.

growing season. As LAI increased above this level, a greater proportion of the water was probably used in the early part of the growing season, or earlier and later in the day. Thus, although the foliage may be inactive during a longer part of the season as LAI increases, water seems to be used during periods of lower evaporative demand, which, in turn, is likely to increase water use efficiency.

By combining the information on the rate of LAI development and water use at a LAI range of 1.5 m²/m² or less (from Oren *et al.*, 1987), the effect of the understory vegetation can be evaluated directly (Fig. 13). In the period before soil water was completely used, the differences in LAI development between stands with and without understories was greater when less water was consumed. Lower water consumption was associated with lower pine LAI, which, in turn, promoted both the development of the understory and its consumption of water. In addition, during the earlier stage of stand recovery from thinning, the roots of the pine did not exploit all the available soil and the pine experienced periodic drought, even though not all the available water was used (Hermann and Petersen, 1969; Youngberg and Cochran, 1982).

As the pine LAI developed, so did its root system. Thus, as recovery from thinning proceeded, the effect of the understory was reduced because the understory was less able to compete with the pine for water and nutrients, and the pine roots accessed a greater proportion of the available moisture. At an LAI of 1.5 m^2/m^2, the amount of water used by *P. ponderosa* stands in the study area was similar, regardless of the understory. Thus, the effect of the understory on stand development was mostly in reducing the rate of LAI development.

E. Causes and Consequences of Leaf Area Senescence

In even-aged stands that develop after a major disturbance such as fire, the numbers of trees per hectare may vary from many thousands at the seedling stage to less than several hundred at the old-growth stage, e.g., *P. menziesii* and *Sequoia sempervirens*. The increase in tree spacing with age is most apparent in even-aged stands, but similar increases occur within individual age groups as young trees become older in stands of mixed ages.

The process of reduced tree number per unit area within age groups of trees involves a divergence in the capacity of individuals to compete favorably for required resources. Individual trees, as is the case for ecosystems, progress over time through periods of efficient capture and utilization of resources, less efficient capture, disturbance and decline, and finally the release of resources with mortality. Within age groups of trees, some trees begin to dominate while others fall behind and become suppressed. A close examination of individual trees reveals a wide variation in leaf area and crown characteristics among trees of similar species and age but differing in their dominance position. Dominant or codominant trees typically have large crowns (both depth and width) and a large amount of foliage on each shoot (a longer shoot length that bears live needles, and perhaps longer needles). In contrast, trees in an intermediate or suppressed crown position generally have much smaller and less dense crowns, with the foliage appearing sparse and often restricted to tufts at the ends of each branch (short length of shoot-bearing live needles) (cf. Jack and Long, 1992). This is particularly true for more shade-intolerant species.

At all but the latest stages of tree growth and development, resource limitation for individual trees appears to be a critical factor in the loss of leaf area, decline of growth, and mortality. Competition for limited resources (light, nutrients, and water) among neighboring trees results in some trees gaining an advantage while other trees become weakened and susceptible to death from various biotic and abiotic agents (Waring, 1987).

Similar competition for resources occurs in old trees, but old trees may reach certain size constraints, strongly affected by genotype and site, that limit them from attaining larger heights and diameters. Thus although younger trees weaken and die as a result of normal competition, older trees may ultimately senesce in the classic sense, limited not strictly by resources but also by other less well-defined morphological features and physiological processes related to old age that prevent them from retaining adequate vigor. These factors may include physical constraints in the distance of material transport, increased resistance for water transport in the bole and branches (see below), a shift in carbon balance and allocation stemming from reduced photosynthate and increased maintenance costs, limited capacity to respond to more favorable resource supplies, and altered susceptibility to biotic and abiotic agents. Studies of leaf area dynamics on older trees appear to have focused primarily on healthy trees with large crowns, with only limited attention to trees in various states of decline.

Kaufmann and Watkins (1990) measured a sample of 20 old-growth *P. contorta* trees growing on one site. These trees were of similar diameter (~30 cm) and age (about 275 years), but had different leaf areas. The results showed that trees that had grown rapidly early in their life had lost leaf area and were growing slowly at the time of measurement, whereas trees that had grown slowly previously had higher leaf areas and growth rates. Large differences existed between neighboring trees, with no explanation in terms of differences in competition for light and availability of water and nutrients. Total leaf areas ranged from 9 to 224 m², and sapwood cross-sectional areas at breast height ranged from 85 to 384 cm². Furthermore, the $A_f:A_s$ ratios varied from as low as 0.09 to as high as 0.88 m²/m² (Fig. 14). These data, taken from a single stand,

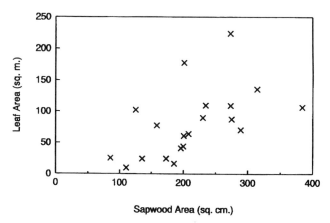

Figure 14 Relationship between leaf area (all surfaces) and sapwood cross-sectional area at 1.37 m for old-growth *Pinus contorta* trees in a single stand.

represent some of the highest and lowest $A_f:A_s$ ratios reported in the literature. Clearly in such trees there is little utility in using sapwood area alone to predict leaf area, but analyses show that sapwood area in conjunction with distance from breast height to the midpoint of the crown (first suggested by Dean and Long, 1986), or simply the number of live branches, provide better predictions of A_f (M. R. Kaufmann, unpublished data).

Recent evidence indicates that the total hydraulic conductance (i.e., the transpiration per unit of water potential gradient between soil and leaves, including the effects of gravity, path length, and hydraulic conductivity through roots, main stem, and branches) in *P. ponderosa* and *P. contorta* trees was considerably lower in old than in young trees (B.J. Yoder, unpublished data, Oregon State University). This was attributed both to differences in crown depth and to longer lengths of smaller branches in the older trees. Other studies have indicated that branches and branch nodes have a higher resistance to water movement and that hydraulic conductance is lower in longer branches (Tyree *et al.*, 1983; Tyree, 1988).

The wide range of $A_f:A_s$ ratios for old *P. contorta* trees suggests that large differences exist either in critical processes or genetic controls governing leaf area development and retention in old trees. The decline in $A_f:A_s$ ratios with increasing distance to the midpoint of the crown suggests that hydraulic effects are significant. Increased flow resistances to the foliage of old trees with long branches may increase water stress, but because of changes in leaf conductance, it is more likely that the *period* of water stress in foliage is increased rather than its *intensity*. The net effect may be that photosynthesis is reduced for longer portions of the day in trees having higher flow resistances than in more vigorous trees having lower flow resistances. This hypothesis is consistent with observations that trees that grew slowly when younger were more vigorous when old in terms of retained leaf area and growth rates, because those trees would not have attained their full branch length and increased their flow resistances until relatively late in the life cycle.

F. Coniferous Leaf Area at Large Spatial Scales

Recent advances in investigating ecophysiological processes of conifer forests using remote-sensing technology have provided some ability to estimate leaf area at large spatial scales. Running *et al.* (1986) and Peterson *et al.* (1987) were the first to measure the LAI of conifer forests using an optical sensor of satellite-level resolution. A scanner measuring red (wavelengths 0.63 to 0.69 μm) and near-infrared (wavelengths 0.76 to 0.90 μm) reflectance bands was placed aboard NASA's ER-2 high-altitude aircraft. The scanner, essentially identical to the sensor in the

Landsat thematic mapper (Landsat-TM) satellite, was flown at 20,000 m altitude and obtained images (24 × 24 m pixel size) for 18 temperate conifer stands in Oregon along a pronounced climatic gradient crossing two mountain ranges.

As is typical for nearly all plant canopies, the conifer canopies strongly absorbed radiation in the red portion of the spectrum (Fig. 15A) but largely scattered or transmitted in the near-infrared (NIR) (Fig. 15B). The ratio of red to NIR reflectance was thus positively correlated with LAI (Fig. 16). Although the ratio partially compensates for differences in topographic position, it does not compensate for atmospheric attenuation, i.e., the scattering of radiation by water vapor and aerosols (Peterson *et al.*, 1987). By collecting simultaneous radiance measurements with a radiometer placed in a helicopter at 100 m above the canopy, algorithms were developed to correct the reflectance data obtained from the ER-2 for atmospheric effects. This correction improved both the precision and the sensitivity of the relationship (Fig. 16).

Spanner *et al.* (1990b) found that the ratio of NIR to red reflectance measured with the Landsat-TM satellite was considerably affected by the degree of canopy closure, understory vegetation, and background reflectance. The ratio of NIR to red reflectance, however, partially compensated for these effects. Gholz *et al.* (1991) also demonstrated that imagery from Landsat-TM could detect seasonal differences in LAI for *P. elliottii* plots to which fertilizer had either been added or not. They found linear relationships between LAI, measured during winter, spring, and autumn, and the normalized difference vegetation index (NDVI) derived from Landsat-TM imagery. NDVI is calculated as (NIR − red)/(NIR + red). The relationship between LAI and NDVI, however, was different at different times of the year.

The size of a Landsat-TM pixel is approximately 30 × 30 m, whereas the Advanced Very High Resolution Radiometer (AVHRR) on the NOAA-9 satellite has a spatial resolution of 1.1 km. Thus, a single AVHRR pixel integrates over much of the fine-scale differences in canopy closure, background reflectance, and understory vegetation that is present in Landsat-TM imagery. Furthermore, by taking the highest value of each pixel recorded for a site over a given period of time (generally 1 week to 1 month), the effects of atmospheric attenuation as well as the effects of scan angle (the angle of the sensor relative to the target) can be minimized (Holben, 1986; Spanner *et al.*, 1990a).

Spanner *et al.* (1990a) evaluated the relationship between the seasonal variation of composite values of NDVI derived from AVHRR data with the LAI for 19 coniferous forest stands in Oregon, Washington, Montana, and California. LAI explained 79% of the variation in the summer maximum of NDVI and was also strongly related to phenological

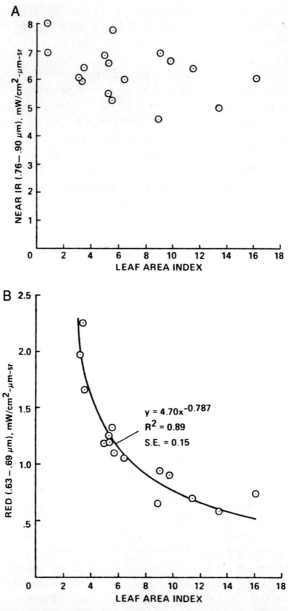

Figure 15 (A) Relationship between near-infrared radiance corrected for atmospheric and topographic effects (measured from the Airborne Thematic Mapper simulator) and LAI for 18 forest stands in Oregon. (B) Relationship between red radiance (corrected as for near infrared) and LAI. (Reprinted by permission of the publisher from Relationship of thematic mapper simulator data to leaf area index of temperate coniferous forests. Peterson *et al.*, 1987), *Remote Sens. Environ.* 22:323–341. Copyright 1987 by Elsevier Science Inc.

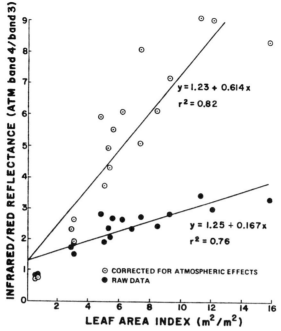

Figure 16 Relationship between near-infrared/red reflectance (measured from the Airborne Thematic Mapper simulator) and LAI for 18 forest stands in Oregon. (From Running et al., 1986.)

changes in LAI over the course of the year (Fig. 17). However, snow cover and low solar zenith angles during the winter tended to reduce NDVI values and need to be considered when estimating LAI from low-resolution, wide field-of-view satellite measurements (Fig. 17). The presence of granitic outcroppings as well as the presence of broadleaf vegetation were also believed to have influenced the NDVI values.

Yoder (1992) studied the relationship between the leaf area, the photosynthetic capacity, and the spectral reflectance of miniature canopies composed of *P. menziesii* seedlings that had been grown under different conditions of light and nitrogen availability. Under controlled laboratory conditions, it appeared that the ratio of visible to NIR reflectance was the better indicator of the amount of leaf area present, whereas NDVI corresponded better to the photosynthetic capacity. These results need to be verified in actual forest conditions.

The relationship between LAI and spectral vegetation indices, such as NDVI, appears to hold reasonably well for certain continuous plant canopies such as grasslands and forests having LAI values less than 3 or 4. In discontinuous forest canopies such as are often found in the boreal

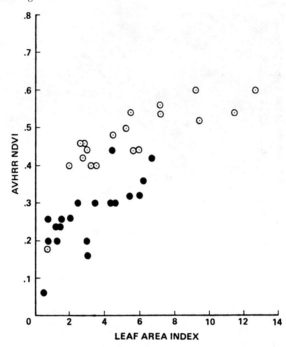

Figure 17 Relationship of maximum summer LAI with normalized difference vegetation index (NDVI) (open circles) and minimum winter LAI with NDVI (filled circles) for 19 forest stands in the western United States. NDVI values were obtained from the NOAA-9 Advanced Very High Resolution Radiometer (AVHRR) sensor having a pixel size of 1.1 km. (Reprinted by permission of the publisher from The seasonality of AVHRK data of temperate coniferous forests: Relationship with leaf area index. Spanner *et al.*, 1990a), *Remote Sens. Environ.* 33:97–112. Copyright 1990 by Elsevier Science Inc.

forest, the effects of the background may be such that the LAI of the overstory may have no obvious relationship at all with the various combinations of red and infrared reflectances that have been successfully applied to other forest types. For example, *Pinus banksiana* is sometimes found in conjunction with an understory comprised of a number of lichen species that have significant reflectance in the blue portion of the spectrum. *Picea mariana*, on the other hand, is often found in conjunction with a number of moss species which have considerable reflectance in the green portion of the spectrum. An interesting possibility for estimating LAI from remotely sensed spectral data in such types of forests (as well as estimating biomass and the percentage of absorbed photosynthetically active radiation) is the use of mixture decomposition algorithms that estimate the fraction of sunlit canopy, shadow, and sunlit background in conifer stands. There is some evidence that indicates these fractions are highly correlated with canopy biophysical properties such as LAI, particularly when there is a high level of contrast in the reflectance characteristics of the sunlit canopy and the sunlit background (Hall *et al.*, 1994).

Direct measurement of the structural and physiological properties of vegetation at large spatial scales is particularly significant to the many urgent questions concerning the role conifer forests in the dynamics of the global climate. Many of these questions can be addressed directly only by measuring forest structure and processes at large spatial scales, i.e., scales considerably larger than the size of the traditional ground-based research plots. Examples include (1) the effect of climate change on forest structure and function at regional levels, (2) the effectiveness of forests as sources or sinks of CO_2 and other trace gases under different climate change scenarios, and (3) the effects of large-scale changes in forest structure on the surface energy balance and the consequent feedback of these changes onto the physical climate system. Obtaining remotely sensed estimates of forest structural properties such as leaf area for use in ecosystem simulation models (e.g., Running *et al.*, 1989; Bonan, 1991) holds promise as a means of addressing many of the important issues in global change research.

IV. Conclusions

Leaf area is critical to understanding the physiological ecology of forests because of its pertinence to biological processes at a wide range of spatial and temporal scales. The close relationship of leaf area to processes such as photosynthesis and transpiration means that an estimate of this single characteristic for a forest stand can be most useful for estimating stand growth as well as fluxes of carbon and water from forest canopies. The leaf area that can be supported by a tree is determined by the equilibrium between the ability of the stem and roots to supply water and nutrients to the foliage and the amount of photosynthetically active radiation intercepted by the crown. Perturbations to this dynamic equilibrium can result in readjustments to various parts of the system such that a new equilibrium is reached. However, for forest stands that are near senescence, additional factors influencing leaf area development also need to be considered.

Although leaf area is often considered as an unchanging state variable in respect to processes operating on the basis of 1 year or less, it should be recognized that LAI can also change considerably over the course of a single growing season. The magnitude of this seasonal change is dependent on both species characteristics and site conditions and has been modeled for *Pinus* spp. Genetic differences in crown architecture can also influence the ability of a tree to capture resources using a given amount of leaf area.

Analysis of leaf area development in forest stands subjected to silvicultural treatment can provide a novel and powerful means of comparing differences in the response of stands to the treatment, e.g., comparing performance at a similar level of leaf area despite the fact that these leaf areas were reached at different times. Silvicultural systems based on the management of stand leaf area hold considerable potential for managing forest stands on a physiological basis.

Remote-sensing technology appears to permit estimates of leaf area at large spatial scales, and these may be used as inputs for process-based forest ecosystem models. However, confounding factors such as snow cover, soil background, changes in solar zenith angle during the year, and atmospheric attenuation of reflected radiation due to haze need to be considered. Furthermore, although estimating leaf area across dramatic climatic gradients with remote-sensing technology has met with some success, a greater understanding of the ways radiation interacts with different types of vegetation is necessary before it will be possible to estimate reliably the range of leaf areas that may occur within a given climatic zone.

Acknowledgments

We thank Richard Waring, Marie Coyea, and John Marshall for their useful comments.

References

Albrektsson, A. (1984). Sapwood basal area and needle mass of Scots pine (*Pinus sylvestris* L.) trees in central Sweden. *Forestry* 57:35–43.

Albrektsson, A., Aronsson, A., and Tamm, A. (1977). The effect of forest fertilization on the primary productivity and nutrient cycling in the forest ecosystem. *Silva Fenn.* 11: 223–239.

Bailey, J. K., Feret, P. P., and Bramlett, D. L. (1974). Crown character differences between well-pruned and poorly-pruned Virginia pine trees and their progeny. *Sylvae Genet.* 6: 181–184.

Baker, T. G., Attiwill, P. M., and Stewart, H. T. L. (1984). Biomass equations for *Pinus radiata* in Gippsland, Victoria. *N.Z. J. For. Sci.* 14:89–96.

Baldocchi, D. D., and Hutchinson, B. A. (1986). On estimating canopy photosynthesis and stomatal conductance in a deciduous forest with clumped foliage. *Tree Physiol.* 2: 155–168.

Barrett, J. W. (1982). Twenty-year growth of ponderosa pine saplings thinned to five spacings in central Oregon. *USDA For. Serv. Pap. PNW* PNW-301.

Beadle, C. L., Talbot, H., and Jarvis, P. G. (1982). Canopy structure and leaf area index in a mature Scots pine forest. *Forestry* 55:105–123.

Bidlake, W. R., and Black, R. A. (1989). Vertical distribution of leaf area in *Larix occidentalis*: A comparison of two estimation methods. *Can. J. For. Res.* 19:1131–1136.
Binkley, D., and Reid, P. (1984). Long-term responses of stem growth and leaf area to thinning and fertilization in a Douglas-fir plantation. *Can. J. For. Res.* 14:656–660.
Blanche, C. A., Hodges, J. D., and Nebeker, T. E. (1985). A leaf area–sapwood area ratio developed to rate loblolly pine tree vigor. *Can. J. For. Res.* 15:1181–1184.
Bonan, G. B. (1991). Atmospheric–biosphere exchange of carbon dioxide in boreal forests. *J. Geophys. Res.* 96:7301–7312.
Booker, R. E., and Kininmonth, J. A. (1978). Variation in longitudinal permeability of green radiata pine wood. *N.Z. J. For. Sci.* 8:295–308.
Borghetti, M., Vendramin, G. G., and Giannini, R. (1986). Specific leaf area and leaf area index ditribution in a young Douglas-fir plantation. *Can. J. For. Res.* 16:1283–1288.
Bormann, B. T. (1990). Diameter-based biomass regression models ignore large sapwood-related variation in Sitka spruce. *Can. J. For. Res.* 20:1098–1104.
Brix, H. (1981). Effects of nitrogen fertilizer source and application rates on foliar nitrogen concentration, photosynthesis and growth of Douglas-fir. *Can. J. For. Res.* 11:775–780.
Brix, H., and Mitchell, K. (1983). Thinning and nitrogen fertilization effects on sapwood development and relationships of foliage quantity to sapwood area and basal area in Douglas-fir. *Can. J. For. Res.* 13:384–389.
Büsgen, M., and Münch, E. (1929). "The Structure and Life of Forest Trees" (T. Thompson, trans.) 3d ed. Chapman & Hall, London.
Campagna, M., and Margolis, H. A. (1989). Effects of short-term atmospheric carbon dioxide enrichment of black spruce seedlings at different stages of development. *Can. J. For. Res.* 19:773–782.
Cannell, M. G. R., and Willett, S. C. (1976). Shoot growth phenology, dry matter distribution and root:shoot ratios of provenances of *Populus trichocarpa*, *Picea sitchensis* and *Pinus contorta* growing in Scotland. *Silvae Genet.* 25:49–58.
Causton, D. R. (1985). Biometrical, structural and physiological relationships among tree parts. *In* "Attributes of Trees as Crop Plants" (M. G. R. Cannell and J. E. Jackson, eds.), pp. 137–159. Institute of Terrestrial Ecology, Huntingdon, UK.
Comstock, G. L. (1970). Directional permeability of softwoods. *Wood Fiber* 1:283–289.
Coyea, M. R., and Margolis, H. A. (1992). Factors affecting the relationship between sapwood area and leaf area of balsam fir. *Can. J. For. Res.* 22:1684–1693.
Dean, T. J., and Long, J. N. (1986). Variation in sapwood area–leaf area relations within two stands of lodgepole pine. *For. Sci.* 32:749–758.
Dean, T. J., Long, J. N., and Smith, F. W. (1988). Bias in leaf area–sapwood area ratios and its impact on growth analysis in *Pinus contorta*. *Trees* 2:104–109.
Drew, A. P., and Ledig, F. T. (1980). Episodic growth and relative shoot:root balance in loblolly pine seedlings. *Ann. Bot. (London)* [N.S.] 45:143–148.
Eamus, D., and Jarvis, P. G. (1989). The direct effects of increase in the global atmospheric CO_2 concentration on natural and commercial temperate trees and forests. *Adv. Ecol. Res.* 19:1–55.
Edwards, W. R. N., and Jarvis, P. G. (1982). Relations between water content, potential, and permeability in stems of conifers. *Plant, Cell Environ.* 5:271–277.
Espinosa-Bancalari, M. A., Perry, D. A., and Marshall, J. D. (1987). Leaf area–sapwood area relationships in adjacent young Douglas-fir stands with different early growth rates. *Can. J. For. Res.* 17:174–180.
Ewers, F. W., and Zimmermann, M. H. (1984). The hydraulic architecture of balsam fir (*Abies balsamea*). *Physiol. Plant.* 60:453–458.
Ford, E. D. (1975). Competition and stand structure in some even-aged plant monocultures. *J. Ecol.* 63:311–333.

Geron, C. D., and Ruark, G. A. (1988). Comparison of constant and variable allometric ratios for predicting foliar biomass of various tree genera. *Can. J. For. Res.* 18:1298–1304.

Gholz, H. L. (1979). Limits on aboveground net primary production, leaf area, and biomass in vegetational zones of the Pacific Northwest. Ph.D. Thesis, Oregon State University, Corvallis.

Gholz, H. L. (1980). Structure and productivity of *Juniperus occidentalis* in central Oregon. *Am. Midl. Nat.* 103:251–261.

Gholz, H. L., Vogel, S. A., Cropper, W. P., Jr., McKelvey, K., Owel, K. C., Teskey, R. O., and Curran, P. J. (1991). Dynamics of canopy structure and light interception in *Pinus elliotti* stands, north Florida. *Ecol. Monogr.* 61:33–51.

Gower, S. T., Grier, C. C., Vogt, D. J., and Vogt, K. A. (1987). Allometric relations of deciduous (*Larix occidentalis*) and evergreen conifers (*Pinus contorta* and *Pseudotsuga menziesii*) of the Cascade Mountains in central Washington. *Can. J. For. Res.* 17:630–634.

Grace, J. G., Rook, D. A., and Lane, P. M. (1987). Modelling canopy photosynthesis in *Pinus radiata* stands. *N.Z. J. For. Sci.* 17:210–228.

Granier, A. (1981). Étude des relations entre la section du bois d'aubier et la masse foliaire chez le Douglas (*Pseudotsuga menziesii* (Mirb.) Franco. *Ann. Sci. For.* 38:503–512.

Grier, C. C., and Logan, R. S. (1977). Old-growth *Pseudotsuga menziesii* communities of a western Oregon watershed: Biomass distribution and protection budgets. *Ecol. Monogr.* 47:373–400.

Grier, C. C., and Running, S. W. (1977). Leaf area of mature northwestern coniferous forests: Relation to site water balance. *Ecology* 58:893–899.

Grier, C. C., and Waring, R. H. (1974). Coniferous foliage mass related to sapwood area. *For. Sci.* 20:205–206.

Grier, C. C., Vogt, K. A., Keyes, M. R., and Edmonds, R. L. (1981). Biomass distribution and above- and below-ground production in young and mature *Abies amabilis* zone ecosystems of the Washington Cascades. *Can. J. For. Res.* 11:155–167.

Hall, F. G., Shimabukuro, Y. E., and Huemmrich, K. F. (1994). Remote sensing of forest biophysical structure in boreal stands of *Picea mariana* using mixture decomposition and geometric reflectance models. *Ecol. Appl.* (in press).

Hashimoto, R. (1991). Canopy development in young sugi *Cryptomeria japonica* stands in relation to changes with age in crown morphology and structure. *Tree Physiol.* 8: 129–143.

Hendrick, R. L., and Pregitzer, K. S. 1993. Patterns of fine root mortality in two sugar maple forests. *Nature (London)* 361:59–61.

Hepp, T. E., and Brister, G. H. (1982). Estimating crown biomass in loblolly pine plantations in the Carolina flatwoods. *For. Sci.* 28:115–127.

Hermann, R. K., and Petersen, R. G. (1969). Root development and height increment of ponderosa pines in pumice soils of central Oregon. *For. Sci.* 15:226–237.

Hinckley, T. M., Braatne, J., Ceulemans, R., Clum, P., Dunlap, J., Newman, D., Smit, B., Scarascia,-Mugnozza, G., and Van Volkenburgh, E. (1992). Growth dynamics and canopy structure. *In* "Ecophysiology of Short Rotation Forest Crops" (C. P. Mitchell, J. B. Ford-Robertson, T. Hinckley, and L. Sennerby-Forsse, eds.), pp. 1–34. Elsevier, London and New York.

Holben, B. N. (1986). Characteristics of maximum-value composite images from temporal AVHRR data. *Int. J. Remote Sens.* 7:1417–1434.

Huber, B. (1928). Weitere quantitative Untersuchungen über das Wasserleitungssytem der Pflanzen. *Jahrb. Wiss. Bot.* 67:877–959.

Hungerford, R. D. (1987). Estimation of foliage area in dense Montana lodgepole pine stands. *Can. J. For. Res.* 17:320–324.

Jack, S. B., and Long, J. N. (1992). Forest production and the organization of foliage within crowns and canopies. *For. Ecol. Manage.* 49:233–245.

Jackson, D. S., and Chittenden, J. (1981). Estimation of dry matter in *Pinus radiata* root systems. 1. Individual trees. *N.Z. J. For. Sci.* 11:164–182.
Jacobs, T. D. (1988). "Differences in Light Interception and Canopy Architecture Between Three Half-Sib Families of Loblolly Pine (*Pinus taeda*)." Duke University Publications, Durham, NC.
Jarvis, P. J. (1975). Water transfer in plants. *In* "Heat and Mass Transfer in the Plant Environment" (D. A. deVries and N. G. Afgan, eds.), Part I, pp. 369–394. Scripta, Washington, DC.
Jarvis, P. J., and Leverenz, J. W. (1983). Productivity of temperate, deciduous, and evergreen forests. *Encyclo. Plant Physiol., New Ser.* 12D:233–280.
Jarvis, P. G., and Stewart, J. B. (1979). Evaporation of water from a plantation forest. *In* "The Ecology of Even-aged Forest Plantations" (E. D. Ford, D. C. Malcolm, and J. Atterson, eds.), pp. 327–350. Institute of Terrestrial Ecology, NERC, Cambridge, UK.
Johnson, J. D., Zedaker, S. M., and Hairston, A. B. (1985). Foliage, stem, and root interrelations in young loblolly pine. *For. Sci.* 31:891–898.
Kaipiainen, L., and Hari, P. (1985). Consistencies in the structure of Scots pine. *In* "Crop Physiology of Forest Trees" (P. M. A. Tigerstedt, P. Puttonen, and V. Koski, eds.), pp. 31–37. Helsinki Univ. Press, Helsinki.
Kärki, L., and Tigerstedt, P. M. A. (1985). Definition and exploitation of forest ideotypes in Finland. *In* "Attributes of Trees as Crop Plants" (M. G. R. Cannell and J. E. Jackson, eds.), pp. 102–109. Institute of Terrestrial Ecology, Huntingdon, UK.
Kaufmann, M. R., and Troendle, C. A. (1981). The relationship of leaf area and foliage biomass to sapwood conducting area in four subalpine forest tree species. *For. Sci.* 27:477–482.
Kaufmann, M. R., and Watkins, R. K. (1990). Characteristics of high- and low-vigor lodgepole pine trees in old-growth stands. *Tree Physiol.* 7:239–246.
Keane, M. G., and Weetman, G. F. (1987). Leaf area–sapwood cross-sectional area relationships in repressed stands of lodgepole pine. *Can. J. For. Res.* 17:205–209.
Kendall-Snell, J. A., and Brown, J. K. (1978). Comparison of tree biomass estimators—DBH and sapwood area. *For. Sci.* 24:455–457.
Keyes, M. R., and Grier, C. C. (1981). Above- and below-ground net production in 10-year old Douglas-fir stands on low and high productivity sites. *Can. J. For. Res.* 11:599–605.
Kinerson, R. S., Higginbotham, K. O., and Chapman, R. C. (1974). The dynamics of foliage distribution within a forest canopy. *J. Appl. Ecol.* 11:347–353.
Kira, T., Shinozaki, K., and Hozumi, K. (1969). Structure of forest canopies as related to their primary productivity. *Plant Cell Physiol.* 10:129–142.
Kittredge, J. (1944). Estimation of the amount of foliage of trees and stands. *J. For.* 42:905–912.
Landsberg, J. J. (1986). "Physiological Ecology of Forest Production." Academic Press, New York.
Lanner, R. M. (1985). On the insensitivity of height growth to spacing. *For. Ecol. Manage.* 13:143–148.
Leverenz, J. W., Deans, J. D., Ford, E. D., Jarvis, P. G., Milne, R., and Whitehead, D. (1982). Systematic spatial variation of stomatal conductance in a Sitka spruce plantation. *J. Appl. Ecol.* 19:835–851.
Lieffers, V. J., and Titus, S. J. (1989). The effects of stem density and nutrient status on size inequality and resource allocation in lodgepole pine and white spruce seedlings. *Can. J. Bot.* 67:2900–2903.
Lindroth, A. (1985). Canopy conductance of coniferous forests related to climate. *Water Resour. Res.* 21:297–304.
Long, J. N., and Smith, R. W. (1988). Leaf area–sapwood area relations of lodgepole pine as influence by stand density and site index. *Can. J. For. Res.* 18:247–250.

Long, J. N., Smith, F. W., and Scott, D. R. M. (1981). The role of Douglas-fir stem sapwood and heartwood in the mechanical and physiological support of crowns and development of stem form. *Can. J. For. Res.* 11:459–464.

Madgwick, H. A. I. (1968). Seasonal changes in biomass and annual production of an old-field *Pinus virginiana* stand. *Ecology* 49:149–152.

Madgwick, H. A. I. (1983a). Seasonal changes in the biomass of a young *Pinus radiata* stand. *N.Z. J. For. Sci.* 13:25–36.

Madgwick, H. A. I. (1983b). Estimation of the oven-dry weight of stems, needles, and branches of individual *Pinus radiata* trees. *N.Z. J. For. Sci.* 13:108–109.

Magnussen, S., Smith, V. G., and Yeatman, C. W. (1986). Foliage and canopy characteristics in relation to above-ground dry matter increment of seven jack pine provenances. *Can. J. For. Res.* 16:464–470.

Maguire, D. A., and Hann, D. W. (1987). Equations for predicting sapwood area at crown base in southwestern Oregon Douglas fir. *Can. J. For. Res.* 17:236–241.

Mäkelä, A. (1986). Implications of the pipe model theory on dry matter partitioning and height growth in trees. *J. Theor. Biol.* 123:103–120.

Marchand, P. J. (1984). Sapwood area as an estimator of foliage biomass and projected leaf area for *Abies balsamea* and *Picea rubens*. *Can. J. For. Res.* 14:85–87.

Margolis, H. A., Gagnon, R. R., Pothier, D., and Pineau, M. (1988). The adjustment of growth, sapwood area, heartwood area and saturated sapwood permeability of balsam fir after different intensities of pruning. *Can. J. For. Res.* 18:723–727.

Marshall, J. D., and Waring, R. H. (1986). Comparison of methods of estimating leaf-area index in old-grown Douglas-fir. *Ecology* 67:975–979.

McNaughton, K. G., and Black, T. A. (1973). Evapotranspiration from a forest: A micrometeorological study. *Water Resour. Res.* 9:1579–1590.

Miller, H. G., and Miller, J. D. (1976). Effect of nitrogen supply on net primary production in Corsican pine. *J. Appl. Ecol.* 13:249–256.

Mori, S., and Hagihara, A. (1991). Crown profile of foliage area characterized with the Weibull distribution in a hinoki (*Chamaecyparis obtusa*) stand. *Trees* 5:149–152.

Myers, B. J. (1988). Water stress integral—A link between short-term stress and long-term growth. *Tree Physiol.* 4:315–324.

Norman, J. M., and Jarvis, P. G. (1975). Photosynthesis in Sitka spruce: Radiation penetration theory and a test case. *J. Appl. Ecol.* 12:839–878.

Oker-Blom, P., and Kellomäki, S. (1982). Theoretical computations on the role of crown shape in the absorption of light by forest trees. *Math. Biosci.* 59:291–311.

Oker-Blom, P., and Kellomäki, S. (1983). Effect of grouping of foliage on the within-stand and within-crown light regime: Comparison of random and grouping canopy models. *Agric. Meteorol.* 28:143–155.

Oren, R., Schulze, E.-D., and Zimmermann, R. (1986a). Estimating photosynthetic rate and annual carbon gain in conifers from specific leaf weight and leaf biomass. *Oecologia* 70:187–193.

Oren, R., Week, K. S., and Schulze, E.-D. (1986b). Relationships between foliage and conducting xylem in *Picea abies* (L.) Karst. *Trees* 1:61–69.

Oren, R., Waring, R. H., Stafford, S. G., and Barrett, J. W. (1987). Twenty-four years of ponderosa pine growth in relation to canopy leaf area and understory competition. *For. Sci.* 33:538–547.

Pearson, J. A., Fahey, T. J., and Knight, D. H. (1984). Biomass and leaf area in contrasting lodgepole pine forest. *Can. J. For. Res.* 14:259–265.

Peterson, D. L., Spanner, M. A., Running, S. W., and Teuber, K. B. (1987). Relationship of thematic mapper simulator data to leaf area index of temperate coniferous forests. *Remote Sens. Environ.* 22:323–341.

Pothier, D., and Margolis, H. A. (1991). Analysis of growth and light interception of bal-

sam fir and white birch saplings following precommercial thinning. *Ann. Sci. For.* 48: 123–132.
Pothier, D., Margolis, H. A., and Waring, R. H. (1989a). Changes in saturated sapwood permeability and total sapwood conductance with stand development. *Can. J. For. Res.* 19:432–439.
Pothier, D., Margolis, H. A., Poliquin, J., and Waring, R. H. (1989b). Relation between the permeability and the anatomy of jack pine with stand development. *Can. J. For. Res.* 19: 1564–1570.
Raich, J. W., and Nadelhoffer, K. J. (1989). Belowground carbon allocation in forest ecosystems: Global trends. *Ecology* 70:1346–1354.
Raison, R. J., Myers, B. J., and Benson, M. L. (1992a). Dynamics of *Pinus radiata* foliage in relation to water and nitrogen stress: I. Needle production and properties. *For. Ecol. Manage.* 52:139–158.
Raison, R. J., Khanna, P. K., Benson, M. L., Myers, B. J., McMurtrie, R. E., and Lang, A. R. G. (1992b). Dynamics of *Pinus radiata* foliage in relation to water and nitrogen stress: II. Needle loss and temporal changes in total foliage mass. *For. Ecol. Manage.* 52: 159–178.
Rogers, D. L., Stettler, R. F., and Heilman, P. E. (1989). Genetic variation and productivity of *Populus trichocarpa* and its hybrids: III. Structure and patterns of variation in a three-year field test. *Can. J. For. Res.* 19:372–377.
Running, S. W., Peterson, D. L., Spanner, M. A., and Teuber, K. B. (1986). Remote sensing of coniferous forest leaf area. *Ecology* 67:273–276.
Running, S. W., Nemani, R. R., Peterson, D. L., Band, L. E., Potts, D. F., Pierce, L. L., and Spanner, M. A. (1989). Mapping regional forest evapotranspiration and photosynthesis by coupling satellite data with ecosystem simulation. *Ecology* 70:1090–1101.
Rutter, A. J. (1966). Studies on the water relations of *Pinus sylvestris* in plantation conditions. IV. Direct observations on the rates of transpiration, evaporation of intercepted water, and evaporation from the soil surface. *J. Appl. Ecol.* 3:393–405.
Santantonio, D. (1989). Stand age and water stress during summer affect efficiency and distribution of production above and below ground in mature stands of Douglas-fir. (*Pseudotsuga menziesii*). *NATO ASI Ser., Ser. E* 166.
Santantonio, D., and Santantonio, E. (1987). Effect of thinning on production and mortality of fine roots in a *Pinus radiata* plantation on a fertile site in New Zealand. *Can. J. For. Res.* 17:919–928.
Schroeder, P. E. (1983). Canopy leaf area and its distribution as an index to stocking and stand growth. M.S. Thesis, Oregon State University, Corvallis.
Schuler, T. M., and Smith, F. W. (1988). Effect of species mix on size/density and leaf-area relations in southwest pinyon/juniper woodlands. *For. Ecol Manage.* 25:211–220.
Schulze, E.-D., Fuchs, M., and Fuchs, M. I. (1977). Spatial distribution of photosynthetic capacity and performance in a mountain spruce forest of northern Germany. *Oecologia* 30:239–248.
Shinozaki, K., Yoda, K., Hozumi, K., and Kira, T. (1964a). A quantitative analysis of plant form: The pipe model theory. I. Basic analysis. *Jpn. J. Ecol.* 14:97–105.
Shinozaki, K., Yoda, K., Hozumi, K., and Kira, T. (1964b). A quantitative analysis of plant form: The pipe model theory. II. Further evidence of the theory and its application in forest ecology. *Jpn. J. Ecol.* 14:133–139.
Sobrado, M. A., Grace, J., and Jarvis, P. G. (1992). The limits of xylem embolism recovery in *Pinus sylvestris* L. *J. Exp. Bot.* 43:831–836.
Spanner, M. A., Pierce, L. L., Running, S. W., and Peterson, D. L. (1990a). The seasonality of AVHRR data of temperate coniferous forests: Relationship with leaf area index. *Remote Sens. Environ.* 33:97–112.
Spanner, M. A., Pierce, L. L., Peterson, D. L., and Running, S. W. (1990b). Remote sensing

of temperate coniferous forest leaf area index: The influence of canopy closure, understory vegetation and background reflectance. *Int. J. Remote Sens.* 11:95–111.

Thompson, D. C. (1989). The effect of stand structure and stand density on the leaf area–sapwood area relationship of lodgepole pine. *Can. J. For. Res.* 19:392–396.

Trousdell, K. B., Dorman, K. W., and Squillace, A. E. (1963). Inheritance of branch length in young loblolly pine progeny. *USDA For. Serv. Res. Note SE* SE-1.

Tyree, M. T. (1988). A dynamic model for water flow in a single tree: Evidence that models must account for hydraulic architecture. *Tree Physiol.* 4:195–217.

Tyree, M. T., and Dixon, M. A. (1986). Water stress induced cavitation and embolism in some woody plants. *Physiol. Plant.* 66:397–405.

Tyree, M. T., Graham, M. E. D., Cooper, K. E., and Bazos, L. J. (1983). The hydraulic architecture of *Thuja occidentalis*. *Can. J. Bot.* 61:2105–2111.

Van Cleve, K., Dyrness, C. T., Viereck, L. A., Fox, J., Chapin, F. S., and Oechel, W. (1983). Tiaga ecosystems in interior Alaska. *BioScience* 33:39–44.

Vose, J. M., and Allen, H. L. (1988). Leaf-area, stemwood growth, and nutrition relationships in loblolly pine. *For. Sci.* 34:547–563.

Vose, J. M., and Swank, W. T. (1990). A conceptual model of forest growth emphasizing stand leaf area. *In* "Process Modeling of Forest Growth Responses to Environmental Stress" (R. K. Dixon, R. S. Meldahl, G. A. Ruark, and W. G. Warren, eds.), pp. 278–287. Timber Hill Press, Portland, OR.

Waring, R. H. (1991). Responses of evergreen trees to multiple stresses. *In* "Responses of Plants to Multiple Stress" (H. A. Mooney, W. E. Winner, and E. J. Pell, eds.), pp. 371–390. Academic Press, San Diego.

Waring, R. H. (1980). Site, leaf area, and phytomass production in trees. *N.Z. For. Serv./For. Res. Inst. Tech. Pap.* 70:125–135.

Waring, R. H. (1982). Estimating forest growth and efficiency in relation to canopy leaf area. *Adv. Ecol. Res.* 13.

Waring, R. H. (1983). Estimating forest growth and efficiency in relation to canopy leaf area. *Adv. Ecol. Res.* 13:327–354.

Waring, R. H. (1987). Characteristics of trees predisposed to die. *BioScience* 37:569–574.

Waring, R. H., and Running, S. W. (1978). Sapwood water storage: Its contribution to transpiration and effect upon water conduction through stems of old-growth Douglas fir. *Plant, Cell Environ.* 1:131–140.

Waring, R. H., Gholz, H. L., Grier, C. C., and Plummer, M. L. (1977). Evaluating stem conducting tissue as an estimator of leaf area in four woody angiosperms. *Can. J. Bot.* 55:1474–1477.

Waring, R. H., Emmingham, W. H., Gholz, H. L., and Grier, C. C. (1978). Variation in maximum leaf area of coniferous forests in Oregon and its ecological significance. *For. Sci.* 24:131–140.

Waring, R. H., Whitehead, D., and Jarvis, P. G. (1979). The contribution of stored water to transpiration in Scots pine. *Plant, Cell Environ.* 2:309–317.

Waring, R. H., Schroeder, P. E., and Oren, R. (1982). Application of the pipe model theory to predict canopy leaf area. *Can. J. For. Res.* 12:556–560.

Whitehead, D. (1978). The estimation of foliage area from sapwood basal area in Scots pine. *Forestry* 51:137–149.

Whitehead, D., and Jarvis, P. G. (1981). Coniferous forest and plantations. *In* "Water Deficits and Plant Growth" (T. T. Kozlowski, ed.), Vol. 6, pp. 49–152. Academic Press, London and New York.

Whitehead, D., Edwards, W. R. N., and Jarvis, P. G. (1984). Relationships between conducting sapwood area, foliage area and permeability in mature *Picea sitchensis* and *Pinus contorta* trees. *Can. J. For. Res.* 14:940–947.

Whitehead, D., Grace, J. C., and Godfrey, M. J. S. (1990). Architectural distribution of foliage in individual *Pinus radiata* D. Don crowns and the effects of clumping on radiation interception. *Tree Physiol.* 7:135–155.

Whitehead, D., Kelliher, F. M., Lane, P. M., and Pollock, D. S. (1993). Seasonal partitioning of evaporation between trees and understory in a widely spaced *Pinus radiata* stand. *J. Appl. Ecol.* 31:528–542.

Worrall, J., Draper, D. A., and Anderson, S. A. (1985). Shoot characteristics of stagnant and vigorous lodgepole pine, and their growth after reciprocal grafting. *Can. J. For. Res.* 15:365–370.

Yoder, B. J. (1992). Photosynthesis of conifers: Influential factors and potentials for remote sensing. Ph.D. Thesis, Oregon State University, Corvallis.

Youngberg, C. T., and Cochran, P. H. (1982). Silviculture on volcanic ash soils in America. *In* "Silviculture Under Extreme Ecological and Economic Conditions," (S. Dafis, ed.), IUFRO Div. 1 Meet., 1980, pp. 331–345. Aristotelian University, Thessaloniki, Greece.

Zimmermann, M. H. (1978). Hydraulic architecture of some diffuse porous trees. *Can. J. Bot.* 56:2286–2295.

Zimmermann, M. H. (1983). "Xylem Structure and the Ascent of Sap." Springer-Verlag, Berlin.

Zobel, B. J., and Talbert, J. T. (1984). "Applied Forest Tree Improvement." Wiley, New York.

8
Causes and Consequences of Variation in Conifer Leaf Life-Span

Peter B. Reich, Takayoshi Koike, Stith T. Gower, and Anna W. Schoettle

I. Introduction

Long-lived foliage of evergreen species (including most conifers) historically has been considered a characteristic feature of this group and symbolic of the evergreen strategy. The two most obvious ways in which conifers often differ from the broad-leafed tree species with which they share much of the temperate and boreal biomes are leaf form (needles versus broad leaves) and habit (evergreen versus deciduous). Exceptions to these rules (e.g., deciduous conifers such as *Larix* and *Taxodium* and evergreen angiosperms in areas dominated by conifers) provide intriguing examples that provoke us to reexamine our interpretation of the advantages and disadvantages of each trait (e.g., Gower and Richards, 1990). Despite our tendency to think of conifers as evergreen species with long-lived needles, there is large variation in leaf life-span among coniferous species, and within species as well. Needle life-span of conifers varies by two orders of magnitude—from less than half a year in European larch (*Larix decidua*) (Benecke *et al.*, 1981) to more than 40 years in Bristlecone pine (*Pinus longaeva*) (Ewers and Schmid, 1981). Within coniferous species, individuals at different sites can also have large differences in needle life-span—from 3 to 9 years for *Pinus sylvestris* (Pravdin, 1969), 4 to 9 years for *Pinus flexilis* (Ewers and Schmid,

1981), 5 to 18 years for *Pinus contorta* (Schoettle, 1990a), and 3 to 8 years for *Pseudotsuga menzesii* (Gower *et al.*, 1992).

Therefore, despite the tendency to think of conifers as a unique group of evergreen species with long-lived foliage, the range of leaf life-span within conifers is broad and overlaps greatly with broad-leafed angiosperms (Reich *et al.*, 1992). Moreover, evergreen species need not have long-lived foliage [as demonstrated by the observations of Coley (1988) and Reich *et al.* (1991a) in tropical forests], and the most consistent distinction between evergreen and deciduous species involves phenology, not leaf life-span. Hence, we begin this chapter by identifying problems with the common terminology that lumps evergreenness and long leaf life-span together, and with the common tendency to think of conifers as a relatively similar group of species with shared traits (evergreen, long-lived foliage) that set them off from other groups with roughly opposite sets of traits. Although these common definitions are accurate in an extremely general sense, they are misleading and obscure the fact that conifers are commonly different from one another, and despite "different-looking" foliage, may be more common physiologically and ecologically with certain angiosperm trees than with other conifers. As we show below, use of leaf life-span as a descriptor of species provides a gradient of identification, rather than an oversimplified lumping of all species into two groups (i.e., evergreen and deciduous).

Leaf life-span has been shown to be correlated with physiology, growth, productivity, and carbon allocation (Sprugel, 1989; Schoettle, 1990a; Reich *et al.*, 1991a, 1992; Gower *et al.*, 1993), and to nutrient use and allocation (Son and Gower, 1991, 1992). Leaf life-span is also an important life-history trait that is strongly related to response to drought, nutrient availability, herbivory, air pollution, and other perturbations (Chapin, 1980; Chabot and Hicks, 1982; Reich, 1987; Coley, 1988; Koike, 1988; Reich *et al.*, 1992). Moreover, as shown below (and to the extent that it has been examined), needle-leafed and broad-leafed species may behave similarly with respect to the relationships between leaf life-span and several plant traits (Son and Gower, 1991; Reich *et al.*, 1992; Gower *et al.*, 1993), suggesting that the behavior of conifers in this regard is not unique.

In this chapter we characterize leaf life-span of conifers largely from the viewpoint of environment, resource availability, and carbon balance, examining both genotypic variation (among and within species) and phenotypic plasticity (within species). We will separate leaf life-span as an adaptation (i.e., genotypically fixed) from variation in leaf life-span that occurs in each species due to acclimation to environment (plasticity). The factors affecting life-span of coniferous leaves are discussed—namely, the physical environment and biological factors related to the growth characteristics of species—as are the implications of variation in leaf life-span for conifers at several levels of hierarchical scale. The implications of leaf life-span are first addressed at the leaf level, and then integrated at the whole-plant and stand level. An important question is whether variation in leaf life-span has similar implications for gymnosperms and angiosperms. A related question is to what extent we can attribute the "uniqueness" of conifers (contrasted to angiosperm trees) to differences in leaf habit (deciduous, evergreen), leaf form (broad-leafed vs. needle-leafed), leaf life-span (which overlaps broadly), or other characteristics.

The geographic distribution of evergreen and deciduous communities has attracted lasting attention (e.g., Axelrod, 1966; Waring and Franklin, 1979; Gower and Richards, 1990). In general, deciduous species are favored where annual variation in temperature (e.g., temperate deciduous forests) or moisture availability (e.g., tropical deciduous forests) results in marked favorable versus unfavorable periods for carbon gain (Kikuzawa, 1991; Reich, 1994). Evergreen communities (either gymnosperm or angiosperm) dominate regions that are aseasonal, are always relatively unfavorable due to low fertility and/or water availability, have short favorable seasons and long unfavorable ones, or have some combination of these or other factors (Reich *et al.*, 1992). For example, Waring and Franklin (1979) hypothesized that the combination of warm, but dry, summers and mild, but wet, winters in the Pacific Northwest of North America results in a climate that favors the evergreen habit. The idea that extended leaf life-span is also a nutrient conservation mechanism that enhances nutrient use efficiency and annual or long-term carbon gain has also been considered as an explanation for geographical patterns of deciduous versus evergreen species (e.g., Monk, 1966; Chapin, 1980; Chabot and Hicks, 1982; but see Son and Gower, 1991). Evergreen conifers occupy a major part of the global evergreen zones outlined above, including dry, infertile and cold sites, but are relatively uncommon in subtropical/tropical regions with little seasonal variation in temperature, where broad-leafed evergreen angiosperms predominate.

II. Variation in Leaf Life-Span

A. Internal and External Factors: Relation to Needle Life-Span

In the sections below we describe how needle life-span varies within and among species in relation to internal and external factors, including temperature, light, water, nutrients, insects, diseases, and air pollution. Although we present and describe a variety of patterns, we argue that these are consistent with two general patterns. First, for conditions that are relatively stable and present during leaf development, a species or genotype will have shorter leaf life-span under conditions conducive to greater carbon gain. Examples include shorter leaf life-span on fertile than on infertile sites, in sunny than in shaded microenvironments, and on cool rather than on cold sites (see below, Ewers and Schmid, 1981; Schoettle and Smith, 1991; Koike *et al.*, 1991; Reich *et al.*, 1991a; Gower *et al.*, 1994). Chabot and Hicks (1982) proposed that leaf life-span could be explained by the balance between photosynthetic rate (benefit) and the costs of producing and maintaining leaves (cost). Based on this idea, Kikuzawa (1991) developed a simple model that estimated the leaf lifespan that would maximize the net gain of a leaf per unit time. Model output was consistent with prior hypotheses and empirical data regarding leaf life-span and other leaf characteristics (Mooney and Gulmon, 1982; Chabot and Hicks, 1982; Reich *et al.*, 1991a, 1992). Using his model, Kikuzawa (1991) obtained two peaks of geographical distribution of evergreeness at higher and lower latitude, consistent with observed patterns; however, the model underestimated the relative proportion of evergreens at high latitudes. Nonetheless, it is encouraging that a simple cost–benefit model consistent with conceptual theory also appears to be qualitatively consistent with empirical observations, suggesting that further modeling may be rewarding.

The second pattern involves conditions that are variable. When variation is temporally regular, some species have adapted with short leaf lifespans and related traits that maximize carbon assimilation during environmentally favorable time windows (e.g., understory seedlings, spring ephemerals, and desert ephemerals). If conditions are also irregular in occurrence and have direct adverse affects on leaf physiology—such as severe drought, disease, or air pollution—these can induce premature leaf senescence before internally regulated leaf senescence would occur under otherwise less stressful, yet common, conditions (e.g., Reich and Borchert, 1982; Reich, 1983).

B. Causes of Variation in Leaf Life-Span

1. Temperature Within a species, leaf life-span of trees is greater for individuals grown at higher than at lower elevations (Weidman, 1939; Ewers and Schmid, 1981; Schoettle, 1990a; Gower *et al.*, 1992). In California, within the genus *Pinus*, Ewers and Schmid (1981) observed that maximum needle life-span increases with elevation (Fig. 1). They also observed that maximum needle life-span within species was significantly greater for native populations growing at higher elevations. For example, maximum needle life-span in *Pinus monophylla* was 10 years at 2270 m elevation versus 8 years at 1950 m elevation and in *Pinus contorta*

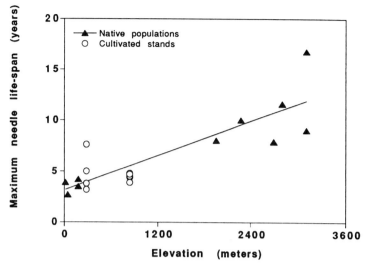

Figure 1 Maximum needle life-span for California *Pinus* species in relation to site elevation. For native populations, this relationship is significant ($P < 0.001$, $r^2 = 0.81$; regression line for native populations only). Needle life-span for populations grown off-site in cultivated plantations falls roughly within the same overall relationship. (Redrawn from data of Ewers and Schmid, 1981.)

was 8 years at 2700 m elevation versus 4 years at 20–180 m elevation (Table I).

Observed patterns such as this result from both adaptation (genetic differences) and acclimation (environmental influences). Significantly greater maximum needle life-spans were observed for individuals of

Table I Differences in Maximum Needle Life-Span of Native Populations of California *Pinus* species[a]

Species	Elevation (meters)	Maximum needle life-span (years)
Pinus monophylla	2274	10.0
	1951	8.0
Pinus contorta	2697	7.9
	182	4.2
	15	3.9
Pinus muricata	182	3.5
	46	2.7

[a] Data from Ewers and Schmid (1981).

Table II Acclimation of Maximum Needle Life-Span within Genotypes of California *Pinus* Species[a]

Species	Elevation (meters)[b]	Maximum needle life-span (years)
Pinus longaeva	3109 (N)	16.8
	287	7.6
Pinus monophylla	1951–2274 (N)	8.0–10.0
	287	7.6
Pinus flexilis	3109 (N)	9.0
	843	4.8
	287	3.8
Pinus muricata	843	4.7
	287	3.8
	46–182 (N)	2.7–3.5

[a] Data from Ewers and Schmid (1981).
[b] The pines were grown at their native (N) elevation and at either lower or higher elevations.

given populations when grown at higher than at lower elevations, indicating acclimation and strong environmental control. Maximum life-span as much as doubled across an elevational gradient as a result of acclimation (Table II). Adaptive (genotypic) differences among California *Pinus* spp. were less clear when species native to different elevations were compared at the same elevations in common garden studies (Table III). However, genotypic variation has been shown in other stud-

Table III Genetically Determined Differences in Maximum Needle Life-Span of California *Pinus* Species[a]

Species	Native elevation (meters)	Max needle life-span at 843 m elevation	Max needle life-span at 287 m elevation
Pinus longaeva	3109	—	7.6
Pinus flexilis	3109	4.8	3.8
Pinus monophylla	1951–2274	—	7.6
Pinus contorta	2697	3.9	3.2
	182	4.4	5.0
	15	4.5	—
Pinus muricata	46–182	4.7	3.8

[a] All pines were grown at the same elevations in the common garden experiments. Data from Ewers and Schmid (1981).

ies, such as that by Zedenbauer (1916), who reported that when grown together in a common garden, low-elevation species of *Picea* retained their needles for about 5 years versus about 9 years for high-elevation species.

Similar to elevation, *Pinus* populations across latitudinal gradients may show differentiation in needle life-span (Pravdin, 1969). *Pinus sylvestris* trees in the central European lowlands commonly retain needles for 3–4 years, whereas in Scandinavia and northwestern Russia needles are commonly retained for as long as 7–8 years. However, to the east, in Asian Russia, *P. sylvestris* populations have long needle life-spans (6–8 years) at all latitudes from 50 to 70°N. A relatively sharp demarcation (across a northwest to southeast transect) between *P. sylvestris* populations of short versus long needle life-spans, and of other differing traits, is attributed by Pravdin (1969) to two distinct races. Clearly, temperature is not the only factor that varies with increasing elevation or latitude. Soil fertility, length of growing season, drought, and other factors tend to follow these same directional patterns, and are likely involved in the establishment of observed patterns of leaf life-span.

Similar to the broad geographic influence of temperature across elevational or latitudinal gradients, at timberline needle life-span in *Pinus cembra*, *Pinus mugo*, and *Picea abies* is greater for trees growing on cooler, shady slopes than on sunny slopes (Koike *et al.*, 1993) (Fig. 2). Within a cost–benefit context, both winter water relations and summer carbon gain may be involved as mechanistic forces for these patterns: (1) because colder winter temperatures require greater cuticles and other structural constituents to minimize desiccation, greater leaf life-span would be required to amortize the greater initial construction costs; (2) greater investment in structural integrity appears to occur at the ex-

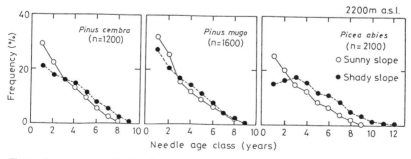

Figure 2 Frequency distribution of needle age classes on relatively sunny vs. shaded slopes on 9 to 10 trees each for three conifers (*n* = 1200–2100 shoots total) at 2200 m elevation near Stillberg, Switzerland. Sunny slopes were east-facing, and trees there received direct sunlight for much of the day. Shaded slopes were north-facing, and trees there received only diffuse light for much of the day. (From Koike *et al.*, 1991.)

pense of carbon assimilation potential (Reich *et al.*, 1991a, 1992), thereby lessening the net assimilation rate and also extending the amortization time; and (3) cooler growing seasons of higher elevations or shadier slopes result in lesser carbon gain even for a given maximum assimilation rate, also requiring greater amortization time and extended life life-span.

2. Light The leaf life-span of trees appears to be inversely related to solar radiation. For instance, in three Japanese conifers, maximum needle life-span of 18-year-old individuals was 15–40% less when grown under full sunlight rather than beneath a hardwood forest canopy (Table IV). Life-span of foliage in sunny parts of the crown, especially the upper part, is shorter than that of shaded crown positions (e.g., Kohyama, 1980; Maillette, 1982; Williams *et al.*, 1989; Schoettle and Smith, 1991). In Wisconsin and Wyoming, maximum needle life-span was 33–75% greater in the lower than upper third of the canopy for four conifer species (Tables V and VI). The larger increase in life-span from the upper to middle (than middle to lower) canopy reflects the exponential decrease in radiation downward in the canopy (Bolstad and Gower, 1990). The reduced leaf life-span of upper canopy foliage is likely correlated with the greater carbon gain and growth in the upper crown, which results from the higher light levels.

Annual shoot growth increment, annual foliar biomass production, and integrated daily photosynthetically active radiation (DPAR) at shoot tips all increase with increased height in the crown of lodgepole pine (Schoettle and Smith, 1991). In contrast to shoot and foliage growth, leaf longevity decreased with increasing height in the crown (Table VI). Within the mature crown, no leaves were retained, regardless of their age, in a DPAR of less than approximately 25% of the available daily

Table IV Needle Longevity of the Lower Branch of Fir and Spruce Grown under Different Light Conditions[a]

	Needle longevity (years ± SD)	
Species	Beneath hardwood canopy[b]	In full sunlight
Abies sachlinensis	6.6 ± 0.2	5.7 ± 0.6
Picea jezoensis	10.2 ± 0.9	8.5 ± 0.7
Picea glehnii	11.8 ± 0.6	7.4 ± 0.5

[a]The trees were 18 years old in 1991 and were growing at 140 m above sea level near Sapporo, Japan. (From T. Koike, unpublished data.)

[b]Relative photosynthetic photon flux density ≈ 12% of full sunlight in July and August.

Table V Maximum Needle Life-Span at Different Canopy Positions[a]

Species	Maximum needle life-span (years)		
	Upper canopy	Middle canopy	Lower canopy
White pine	2.4	3.6	4.2
Red pine	3.8	5.0	5.2
Norway spruce	4.4	7.0	7.4
Lodgepole pine	9.2	11.2	12.2

[a] Data are for three species in adjacent 27-year-old plantations in a common garden experiment in southwestern Wisconsin (S. T. Gower, unpublished data) and for lodgepole pine in southeastern Wyoming (Schoettle and Smith, 1991).

PAR above the crown (Fig. 3). Therefore, within this species, shoot structure was regulated in such a way that foliage was produced and retained in high light areas of the canopy, whereas foliage was shed in regions of the canopy that had become heavily shaded (Schoettle and Smith, 1991). The degree of shading that cannot support leaves may be the light environment for which leaves cannot maintain a positive carbon balance (McMurtrie *et al.*, 1986; Schoettle, 1990b).

3. Water and Nutrients Deciduous trees may have originated from tropical regions, with the deciduous habit an adaptation to annual dry seasons (Axelrod, 1966). Deciduousness may be partially facultative in

Table VI Summary of Shoot Structural Characteristics Averaged over Twenty Lodgepole Pine Trees[a]

Crown location	Sample size (no. of shoots)	Shoot increment (cm/yr)	Foliage production (g/yr)	Leaf longevity (years)	Foliage biomass (g/shoot)	DPAR at the shoot tip (% full sun)
Crown third						
Top	120	1.9a (0.1)	1.27a (0.09)	9.2a (0.3)	8.79a (0.54)	66a
Middle	120	1.4b (0.1)	0.82b (0.05)	11.2b (0.3)	7.05b (0.49)	37b
Bottom	120	0.9c (0.1)	0.43c (0.04)	12.2c (0.3)	4.88c (0.37)	30c

[a] The trees were from five sites in southeastern Wyoming. Values followed by different letters were significantly different at $\alpha = 0.05$. Means are presented with the standard errors in parentheses (Schoettle and Smith, 1991).

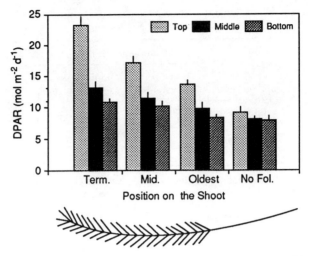

Figure 3 Measured daily integrated quantum flux (DPAR) at four positions along the shoot at three locations within the crown of lodgepole pine. There was a statistically significant ($P < 0.05$) increase in DPAR with increasing crown height at each foliated position along the shoot. However, the DPAR above the first increment of stem without needles, beyond the oldest leaves, was not different at different crown heights. Each bar for the upper and middle crown thirds is an average of 20 shoots and values from 14 shoots were used to generate the bars for the lower crown third (Schoettle and Smith, 1991).

some tree species. For instance, *Tabebuia neochrysantha*, a tropical deciduous angiosperm, may be evergreen on moist sites (Reich and Borchert, 1982). Conversely, seedlings of *Larix lyallii*, normally a deciduous conifer, can remain wintergreen if snow depth is adequate to prevent winter desiccation—however, more typically, by being deciduous, larch avoids winter damage (Richards and Bliss, 1986) from winds and blowing ice crystals (Hadley and Smith, 1986).

In natural ecosystems, leaf life-span of conifers and other vegetation appears to be evolutionarily modulated in relationship with the nutrient availability of their habitat. Species with long-lived foliage are most common on infertile soils in many ecosystems and biomes around the world (Monk, 1966; Chapin, 1980; Reich *et al.*, 1992). These species also have lower leaf N concentrations and lower photosynthetic capacities (Field and Mooney, 1986; Williams *et al.*, 1989; Reich *et al.*, 1991a). In many instances, other resources and conditions are also suboptimal in such environments, which may be cold, dry, shaded, or otherwise not conducive to rapid growth. Thus, although the validity of this pattern is widely accepted, the relative role of genetic versus environmental influence on this pattern is not well understood. Advances will require further manipulative and/or common garden experiments such as those by Miller and Miller (1976), Shaver (1983), Ewers and Schmid (1981), and Gower et al. (1993b).

In seedlings of three broadleaf species, leaf life-span decreased with increased leaf nitrogen that resulted from nitrogen fertilization (Koike and Sanada, 1989). Similar results have been reported elsewhere (Shaver, 1983; Linder and Rook, 1984; Linder, 1987; Stow and Gower, unpublished data). A summary of experimental studies demonstrates that leaf life-span in evergreen conifers is inversely related to resource availability (Table VII) (but it should be remembered that patterns of

Table VII Influence of Water and Nutrient Availability on Leaf Life-Span for Evergreen Conifers

Species	Location	Treatment[a]	Leaf life-span	Change relative to C (%)	Source
Pinus elliottii	Florida	C	2.6	−31	Gholz et al. (1991)
		F (180/2 to 4/2)	1.8		
Pinus sylvestris	Sweden	C	1.8		Linder and Axelsson (1981)
		FI (174/d/5; 3 mm/d/5)	1.4	−22	
Pinus resinosa	Wisconsin	C	4.1	+12	Bockheim et al. (1986)
		F (100/1/1)	4.6		
Pinus radiata	New Zealand	C[b]	2.4	−29	Beets and Madgwick (1988)
		F[b] (96/2/10)	1.7		
		C[c]	2.3	−17	
		F[c] (96/2/10)	1.9		
		C[d]	2.4	−17	
		F[d] (96/2/10)	2.0		
Pinus nigra	Scotland	C	2.5	0	Miller and Miller (1976)
		F (84/1/2)	2.5		
		F (168/1/2)	2.4	−4	
		F (336/1/2)	2.3	−8	
		F (504/1/2)	2.3	−8	
Pseudotsuga menziesii	Washington	WC	4.7	+9	Turner (1977)
		C	4.3		
		F	3.9	−9	
		C	4.0	−7	
Pseudotsuga menziesii	New Mexico	WC	6.9	+3	Gower et al. (1992)
		C	7.1		
		F (200/1/2)	7.0	−1	
		I	7.1	−3	
		WC/I	6.7	−6	

[a] C, Control; F, fertilized; WC, wood chips added; WC/I, wood chips and irrigated. Values in parentheses following F symbol represent amount of N added (Kg/ha/yr), application frequency per year, and number of years applied, respectively; d, daily.
[b] Density of 2224 trees/ha.
[c] 1483 trees/ha.
[d] 741 trees/ha.

leaf area duration at the canopy level do not follow this same pattern). In studies where fertilization was applied for 2 or more years, needle life-span decreased on average by 7% in *Pinus nigra* (Miller and Miller, 1976), 1–8% in *P. menziesii* (Turner, 1977; Gower *et al.*, 1992), 21% in *Pinus radiata* (Beets and Madgwick, 1988), 22% in *P. sylvestris* (Linder and Axelsson, 1981), and 31% in *Pinus elliottii* (Gholz *et al.*, 1991). The addition of wood chips, which act to increase C/N ratios and thus decrease nutrient availability, resulted in 3–9% increases in *P. menziesii* (Turner, 1977; Gower *et al.*, 1992). Increased water availability had effects similar to those of increased nutrient availability, resulting in shortened needle life-span (Gower *et al.*, 1992). In contrast, *Picea glehnii* growing in alpine swamps keeps needles for 2 years longer than at more typical sites (Koike *et al.*, 1991).

4. Pollutants and Elevated CO_2 A large number of reports linking leaf loss, declining forests, and air pollutants (SO_2, O_3, NO_x, acid deposition, etc.) exist from around the world. When pollutant stress is sufficient that acute injury results in visible damage, entire needles or leaves, or parts thereof, may be killed outright (obviously reducing the life-span). Such damage can occur for foliage of any age. Chronic low levels of pollutants (a more common situation) may not produce any visible symptoms of damage, but such stress from air pollutants such as SO_2, O_3, or acid deposition often accelerates leaf aging and premature senescence (Reich, 1983; Schulze *et al.*, 1989). Under high CO_2 levels, life-span of older leaves of *Taxus cuspidata, Picea jezoensis,* and *Betula maximowicziana* is diminished (Koike *et al.*, 1991), consistent with the principle that greater carbon gain is related to shorter leaf duration.

However, at the whole-plant level, reduced carbon gain and leaf area resulting from pollutant stress can lead to compensatory responses such as increased production of new leaf area (Reich and Lassoie, 1985), increased needle length, and/or increased life-span of surviving foliage (J. Oleksyn and P. Reich, unpublished data). These responses (as observed in the cited studies) were minimally effective at compensating for stress, but nonetheless they demonstrate important biological interactions. For example, on a severely polluted site in Poland, several *P. sylvestris* provenances had greater needle life-span than they did on a control site (J. Oleksyn and P. Reich, unpublished data). Slow growth and severe branch dieback at the polluted site resulted in mimimal self-shading of surviving needles and a low total tree leaf area. If conifer needles are shed when their branch reaches a state of negative rather than positive net carbon balance (Sprugel, 1989), usually due to shading (Schoettle and Smith, 1991), then the continued high light availability to the surviving older foliage in these *P. sylvestris* may have resulted in their greater needle life-span. Slow growth rate correlates with long leaf life-span (Schoettle, 1990a) and thus also may be a factor increasing the life-span of these *P. sylvestris* needles. Apparently, the influence of air pollution on leaf life-span can be complex—stress at the tissue level can accelerate aging and diminish life-span; dieback elsewhere on the tree can influence the microenvironment in a way as to alter carbon balance and the timing of senescence; and at the branch or whole-tree level, longer term reductions in growth may increase tissue life-span.

5. Insects and Diseases Because leaf properties are constrained by unavoidable trade-offs in morphology and chemistry that maximize either productivity or persistence (Reich *et al.*, 1991a, 1992), it is not surprising that adaptations to insects and other animals would be related to habitat resource availability and leaf life-span (Coley *et al.*, 1985). Although no parallel work with multiple species has been made in systems with coniferous species, work by Coley (1988) in a tropical forest demonstrated that species with longer leaf life-spans had greater investments in defenses than did those with shorter leaf life-spans. Gower *et al.* (1989) found the same pattern in two North American conifers, where greater lignin and lower %N were associated with greater needle life-span. In tropical trees, long-lived leaves have greater levels of carbon-based chemical defenses (Coley, 1988) and are physically tougher (Coley, 1988; Reich *et al.*, 1991a). Species with greater defenses and longer leaf life-spans also grew more slowly than their opposites (Coley, 1988), which is consistent with the lower N concentrations, photosynthetic rates, and growth rates of species of long leaf life-span (Reich *et al.*, 1991a, 1992). Although we know of no large comparative study of variation in defensive chemicals among species of differing leaf life-span that includes conifers, there are data indicating that increasing leaf life-span is related to both increasing leaf toughness within conifers and across all species types (P. B. Reich, unpublished data) and to an increasing C/N ratio (decreasing leaf N concentration). The hypotheses of Coley *et al.* (1985) are consistent with cost–benefit hypotheses as qualitatively described above, and as modeled, such as by Kikuzawa (1991).

Based on defoliation experiments with conifers (Reich *et al.*, 1993) and hardwoods (Heichel and Turner, 1983), the photosynthetic rate of residual leaves may be increased by simulated grazing. Leaf life-span might theoretically be shortened by this photosynthetic stimulation, based on the negative relationship between leaf carbon gain and life-span. However (similar to the scenario for pollutants), because branch or whole plant carbon gain may be reduced by losses of leaf area, this could promote extended leaf life-span, and the actual result may depend on the interaction between individual leaf and higher scale (branch, whole plant) carbon balances. For instance, leaf life-span in remaining healthy leaves was extended in defoliation experiments (Fig. 4) (Kiku-

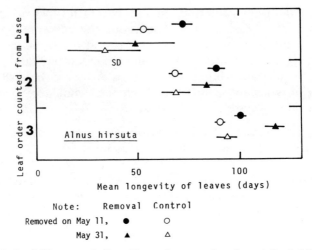

Figure 4 Leaf life-span of *Alnus hirsuta* for control and partially defoliated plants. (Redrawn from data of Kikuzawa, 1988.)

zawa, 1988; Nowak and Caldwell, 1984). Following leaf loss, compensation may also occur in the form of secondary shoot and leaf growth (Furuno and Shidei, 1960; Potter and Redmond, 1989) and this foliage may have a shorter life-span than the original foliage would have had without herbivory. The relationships of leaf life-span to defoliation and plant defenses are covered in Chapter 6, this volume.

III. Relationship of Leaf Life-Span to Leaf, Plant, and Ecosystem Traits

We have shown that leaf life-span varies widely in a relatively systematic fashion in relation to many environmental conditions known to affect carbon gain potential. Do leaf physiological traits vary in relation to leaf life-span in a pattern consistent with the idea that leaf life-span and carbon gain are linked? We address this question below, and also ask whether and how conifers might differ from broad-leafed trees in this respect.

A. Variation among Species in Leaf and Plant Traits vis-à-vis Leaf Life-Span

For plants surveyed at many sites around the world, mass-based foliar N concentration (leaf N_{mass}) in conifers decreases from about 30 mg/g to less than 10 mg/g with increasing leaf life-span (Reich *et al.*, 1992) (Fig. 5). The regression relationship between leaf N_{mass} and life-span was significant and did not differ between broad-leafed and needle-leafed

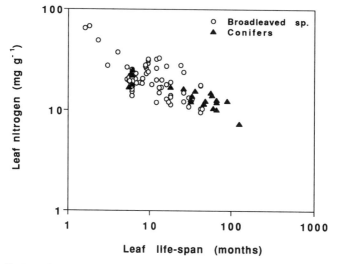

Figure 5 Mass-based leaf nitrogen concentration in relation to leaf life-span (months). Data are for species from diverse ecosystems and biomes. Regression relationships for conifers and broad-leafed species were not statistically different (all data pooled, $P < 0.001$, $r^2 = 0.52$). (Redrawn from Reich et al., 1992.)

groups, indicating that variation in leaf traits in relation to leaf life-span is similar for species with both leaf types. Area-based leaf N (projected area, leaf N_{area}) was not significantly correlated with leaf life-span in conifers, broad-leafed species, or all groups combined. The broad survey of Reich et al. (1992) included data for species in different and largely native environments. Thus, the differences observed among species may be due to site differences, genotypic differences, or both. In a study by Gower et al. (1993), five tree species of differing leaf life-spans were compared in a common garden experiment, with results consistent (i.e., N_{mass}, but not N_{area}, decreases with leaf life-span) (Fig. 5) with observations across sites as reported by Reich et al. (1992), clearly demonstrating a strong genetic component for ecological differences in leaf traits.

Leaf structure and function are closely related to leaf nutrient status (Field and Mooney, 1986; Reich et al., 1992; Gower et al., 1993). With increasing nitrogen content in a leaf, leaf cells show increasing elongation, which may reduce leaf anatomical (Koike and Sanada, 1989; Koike, 1988) and mechanical (Reich et al., 1991a) strength, and thereby limit durability. With increasing leaf nitrogen concentration, photosynthetic rates increase across species (Field and Mooney, 1986; Reich et al., 1991a, 1992; Gower et al., 1993) and usually increase within a species as well, but less consistently in conifers than in other species (Brix, 1971; Keller, 1972; Sheriff et al., 1986; Field and Mooney, 1986; Reich and

Schoettle, 1988; Schoettle, 1990b; Reich et al., 1991b). Variation in mass-based net photosynthetic rate (A_{mass}) among conifer species was a linear function of N_{mass} (Reich et al., 1992; Gower et al., 1993), largely as a result of high N and photosynthetic rates in larch. When only conifers with long leaf life-spans are compared, there is little overall relationship between A and N, despite the data still fitting within the framework of a global mass-based A-to-N relationship (Field and Mooney, 1986; Reich et al., 1992). Specific leaf area (SLA; cm^2 projected leaf area/gram of leaf) decreases significantly in conifers from ≈140 to 40 cm^2/g as leaf life-span varies from 0.5 months to 8 years (Reich et al., 1992; Gower et al., 1993). This relationship also did not differ significantly between broad-leafed and needle-leafed species.

Maximum mass-based net photosynthetic rate (A_{mass}) of conifers decreases by roughly an order of magnitude (≈100 nmol/g/sec to ≈ 10 nmol/g/sec) as leaf life-span increases from 0.5 year to many years, as shown in both a broad species–site comparison (Reich et al., 1992) and in a common garden (plantation) study (Gower et al., 1993) (Fig. 6). These results are consistent with reports of higher net photosynthesis in larch than in cooccurring evergreen conifers (Matyssek, 1986; Kloeppel et al., 1993). High A_{mass} in species with short-lived foliage results from their high N_{mass} and SLA. How does the relationship of photosynthesis to leaf life-span in conifers compare to the relation for broadleafed spe-

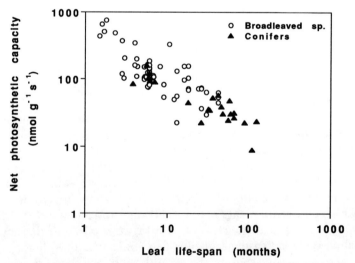

Figure 6 Mass-based net photosynthetic rates in relation to leaf life-span (months). Data are for species from diverse ecosystems and biomes. Regression relationships for conifers and broad-leafed species were not statistically different (all data pooled, $P < 0.001$, $r^2 = 0.70$). (Redrawn from Reich et al., 1992.)

cies, which range from 0.1 year to many years in leaf life-span? When analyzed separately, the regressions between A_{mass} and leaf life-span were significant and similar in these two groups (Fig. 6), suggesting that the relationship of leaf life-span and photosynthetic capacity does not differ among species of differing leaf forms. The lack of difference in these relationships highlights why one should use leaf life-span and discourage continuing use of evergreen–deciduous contrasts—the relationships clearly fall along a continuum within conifers or any other groups, and there is substantial overlap in leaf life-span between conifers and broadleafed tree species.

In contrast, area-based net photosynthesis (A_{area}) was not significantly related to leaf life-span among coniferous species using either data pooled from many ecosystems (Reich et al., 1992) or for species from a common environment (Gower et al., 1993). The reason why there is no relationship between A_{area} and leaf life-span among conifers is that increasing needle "thickness" ($\approx$decreasing SLA; see below) offsets decreasing photosynthesis per unit dry mass as leaf life-span increases, such that there is much less variation in A_{area} than in A_{mass} across coniferous species of differing leaf life-span. Despite the fact that the relationship of A_{area} to leaf life-span does not hold within coniferous species, the values do not "deviate" substantially from a significant overall relationship among all species (Reich et al., 1992), which is, however, much weaker than the mass-based relationship, again because of the offsetting influence of thick leaves and low A_{mass} on A_{area}. It seems that the area-based photosynthesis–leaf life-span relationship may be one that holds true more often among angiosperms (e.g., Reich et al., 1991a; Reich, 1993) than among conifers (Reich et al., 1992; Gower et al., 1993).

Variation among species in maximum relative growth rate of seedlings is well correlated with leaf life-span (Reich et al., 1992). In general, species with long leaf life-span tend to have thick leaves (low SLA) with low nitrogen concentrations and low photosynthetic rates, and low leaf area per whole plant mass (leaf area ratio). In combination, these traits limit maximum relative growth rate (Poorter et al., 1990; Reich et al., 1992; Walters et al., 1993a,b). Many conifers have long leaf life-span and possess this combination of traits—however, so do many other species (Grime and Hunt, 1975; Shipley and Peters, 1990)—and it is the set of traits, rather than any other attribute of a species, that appears to determine growth capacity. Similarly, height growth of young open-grown trees is greater in species with shorter than longer leaf life-spans (Coley, 1988; Reich et al., 1992) (Fig. 7). Again, this pattern appears to hold true regardless of whether a species with long-lived leaves is a conifer or a broadleaf angiosperm.

Within a species, growth rate and leaf life-span are also negatively

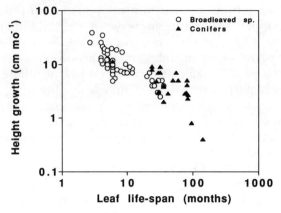

Figure 7 Height growth rates of tree saplings in relation to leaf life-span (months). Data are for species from diverse ecosystems and biomes. Regression relationships for conifers and broad-leafed species were not statistically different (all data pooled, $P < 0.001$, $r^2 = 0.72$). (Redrawn from Reich et al., 1992.)

correlated. For lodgepole pine in the central Rocky Mountains, leaf life-span was 38% greater and annual shoot growth increment was 33% less on shoots at 3200 m elevation, relative to those at 2800 m elevation (Table VIII). The negative correlation between leaf life-span and shoot growth observed among elevations is consistent with the relationship observed within individual crowns of lodgepole pine (Table VI). The wide range in leaf life-spans within conifer species (e.g., 5–18 years in *P. contorta*; see above also) enables the species to maintain a constant foliar biomass per shoot in different environments (Table VIII). If a species had a fixed leaf life-span, an individual with low shoot growth would

Table VIII Effect of Elevation on Shoot Characteristics of Lodgepole Pine Grown at Six Sites in the Central Rocky Mountains[a]

	Site elevation	
Trait	2800 m	3200 m
Leaf longevity (years)**	9.5 (0.6)	13.1 (0.7)
Foliar biomass per shoot (grams)	4.6 (0.6)	5.9 (0.8)
Foliated length (cm)	19.9 (2.1)	20.7 (2.5)
Annual shoot increment (cm)*	2.1 (0.3)	1.4 (0.2)

[a] Approximately 10 shoots were characterized from the lowest crown third from 4 trees per site, for a total of 12 trees per elevation. The means are given, plus or minus standard errors in parentheses (Schoettle, 1990a).
 *$P < 0.05$.
 **$P < 0.001$.

support little foliage per shoot. Conversely, an individual with a greater annual shoot growth capacity would retain many leaves, some of which may be in a microenvironment that could not support a positive carbon balance due to inadequate sunlight (Schoettle, 1990b). Therefore, the plasticity of leaf life-span in pines may enable a consistency in crown architecture among environments (Schoettle and Fahey, 1994).

B. Changes over Leaf Life-Span

The maximum photosynthetic rate and the nitrogen concentration of a leaf decrease with age (e.g., Hom and Oechel, 1983; Teskey *et al.*, 1984; Reich *et al.*, 1991b). Leaf nitrogen is retranslocated from older to newer leaves (Field, 1983; Carlyle and Malcolm, 1986; Tyrrell and Boerner, 1987; Son and Gower, 1992). As a result, the photosynthetic rate per unit nitrogen does not vary with leaf age as much as does photosynthesis per unit leaf mass (Hom and Oechel, 1983; Schoettle, 1990b). Although the general pattern of decline in photosynthesis with leaf age occurs in all species, the pattern differs among species of differing leaf life-span. After reaching the maximum photosynthetic rate, early successional species with shorter leaf life-spans decrease their photosynthetic rate more rapidly in relation to absolute leaf age than do late successional species with longer leaf life-spans (Fig. 8). However, the rate of

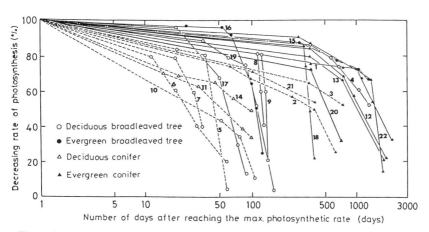

Figure 8 Decrease in photosynthetic rate after reaching its maximum value for several broad species groups. Species: (1) *Picea glauca*, (2) *Pinus thunbergii*, (3) *Pinus ponderosa*, (4) *Pinus pumila*, (5) *Alnus hirsuta*, (6) *Populus maximowiczii*, (7) *Betula maximowicziana*, (8) *Quercus mongolica* var. *grosserata*, (9) *Fagus crenata*, (10) *Rubus*, (11) *Larix kampferi*, (12) *Abies mariesii*, (13) *Abies veitchii*, (14) *Larix decidua*, (15) *Cleyera japonica*, (16) *Aucuba japonica* var. *borealis*, (17) *Acer saccharum*, (18) *Abies sacharinensis*, (19) *Fagus sylvatica*, (20) *Cryptomeria japonica*, (21) *Pinus sylvestris*, and (22) *Picea abies*. (Redrawn from Koike *et al.*, 1991.)

change in photosynthesis with leaf age expressed as a proportion of the leaf life-span does not differ appreciably among species of differing leaf life-span, to the extent that data are available to make such assessments.

C. Relationship to Stand and Ecosystem Traits

At the mature tree and stand level, long-lived foliage may confer advantages that do not occur at the seedling or sapling stage. As shown above, whole-plant relative growth rate and height growth rate of evergreen conifer seedlings and saplings, respectively, are low compared to other species—consistent with their long leaf life-span and low A_{mass} and LAR (Assmann, 1970; Grime and Hunt, 1975; Gowin et al., 1980; Reich et al., 1992). However, once saplings have grown into trees, those with long leaf life-spans support a greater total foliage mass per tree, for either mean (and thus different) or standardized (e.g., at a given diameter) stem diameter (Gower et al., 1987, 1993) (Fig. 9). This trait may enable trees with long leaf life-spans to maintain growth or annual aboveground net primary production (ANPP) similar to that of trees with short leaf life-spans (see below; Gower et al., 1989, 1993; Gower and Richards, 1990; Reich et al., 1992), yet due to differences in allocation and allometry, trees with shorter leaf life-spans are still taller than those with long-lived foliage (Gower and Richards, 1990).

Net primary production is determined by the photosynthetic rate, photosynthetic surface and its duration, and length of growing season (Ondak, 1970). Theoretically, a tree with a large photosynthetic surface

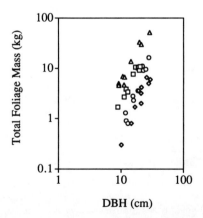

Figure 9 Total foliage mass per tree in relation to stem diameter at breast height (DBH) for four coniferous tree species with different needle life-spans growing in a common garden experiment in southwestern Wisconsin. Species and respective needle life-spans were (◊) *Larix decidua* (5 months), (○) *Pinus strobus* (36 months), (□) *Pinus resinosa* (46 months), and (△) *Picea abies* (66 months). (From Gower et al., 1993.)

area (i.e., high leaf area or mass) and high photosynthetic rate and a long life-span should exhibit the highest production rates. However, we know of no trees that exhibit this suite of characteristics. Physiologically, it may be impossible for a tree to have high photosynthetic rate, a large canopy, and extended leaf duration because of mutual shading (Sprugel, 1989; Schoettle and Smith, 1991) and unavoidable leaf-level trade-offs between productivity and defense/stress tolerance characteristics (Coley, 1988; Reich *et al.*, 1991a, 1992), because leaf morphological and biochemical characteristics that are positively related to maximum net photosynthetic rates are negatively related to antiherbivore defense, desiccation, and physical integrity.

In closed-canopy forest stands, total foliage mass (Mg/ha) increases sharply with increasing leaf life-span among species, and this is true whether species are examined across a range of sites (Reich *et al.*, 1992) or in a common garden (Gower *et al.*, 1993) (Fig. 10). This trend was not seen however, within lodgepole stands across an elevational gradient (Schoettle, 1990b). Total leaf area (leaf area index, LAI) also increases with leaf life-span among species, but not as steeply as foliage mass, as a result of offsetting trends in total foliage mass (increasing) and SLA (decreasing) in relation to leaf life-span. Annual aboveground net primary production of forest stands was not significantly correlated with leaf life-span (Reich *et al.*, 1992; Gower *et al.*, 1993) when all data were examined together. For conifers with leaf life-spans greater than 1 year, however, ANPP decreased significantly ($p < 0.001$, $r^2 = 0.48$) (Reich *et al.*, 1992) with increasing leaf life-span, when data from diverse ecosystems

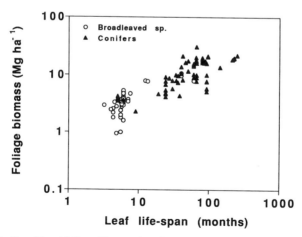

Figure 10 Stand level foliage biomass in relation to leaf life-span (months) for diverse forest stands (all data pooled, $P < 0.001$, $r^2 = 0.74$). (Redrawn from data of Reich *et al.*, 1992, and Gower *et al.*, 1993.)

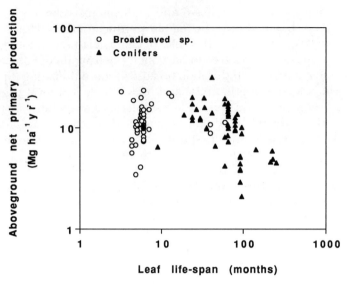

Figure 11 Stand level aboveground net primary production (ANPP) in relation to leaf life-span (months) for diverse forest stands. (Redrawn from data of Reich et al., 1992, and Gower et al., 1993.) No significant relationship ($P > 0.05$) observed for all data pooled. Using data for conifer species with leaf life-span greater than 1 year, ANPP was significantly related to needle life-span ($r^2 = 0.48$, $P < 0.001$).

were compared (Fig. 11). This may result from species with longest lived foliage inhabiting sites of low potential productivity (as opposed to having lower potential productive capacities), because no decrease in ANPP was observed with increasing needle life-span among evergreen conifers grown in a common garden experiment (Gower et al., 1993).

Stand level production efficiency of coniferous forest canopies (expressed on a mass basis, ANPP/foliage biomass) decreases with increasing leaf life-span ($p < 0.001$, $r^2 = 0.78$) (Fig. 12), whether evaluated across different (Reich et al., 1992) or shared environments (Gower et al., 1993). When these data are compared to data for broad-leafed species, one general pattern is found for all species. Canopy production efficiency expressed on a leaf area basis also decreases with increasing leaf life-span (Fig. 12). There is thus an offsetting inverse relationship between leaf production efficiency and stand foliage biomass (Fig. 13) or area as noted previously (Tadaki, 1966, 1986; Saito, 1981; Oohata, 1986; Gower et al., 1989, 1993; Reich et al., 1992). Conifers and broad-leafed tree species follow the same relationships between foliage mass, production efficiency, and leaf life-span (Figs. 10–13), suggesting a fundamental interdependence among these traits.

Gower et al. (1993) and Reich et al. (1992) suggest that the large ac-

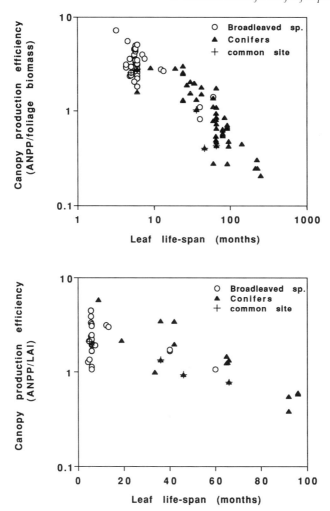

Figure 12 Stand level canopy production efficiency on a mass (kg ANPP/kg foliage biomass) and area basis (kg ANPP/LAI) in relation to leaf life-span (months) for diverse forest stands (all data pooled, $P < 0.001$, $r^2 = 0.78$ and 0.63, respectively). Species examined on a common site (in Wisconsin) are identified by a cross. (Redrawn from data of Reich *et al.*, 1992, and Gower *et al.*, 1993.)

cumulation of foliage mass by species with long-lived foliage compensates for their intrinsically low A_{mass}, N_{mass}, and SLA, resulting in the general lack of relationship between ANPP and leaf life-span. This accumulation of foliage mass, which takes many years to develop, enables conifers (and other species) with long-lived leaves to have a productivity at the mature tree or stand stage equal to that of species with shorter

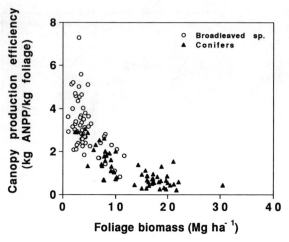

Figure 13 Canopy production efficiency (ANPP/foliage biomass) in relation to foliage biomass for diverse forest stands (all data pooled, $P < 0.001$, $r^2 = 0.73$). (Drawn from data of Tadaki, 1986, Reich et al., 1992, and Gower et al., 1993.)

leaf life-spans, despite lower per leaf productivity and seedling and sapling growth rates.

D. Leaf Life-Span and Nutrient Use Efficiency

Because carbon assimilation and biomass accumulation of conifer forests are commonly limited by nutrient availability, efficient nutrient use should enhance the productivity of conifers. In general, nitrogen use efficiency (NUE) is defined as carbon fixed per unit of nitrogen invested, although the exact definition used by different researchers ranges from an instantaneous time scale for needles to an annual (or longer) time scale for all tissues. The different temporal and spatial scales of NUE reflect, in part, different processes that influence NUE. For example, high photosynthetic rates per unit foliage N, large withdrawal of N from senescing tissue, and greater life-span of tissues are all processes that increase whole-tree NUE.

Among species, net photosyntheis per unit foliage N (i.e., PPNUE) (Field and Mooney, 1986) is positively correlated with foliage N concentration, which is often inversely correlated with leaf life-span (Reich et al., 1991a, 1992; Gower et al., 1993). PPNUE is also inversely correlated to leaf life-span in natural environments; however, it often does not differ for control and fertilized conifers, despite the fact that fertilization often decreases leaf life-span (S. T. Gower, unpublished data). Fertilization has been shown to increase nitrogen in storage compounds such as amino acids, but does not appear to increase enzymes responsible for photosynthesis (Billow et al., 1994).

Withdrawal of nutrients from senescing or aging tissue is another physiological mechanism that increases overall efficient nutrient use. Comparison of the percentage withdrawal of nitrogen from senescing foliage has shown that the percentage of foliar retranslocation of N does not differ for evergreen and deciduous species (Small, 1972; Chapin and Kedrowski, 1983), nor is it correlated with leaf life-span (Reich et al., 1992). Percentage withdrawal of N from foliage or stemwood was not correlated with leaf life-span for four conifers grown on adjacent plots with similar soils (Son and Gower, 1991). The deciduous conifer, *Larix* spp., however, withdraws N from senescing foliage more efficiently than do broadleaf deciduous hardwoods or evergreen conifers (Tyrrell and Boerner, 1987; Gower and Richards, 1990; Son and Gower, 1991).

Vitousek (1982) reported that the ratio of litterfall mass:litterfall N (considered an index of N use efficiency) was greater for conifers than for deciduous broadleaf trees; however, the same data also revealed that nitrogen availability, approximated by annual litterfall N content, was greater for deciduous (largely short leaf life-span) than for evergreen (largely long leaf life-span) forests. Differences in N availability may thus confound the interpretation of litterfall-based N use efficiency patterns (Turner, 1977; Shaver and Melillo, 1984; but see Birk and Vitousek, 1986), making it difficult to determine if N use efficiency is related to leaf life-span. However, both instantaneous, leaf level (PPNUE) and annual, stand-level indices of N use efficiency have been shown to be inversely correlated with leaf life-span, when conifers with different leaf life-spans were planted on the same soil (Son and Gower, 1991; Gower et al., 1993). [For more detailed analysis of nutrient use efficiency, see Sheriff et al. (1994).]

IV. Summary

Species with mutually supporting traits, such as high N_{mass}, SLA, and A_{mass} and short leaf life-span, tend to inhabit either generally resource-rich environments or spatial and/or temporal microhabitats that are resource-rich in otherwise more limited habitats (e.g., "precipitation" ephemerals in warm deserts or spring ephemerals in the understory of temperate deciduous forests) (Reich et al., 1992). In contrast, species with long leaf life-span often support foliage with low SLA, N_{mass}, and A_{mass}, and often grow in low-temperature limited, dry, and/or nutrient-poor environments (e.g., Monk, 1966; Small, 1972; Chapin, 1980). The contrast between evergreen and deciduous species, and the implications that emerge from such comparisons, can be considered a paradigm of modern ecological theory (e.g., Monk, 1966; Schulze et al., 1977; Chapin, 1980; Chabot and Hicks, 1982; Sprugel, 1989). However, based on the results of Reich et al. (1992) and Gower et al. (1993), coniferous species with foliage that persists for 9–10 years are likely to assimilate and allocate carbon and nutrients differently than other evergreen conifers that retain foliage for 2–3 years. Thus, attempts to contrast ecophysiological or ecosystem characteristics of evergreen versus deciduous life forms may be misleading, and pronounced differences among evergreen conifers may be ignored. Clearly, the deciduous–evergreen contrast, although useful in several ways, should be viewed from the broader perspective of a gradient in leaf life-span.

References

Assmann, E. (1970). "The Principles of Forest Yield Study." Pergamon, New York.

Axelrod, D. I. (1966). Origin of deciduous and evergreen habits in temperate forests. *Evolution (Lawrence, Kans.)* 20:1–15.

Beets, P. N., and Madgwick, H. A. I. (1988). Aboveground dry matter and nutrient content of *Pinus radiata* as affected by lupin, fertilizer, thinning, and stand age. *N.Z. J. For. Sci.* 18:43–64.

Benecke, Y., Schulze, E.-D., Matyssek, R., and Havranek, W. M. (1981). Environmental control of CO_2 assimilation and leaf conductance in *Larix decidua* Mill. I. A comparison of contrasting natural environments. *Oecologia* 50:54–61.

Billow, C., Matson, P. A., and Yoder, B. (1994). Seasonal biochemical changes in coniferous forest canopies and their response to fertilization. *Tree Physiol.* (submitted for publication).

Birk, E. M., and Vitousek, P. M. (1986). Nitrogen availability and nitrogen use efficiency in loblolly pine stands. *Ecology* 67:69–79.

Bockheim, J. A., Leide, J. E., and Tavella, D. S. (1986). Distribution and cycling of macronutrients in a *Pinus resinosa* plantation fertilized with nitrogen and potassium. *Can. J. For. Res.* 16:778–785.

Bolstad, P., and Gower, S. T. (1990). Estimation of leaf area index in fourteen southern Wisconsin forest stands using a portable radiometer. *Tree Physiol.* 4:115–24.

Brix, H. (1971). Effects of nitrogen fertilization on photosynthesis and respiration in Douglas-fir. *For. Sci.* 17:407–414.

Carlyle, J. C., and Malcolm, D. C. (1986). Larch litter and nutrient availability in mixed larch–spruce stands. I. Nutrient withdrawal, redistribution, and leaching losses from larch foliage and senescence. *Can. J. For. Res.* 16:321–326.

Chabot, B. F., and Hicks, D. J. (1982). The ecology of leaf life spans. *Annu. Rev. Ecol. Syst.* 13:229–259.

Chapin, F. S., III (1980). The mineral nutrition of wild plants. *Annu. Rev. Ecol. Syst.* 11:233–260.

Chapin, F. S., III, and Kedrowski, R. A. (1983). Seasonal changes in nitrogen and phosphorus fractions and autumnal retranslocation in evergreen and deciduous taiga trees. *Ecology* 64:376–391.

Coley, P. D. (1988). Effects of plant growth rate and leaf lifetime on the amount and type of antiherbivore defense. *Oecologia* 74:531–536.

Coley, P. D., Bryant, J. P., and Chapin, F. S., III (1985). Resource availability and plant antiherbivore defense. *Science* 230:895–899.

Ewers, F. W., and Schmid, R. (1981). Longevity of needle fascicles of *Pinus longaeva* (Bristlecone Pine) and other North American pines. *Oecologia* 51:107–115.

Field, C. (1983). Allocating leaf nitrogen for the maximization of carbon gain: Leaf age as a control on the allocation program. *Oecologia* 56:341–347.
Field, C., and Mooney, H. A. (1986). The photosynthesis–nitrogen relationship in wild plants. *In* "On the Economy of Plant Form and Function" (T. J. Givnish, ed.), pp. 25–55. Cambridge Univ. Press, Cambridge, UK.
Furuno, T., and Shidei, T. (1960). Effect of leaf-cutting on elongation of needle-leaf of *Pinus densiflora* Sieb. et Zucc. and *P. thunbergii*. *J. Jpn. For. Soc.* 42:435–440.
Gholz, H. L., Vogel, S. A., Cropper, W. P., Jr., McKelvey, K., Ewel, K. C., Teskey, R. O., and Curan, P. J. (1991). Dynamics of canopy structure and light interception in *Pinus elliottii* stands, north Florida. *Ecol. Monogr.* 61:33–51.
Gower, S. T., and Richards, J. H. (1990). Larches: Deciduous conifers in an evergreen world. *BioScience* 40:818–826.
Gower, S. T., Grier, C. C., Vogt, D. J., and Vogt, K. A. (1987). Allometric relations of deciduous (*Larix occidentalis*) and evergreen (*Pinus contorta*) and (*Pseudotsuga menziesii*) trees of the Cascade Mountains in central Washington. *Can. J. For. Res.* 17:640–646.
Gower, S. T., Grier, C. C., and Vogt, K. A. (1989). Aboveground production and N and P use by *Larix occidentalis* and *Pinus contorta* in the Washington Cascades, USA. *Tree Physiol.* 5:1–11.
Gower, S. T., Vogt, K. A., and Grier, C. C. (1992). Carbon dynamics of Rocky Mountain Douglas-fir: Influence of water and nutrient availability. *Ecol. Monogr.* 62:43–65.
Gower, S. T., Reich, P. B., and Son, Y. (1993). Canopy dynamics and aboveground production of five tree species with different leaf longevities. *Tree Physiol.* 12:327–345.
Gower, S. T., Haynes, B. E., Fassnacht, K. S., Running, S. W., and Hunt, E. R., Jr. (1994). Influence of fertilization on the allometric relations for two pines in contrasting environments. *Can. J. For. Res.* (in press).
Gowin, T., Lourtioux, A., and Mousseau, M. (1980). Influence of constant growth temperature upon the productivity and gas exchange of seedlings of Scots pine and European larch. *For. Sci.* 26:301–309.
Grime, J. P., and Hunt, R. (1975). Relative growth rate: Its range and adaptive significance in a local flora. *J. Ecol.* 63:393–422.
Hadley, J. L., and Smith, W. K. (1986). Wind effects on needles of timberline conifers: Seasonal influence on mortality. *Ecology* 67:12–19.
Heichel, G. H., and Turner, N. C. (1983). CO_2 assimilation of primary and regrowth foliage of red maple (*Acer rubrum* L.) and red oak (*Quercus rubra* L.): Response to defoliation. *Oecologia* 57:14–19.
Hom, J. L., and Oechel, W. C. (1983). The photosynthetic capacity, nutrient content, and nutrient use efficiency of different needle age-classes of black spruce (*Picea mariana*) found in interior Alaska. *Can. J. For. Res.* 13:834–839.
Keller, T. (1972). Gaseous exchange of forest trees in relation to some edaphic factors. *Photosynthetica* 6:197–206.
Kikuzawa, K. (1988). Effect of defoliation of alder (*Alnus hirsuta* Turcz.) on leaf lifespan. *Proc. 35th Ecol. Soc. Jpn.*, p. 210 (in Japanese).
Kikuzawa, K. (1991). A cost–benefit analysis of leaf habit and leaf longevity of trees and their geographical pattern. *Am. Nat.* 138:1250–1263.
Kloeppel, B. D., Gower, S. T., and Reich, P. B. (1993). Net photosynthesis of western larch and sympatric evergreen conifers along a precipitation gradient in western Montana. *Proc. Symp. Ecol. Manage. Larix For.: A Look Ahead*, Whitefish, MT, *1992* (in press).
Kohyama, T. (1980). Growth pattern of *Abies mariesii* saplings under conditions of opengrowth and suppression. *Bot. Mag. (Tokyo)* 93:13–24.
Koike, T. (1988). Leaf structure and photosynthetic performance as related to the forest succession of deciduous broad-leaved trees. *Plant Species Biol.* 3:77–87.
Koike, T., and Sanada, M. (1989). Photosynthesis and leaf longevity in alder, birch, and ash seedlings grown under different nitrogen levels. *Ann. Sci. For.* 46S:476–478.

Koike, T., Hasler, R., and Matyssek, R. (1991). Leaf longevity and growth characteristics of trees. *Hoppo Ringyo (North. For.)* 43:358–363.

Koike, T., Hasler, R., and Item, H. (1993). Needle longevity and photosynthetic performance in Cembran pine and Norway spruce growing on the north- and east-facing slopes at the timberline of Stillberg in the Swiss Alps. *Trees* (in press).

Linder, S. (1987). Responses to water and nutrients in coniferous ecosystems. *Ecol. Stud.* 61:180–202.

Linder, S., and Axelsson, B. (1981). Changes in carbon uptake and allocation patterns as a result of irrigation and fertilization in a young *Pinus sylvestris* stand. *In* "Carbon Uptake and Allocation Key to Management of Subalpine Forest Ecosystems" (R. Waring, ed.), pp. 38–44. Int. Union For. Res. Organ. Workshop, For. Res. Lab., Oregon State University, Corvallis.

Linder, S., and Rook, D. A. (1984). Effects of mineral nutrition on carbon dioxide exchange and partitioning of carbon in trees. *In* "Nutrition of Plantation Forests" (G. D. Bowen and E. K. S. Nambiar, eds.), pp. 211–236. Academic Press, Orlando, FL.

Maillette, L. (1982). Needle demography and growth pattern of Corsican pine. *Can. J. Bot.* 60:15–116.

Matyssek, R. (1986). Carbon, water, and nitrogen relations in evergreen and deciduous conifers. *Tree Physiol.* 2:177–187.

McMurtrie, R. E., Linder, S., Benson, M. L., and Wolf, L. (1986). A model of leaf area development for pine stands. *In* "Crown and Canopy Structure in Relation to Productivity" (T. Fujimori and D. Whitehead, eds.), pp. 284–307. Forestry and Forest Products Research Institute, Ibaraki, Japan.

Miller, H. G., and Miller, J. D. (1976). Effect of nitrogen supply on net primary production in Corsican pine. *J. Appl. Ecol.* 13:249–256.

Monk, C. D. (1966). An ecological significance of evergreeness. *Ecology* 47:504–505.

Mooney, H. A., and Gulmon, S. L. (1982). Constraints on leaf structure and function in reference to herbivory. *BioScience* 32:198–206.

Nowak, R. S., and Caldwell, M. M. (1984). A test of compensation photosynthesis in the field: Implications for herbivory tolerance. *Oecologia* 61:311–318.

Ondak, S. (1970). Growth analysis applied to the estimation of gross assimilation and respiration rate. *Photosynthetica* 4:214–222.

Oohata, S. (1986). Net production and matter distribution to photosynthetic organ in forest trees. *Bull. Kyoto Univ. For.* 57:46–59.

Poorter, H., Remkes, C., and Lambers, H. (1990). Carbon and nitrogen economy of 24 wild species differing in relative growth rate. *Plant Physiol.* 94:621–627.

Potter, D. A., and Redmond, C. T. (1989). Early spring defoliation, secondary leaf flush, and leafminer outbreaks on American holly. *Oecologia* 81:192–197.

Pravdin, L. F. (1969). "Scots Pine Variation, Intraspecific Taxonomy and Selection." Izd. Nauka, Moskva, 1964 (translated from Russian by Israel Program for Scientific Translations, Jerusalem).

Reich, P. B. (1983). Effects of low concentrations of O_3 on net photosynthesis, dark respiration, and chlorophyll contents in aging hybrid poplar leaves. *Plant Physiol.* 73:291–296.

Reich, P. B. (1987). Quantifying plant response to ozone: A unifying theory. *Tree Physiol.* 3:63–91.

Reich, P. B. (1993). Reconciling apparent discrepancies among studies relating life-span, structure and function of leaves in contrasting life forms and climates: "The blind men and the elephant retold." *Funct. Ecol.* 7:721–725.

Reich, P. B. (1994). Phenology of tropical forests: Patterns, causes and consequences. *Can. J. Bot.* (in press).

Reich, P. B., and Borchert, R. (1982). Phenology and ecophysiology of the tropical tree, *Tabebuia neochrysantha* (Bignoniaceae). *Ecology* 63:294–299.

Reich, P. B., and Lassoie, J. P. (1985). Influence of low concentrations of ozone on growth, biomass partitioning, and leaf senescence in young hybrid poplar plans. *Environ. Pollut., Ser. A* 39:39–51.

Reich, P. B., and Schoettle, A. W. (1988). Role of phosphorus and nitrogen in photosynthetic and whole plant carbon gain and nutrient-use efficiency in eastern white pine. *Oecologia* 77:25–33.

Reich, P. B., Uhl, C., Walters, M. B., and Ellsworth, D. S. (1991a). Leaf lifespan as a determinant of leaf structure and function among 23 tree species in Amazonian forest communities. *Oceologia* 86:16–24.

Reich, P. B., Walters, M. B., and Ellsworth, D. S. (1991b). Leaf age and season influence the relationships between leaf nitrogen, leaf mass per area and photosynthesis in maple and oak trees. *Plant, Cell Environ.* 14:251–259.

Reich, P. B., Walters, M. B., and Ellsworth, D. S. (1992). Leaf lifespan in relation to leaf, plant, and stand characteristics among diverse ecosystems. *Ecol. Monogr.* 62:365–392.

Reich, P. B., Walters, M. B., Krause, S. C., Vanderklein, D. W., Raffa, K. F., and Tabone, T. (1993). Growth, nutrition and gas exchange of *Pinus resinosa* following artificial defoliation. *Trees* 7:67–77.

Richards, J. H., and Bliss, L. C. (1986). Winter water relations of a deciduous timberline conifer, *Larix lyallii* Parl. *Oecologia* 69:16–24.

Saito, H. (1981). Materials for the studies of litterfall in forest stands. *Bull. Kyoto Pref. Univ. For.* 25:78–89.

Schoettle, A. W. (1990a). The interaction between leaf longevity, shoot growth, and foliar biomass per shoot in *Pinus contorta* at two elevations. *Tree Physiol.* 7:209–214.

Schoettle, A. W. (1990b). Importance of shoot structure to sunlight interception and photosynthetic carbon gain in *Pinus contorta* crowns. Ph.D. Dissertation, University of Wyoming, Laramie.

Schoettle, A. W., and Fahey, T. J. (1994). Foliage and fine root longevity in pines. *Ecol. Bull.* (in press).

Schoettle, A. W., and Smith, W. K. (1991). Interrelation between shoot characteristics and solar irradiance in the crown of *Pinus contorta* spp. *latifolia*. *Tree Physiol.* 9:245–254.

Schulze, E.-D., Fuchs, M., and Fuchs, M. I. (1977). Spatial distribution of photosynthetic capacity and performance in a mountain spruce forest of northern Germany. III. The ecological significance of the evergreen habit. *Oecologia* 30:239–248.

Schulze, E.-D., Lange, O., and Oren, R. (1989). Forest decline and air pollution. *Ecol. Stud.* 77:475.

Shaver, G. R. (1983). Mineral nutrition and leaf longevity in *Ledum palustre*: The role of individual nutrients and timing of leaf mortality. *Oecologia* 56:160–165.

Shaver, G. R., and Melillo, J. M. (1984). Nutrient budgets of marsh plants: Efficiency concepts and relation to availability. *Ecology* 65:1491–1510.

Sheriff, D. W., Nambiar, E. K. S., and Fife, D. N. (1986). Relationships between nutrient status, carbon assimilation, and water use efficiency in *Pinus radiata* (D. Dan) needles. *Tree Physiol.* 2:73–88.

Sheriff, D. W., Margolis, H. A., Kaufmann, M. R., and Reich, P. B. (1994). Resource use efficiency. *In* "Resource Physiology of Conifers: Acquisition, Allocation, and Utilization" (W. K. Smith and T. M. Hinckley, eds.), pp. 143–178. Academic Press, San Diego.

Shipley, B., and Peters, R. H. (1990). A test of the Tilman model of plant strategies: Relative growth rate and biomass partitioning. *Am. Nat.* 136:139–153.

Small, E. (1972). Photosynthetic rates in relation to nitrogen recycling as an adaptation to nutrient deficiency in peat bog plants. *Can. J. Bot.* 50:2227–2233.

Son, Y., and Gower, S. T. (1991). Aboveground nitrogen and phosphorus use by five plantation-grown trees with different leaf longevities. *Biogeochemistry* 14:167–191.

Son, Y., and Gower, S. T. (1992). Nitrogen and phosphorus distribution for five plantation species in southwestern Wisconsin. *For. Ecol. Manage.* 53:175–193.

Sprugel, D. G. (1989). The relationship of evergreeness, crown architecture, and leaf size. *Am. Nat.* 133:465–479.

Tadaki, Y. (1966). Some discussions on the leaf biomass of forest stands and trees. *Bull. Gov. For. Exp. Sta. (Jpn.)* 184:135–161.

Tadaki, Y. (1986). Productivity of forests in Japan. *In* "Crown and Canopy Structure in Relation to Productivity" (T. Fujimori and D. Whitehead, eds.), pp. 7–25. Forestry and Forest Products Research Institute, Ibaraki, Japan.

Teskey, R. O., Grier, C. C., and Hinckley, T. M. (1984). Change in photosynthesis and water relations with age and season in *Abies amabilis*. *Can. J. For. Res.* 14:77–84.

Turner, J. (1977). Effect of nitrogen availability on nitrogen cycling in a Douglas-fir stand. *For. Sci.* 23:307–316.

Tyrrell, L. E., and Boerner, R. E. (1987). *Larix laricina* and *Picea mariana*: Relationships among leaf lifespan, foliar nutrient patterns, nutrient conservation, and growth efficiency. *Can. J. Bot.* 65:1570–1577.

Vitousek, P. M. (1982). Nutrient cycling and nutrient use efficiency. *Am. Nat.* 119:553–572.

Walters, M. B., Kruger, E. L., and Reich, P. B. (1993a). Growth, biomass distribution and CO_2 exchange of northern hardwood seedlings in high and low light: Relationships with successional status and shade tolerance. *Oecologia* 94:7–16.

Walters, M. B., Kruger, E. L., and Reich, P. B. (1993b). Relative growth rate in relation to physiological and morphological traits for northern hardwood seedlings: Species, light, and ontogenetic considerations. *Oecologia* 96:219–231.

Waring, R. H., and Franklin, J. F. (1979). Evergreen coniferous forests of the Pacific Northwest. *Science* 204:1380–1386.

Weidman, P. H. (1939). Evidences of racial influence in a 25-year test of ponderosa pine. *J. Agric. Res.* 59:855–887.

Williams, K., Field, C. B., and Mooney, H. A. (1989). Relationships among leaf construction cost, leaf longevity, and light environment in ran forest plants of the genus *Piper*. *Am. Nat.* 133:198–211.

Zedenbauer, E. (1916). Beitrage zur Biologie der Waldbaume. *Zentralbl. Gesamte Forstwes.* 42:233–247.

9

Response Mechanisms of Conifers to Air Pollutants

Rainer Matyssek, Peter Reich, Ram Oren, and William E. Winner

I. Introduction

A. Assessing Risk to Conifers from Air Pollution: Need and Approach

Gases such as ozone (O_3), sulfur dioxide (SO_2), and nitrogen oxides (NO_x) have always been part of the environment, occurring only locally or in trace concentrations (e.g., O_3 formation under high solar radiation, SO_2 set free by volcanic activity, and NO_x by N cycling in ecosystems). Over the past 50–100 years, concentrations of these and other constituents of the atmosphere have increased. Scientists now recognize that this increase represents anthropogenic pollution, which can have both local and widespread distributions with concentrations above natural levels, and has the capacity to affect vegetation directly and indirectly. There is evidence that some species with short generation times may be evolving in response to these relatively new selective forces, as air pollution alters plant productivity and, ultimately, fitness (Taylor et al., 1991). However, the postindustrial appearance of air pollutants has occurred within such a small part of the lives of long-lived perennial plants that some species may not be able to adapt to this new, anthropogenic stress. Nevertheless, resource acquisition, allocation, and organ differentiation proceed in the presence of, and are affected by, air pollutants. A number of texts have reviewed responses of vegetation to SO_2 (Winner et al., 1985), O_3 (Guderian, 1985), and acid deposition (Schulze et al., 1989), as well as the effects of air pollutants on forests (Smith, 1990).

The two main goals of this chapter are to provide information highlighting the potential impacts, or risks, that air pollutants pose for coniferous trees, and to characterize the way in which coniferous trees, as integrated systems, respond to these stress agents. In order to manage forest ecosystems efficiently and to predict the long-term changes in ecosystem processes, scientists and resource managers need to understand the mechanisms of pollutant impact and plant response to current pollutant regimes. Yet, as this review will show, the understanding of conifer responses to air pollution is insufficient to manage forest ecosystems with confidence. There is, however, sufficient knowledge to begin the process of assessing the risk to coniferous forests from air pollution.

Ecological risk assessment provides such a framework for arranging information about the biology of conifers and the atmospheric chemistry of air pollution so that forests at highest risk from air pollution are identified (Suter, 1990). One approach to ecological risk assessment is to use a geographical information system (GIS) for developing sequential overlay maps of biological resources and environmental stresses. Keys to effective analysis include the availability of data on the distributions of environmental factors and biological resources, clear understanding of biological responses to stress, and effective mathematical links between scales of space and time (Fig. 1).

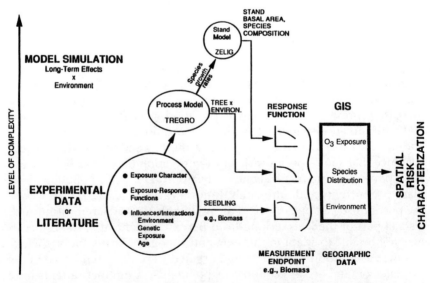

Figure 1 Approach used to develop an ecological risk assessment, such as determining the risk posed by O_3 to conifers. GIS, Geographical information system. (From Hogsett et al., 1993.)

This chapter will focus on responses of conifers to O_3, because this pollutant is known to occur at biologically significant levels across the many parts of North America and Europe where conifers are economically and ecologically important. GIS techniques for ecological risk assessment have been used to show that ambient O_3 in the eastern United States could reduce growth of some tree species, including two coniferous species (Fig. 2), from 5 to 35% (Hogsett et al., 1993). Predicted O_3-caused biomass losses for *Pinus taeda* in the southeastern United States range from 10 to 15%, whereas predicted biomass losses for *Pinus strobus* in the northeastern and Atlantic states ranges from 3 to 6%. Four kinds of information were mapped or integrated to make this assessment:

1. Estimation of the concentrations and aerial extent of O_3 across the United States.
2. Estimation of the distribution of conifers in the United States.
3. Estimation of the biological and physical factors that affect responses of conifers to O_3.
4. Estimation of O_3 response functions for conifers.

Ecological risk assessment to estimate impacts of air pollution reveals the need for basic biological information about O_3 responses of trees in order to determine the magnitude of lost forest productivity, the geographical regions where such losses are highest, and differences in risk for ecologically and economically important tree species. The ecological risk assessment framework is also important because it reveals the lack of important information on the distribution of air pollution, including O_3, and shows the difficulty of dealing with issues of multiple stresses at the stand, forest, and regional level. Currently, as part of the European Community "Critical Levels" program, an international group of scientists is developing a GIS-based system for all of Europe that incorporates O_3 levels and vegetation communities and their responses to ozone, in order to evaluate the risks of ozone damage to vegetation in a large regional context.

B. Objectives and Focus

Recognizing that O_3-caused reductions in the growth of conifers may range between 3 and 15% presents compelling reasons for the most general goal of this chapter, which is to review information on conifers and air pollution. The chapter will highlight analysis of tree responses to O_3 and interactive stresses, for the following specific reasons:

1. To provide a framework for understanding uptake of air pollutants by conifers that is based on principles of gas exchange, explains processes for single needles and seedlings, and can be applied to risk

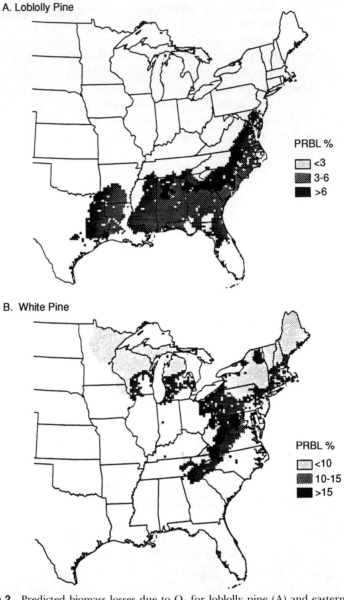

Figure 2 Predicted biomass losses due to O_3 for loblolly pine (A) and eastern white pine (B) in eastern North America. PRBL, Predicted biomass losses. (From Hogsett et al., 1993.)

assessments that are founded on calculations of air pollution uptake by forests.
2. To characterize the effects of air pollution on the development and senescence of trees.
3. To summarize attempts to scale from leaf-level responses of conifers to air pollution to responses of whole trees.
4. To show the need for and difficulty of placing air pollution stress in a multiple-stress framework.
5. To summarize attempts to scale from tree-level responses of conifers to air pollution to responses of conifer stands and regions.
6. To show the potential for modeling to contribute to the assessment of conifer responses to air pollution.

Emphasis in this chapter is placed on developing a mechanistic understanding of plant–environment relationships. To do so often requires controlled experiments that analyze air pollution uptake processes by plants, physiological and growth responses of seedlings and small trees to air pollutants, and plant responses to air pollution combined with co-occurring stresses. Controlled experiments, whether in the laboratory or the field, may involve environmental conditions that differ from field conditions. For example, pollutant doses and the numerous other environmental interactions that occur naturally may be absent from controlled environment studies. Work discussed in this chapter emphasizes studies in which fumigation regimes were at least roughly comparable to those on natural sites (see Reich, 1987).

This chapter will also focus on issues related to scale. Scale is important because the effects of air pollution are better known at the short-term, tissue-cellular level than at the long-term, whole-plant/forest stand levels of biological organization. Unfortunately, the long-term, plant/forest-level analysis is most relevant to conifer risk assessment. In this analysis plants are viewed as a system consisting of several levels of biological organization, each of them integrating functional and structural features (Fig. 3). Processes occur within and between each level. As a consequence, pollutants (or any environmental factor), may act on the plant at any level, and thus indirectly affect all other parts of the system. A mechanistic understanding of plant responses to air pollution stress can be achieved by gathering observations from each of the levels and by attempting to analyze their interactions. This "holistic" understanding is difficult to achieve because such thorough and comprehensive analysis does not yet exist for a single species. However, such an approach offers a promising avenue for improving our knowledge of conifer responses to pollutant stress.

In this chapter we focus on O_3 because this pollutant occurs across

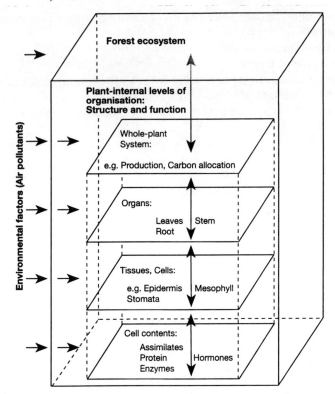

Figure 3 Hierarchical levels of biological organization. The interactions between the levels represent the causal basis of the processes occurring in each level. Environmental factors may act on each of these levels, whereas impact on any level may indirectly influence processes on all other levels.

much of North America and Europe where conifers are economically and ecologically important. In addition, there is an extensive literature (e.g., Reich, 1987; Pye, 1988) documenting the sensitivity of coniferous trees, and other plants, to O_3. As compared with SO_2, NO_x, or acid precipitation, O_3 is also of interest because of its rather singular action on plants: O_3 primarily acts at the leaf level and thus indirectly influences other plant organs, whereas other pollutants can exert impact on foliage or on the soil–root system. Because O_3 acts directly on leaves, and indirectly on the whole plant, studies with this pollutant may facilitate the development of a conceptual understanding of whole-tree physiology under pollutant stress. O_3 has also been hypothesized as a contributing agent in several forest decline problems of the Northern Hemisphere (e.g., Miller, 1973; Peterson *et al.*, 1987; Schulze *et al.*, 1989). Finally, O_3 is important because it will remain a significant air pollutant as long as the emissions of nitrogen oxides (which are part of the tropospheric formation processes of O_3) remain high.

C. Unique Features of Past Research with Conifers

Piecing together a mechanistic, holistic analysis of conifer responses to O_3 is difficult for several reasons:

1. Experiments in the 1960s and 1970s often included "unrealistically" high concentrations compared with current ambient pollution levels (Reich, 1987).
2. Single O_3 effects on plants were often described from a statistical perspective, rather than emphasizing a biological perspective, and often without considering plant ontogeny (see Mooney and Winner, 1988).
3. Research results may be confounded by fumigation artifacts. For example, generating O_3 for fumigation experiments from air rather than from O_2 may also result in the production of other pollutants, such as NO_x (Brown and Roberts, 1988).
4. Studies have usually been short with respect to needle longevity (Reich, 1987).
5. Experiments for trees larger than seedlings are rare.

Even though the useful data base is less than ideal, the process of analyzing mechanisms of conifer responses to O_3 should begin in order to summarize current understanding, to help prioritize future research needs, and to further thinking about ecological risk assessments applied to air pollution and conifers.

II. O_3 Uptake and Impacts at Leaf to Cellular Tissue Scales

The analysis of O_3 uptake by vegetation is a starting point for understanding the environmental issues surrounding gaseous air pollutants. The vegetation across the landscape is the largest sink for O_3, and therefore uptake by plants affects ambient O_3 concentrations. In addition, O_3 that is absorbed into leaves has the capacity to alter biochemical, physiological, and growth processes. Thus efforts to understand O_3 and its impacts on coniferous trees must begin with analysis of O_3 uptake by leaves in order to analyze metabolic responses to specific quantities of absorbed O_3 (Reich and Amundson, 1985; Reich, 1987). In addition, leaf-level studies of air pollution uptake are necessary to identify principles useful for analysis of O_3 uptake at the landscape level. Such landscape-level analyses have the potential to show regional scale rates of O_3 flux, seasonal trends in uptake, and differences in O_3 sink potential for important ecosystem components.

A. The Process of O_3 Uptake at the Leaf Level

Analyzing O_3 uptake by vegetation is simpler than analysis of other pollutants, such as SO_2 and acid deposition. O_3 can enter plants through foliage only by diffusing through stomata, whereas some gaseous pollutants, such as N_2O, can also diffuse through the cuticle. In addition, chemicals deposited in the form of aerosols and precipitation can enter plants through foliage and can also infiltrate into soils and be absorbed through roots. Another important feature of O_3 is that it reacts first to form phytotoxic free radicals and then its ultimate reaction product is O_2, which is biologically harmless in low concentrations. On the other hand, the reaction products of SO_2 and compounds in atmospheric deposition include free radicals as well as oxidized and reduced forms of sulfur, nitrogen, and other elements, all of which can alter metabolism to some degree.

For technical reasons, most air pollution uptake studies have focused on SO_2, which can serve as a model for understanding O_3 uptake (Black and Unsworth, 1980; Winner and Mooney, 1980). Another difficulty is that concepts needed for analyzing O_3 absorption by conifers rely on air pollution uptake experiments done with broad-leafed plants. Nonetheless, we will attempt to establish some general principles of O_3 uptake by conifers. One principle to emerge is that the uptake of SO_2 and O_3 through the stomata is large, and diffusion of these pollutants across the cuticle is slow. The central role of stomata in regulating O_3 uptake is unavoidable, because biological and environmental factors that account for differences in stomatal conductance will result in differences in absorption of air pollution into stomata. Therefore research with conifers and O_3 should emphasize stomatal uptake processes.

Even though the analysis of O_3 uptake by plants may be simpler than that for other pollutants, it is sufficiently complex that research on the topic has been scarce and scientists needing to know O_3 uptake rates by vegetation have resorted to crude approximations.

The difficulties with quantifying O_3 absorption are (1) that only a portion of the total uptake of O_3 by foliage is absorbed through stomata into the mesophyll, and the other portion absorbs to external leaf features; (2) that approaches for using stomatal conductance for calculating O_3 flux rates have not been verified; and (3) that biological and environmental factors, and limits to current technologies, make it difficult to extend measures of O_3 absorption for single leaves to estimates of O_3 absorbed by whole plants or forests.

B. O_3 Absorption and Adsorption

Total uptake rates of O_3 by conifers is partitioned between that which is adsorbed onto external surfaces of needles, stems, and branches and that which is absorbed through stomata into the needle mesophyll (Winner and Greitner, 1989). Both O_3 uptake processes must be understood in order to quantify the role of vegetation as an O_3 sink and to define O_3-caused changes in plants relative to quantities of O_3 absorbed.

O_3 adsorption can be a significant sink for atmospheric O_3 because adsorption rates may be nearly as high as rates of absorption. Adsorption rates are affected by needle size and shape, canopy architecture, relative humidity, the texture of plant surfaces, and the factors that affect the half-life of O_3 occupying adsorption sites. Rates of O_3 absorption through stomata can be highly variable, reflecting those factors that affect stomatal conductance.

Analysis of total O_3 uptake fits the simple summation:

$$J_{O_3} = O_{3ads} + O_{3abs},$$

where J_{O_3} is the total O_3 uptake rate, O_{3ads} is the rate of O_3 adsorption to exterior leaf/plant surfaces, and O_{3abs} is the rate of O_3 absorption into the mesophyll via stomata. Fortunately, solving this equation requires measurement of only two of the three elements. Total O_3 uptake is the easiest element to measure. Typically this is done in a cuvette, or fumigation chamber, where total O_3 uptake is calculated from the following equation:

$$J_{O_3} = (O_{3in} - O_{3out})FL^{-1},$$

where J_{O_3} is the O_3 absorption rate, O_{3in} is the O_3 concentration entering a fumigation chamber, O_{3out} is the O_3 concentration leaving a fumigation chamber, L is leaf area, and F is flow rate. This approach calculates O_3 uptake by both the leaf and the cuvette. Because the leaf and the cuvette both absorb O_3, blank cuvette fumigations are necessary to define the cuvette sink strength for O_3.

C. Calculating O_3 Absorption from Stomatal Conductance

Partitioning total O_3 uptake between that adsorbed and that absorbed requires further analysis. Because O_{3ads} for leaf and plant surfaces cannot be measured directly, approaches have been developed for measuring O_3 absorption. One such approach, the water vapor surrogate method, has enormous value because absorption can be calculated simply from measurements of stomatal conductance and ambient air pollution concentrations. Both of these types of data are relatively easy to obtain and offer the potential for estimating air pollution uptake at the landscape level. There are assumptions underlying the water vapor method for calculating air pollution fluxes, including that (1) stomatal conductance for any gas can be calculated from conductance values for water vapor and (2) the air pollution concentration inside the leaf mesophyll is nil.

The water surrogate approach for calculating air pollution absorption is based on the idea that measuring stomatal conductance for water vapor provides an accurate measurement of the potential for the leaf to exchange gases between the air and the leaf mesophyll. Thus once the stomatal conductance for water vapor is known, conductance for any gas can be calculated by

$$g_{xgas} = g_{H_2O}(D_{xgas}D_{H_2O}^{-1}),$$

where g_{xgas} is stomatal conductance for any gas, g_{H_2O} is stomatal conductance for water vapor, D_{xgas} is diffusivity for any gas, and D_{H_2O} is diffusivity for H_2O.

Once stomatal conductance for a gas is known, gas flux rates between the leaf and air are calculated as $J = gd$, where J is the flux rate between the leaf and air, g is stomatal conductance, and d is the gas concentration gradient between the leaf mesophyll and the air. An assumption for calculating flux rates for air pollutants is that their concentration in the leaf mesophyll is nil. The importance of this assumption is that, if true, pollution concentration gradients can be measured simply by measuring ambient air pollution concentrations. If pollution concentrations in the leaf mesophyll are not nil, then the concentration gradient between the leaf and air cannot be accurately determined and the water vapor surrogate method will be inaccurate.

Analysis has shown that O_3 concentrations in the leaf mesophyll are undetectably low (Laisk et al., 1989), which may also be true for other air pollutants. Directly measuring, or indirectly calculating, the levels of SO_2 and O_3 in leaves is technically difficult because air pollution concentrations in air around leaves is low (in the parts per million or billion range) relative to other gases. Thus failing to measure concentrations of air pollution in the mesophyll does not rule out the possibility that it does exist. Nonetheless, existing evidence suggests that such concentrations are low and that their resistance to O_3 flux into leaves is also low.

D. Scaling from the Leaf to the Cellular Level

1. Mechanistic Gas Exchange Studies Because leaves are the main uptake site of gaseous pollutants, they are potentially the first organs to be injured. Photosynthesis and transpiration are leaf functions that are affected by O_3 and that represent an integrated response of an array of biochemical pathways that account for leaf metabolism. The O_3 molecules that are absorbed through stomata become toxic by setting in motion the processes of free radical production. The radicals attack the cell walls and plasmalemma and lead to cell and organelle deterioration, the activation of the antioxidant systems in the foliage, accelerated senescence of leaves, loss of capacity to tolerate other stresses, and growth reductions.

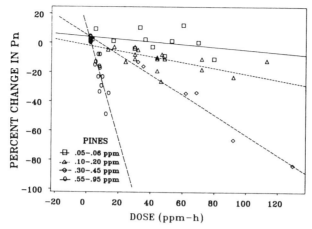

Figure 4 Percentage reduction in net photosynthesis (Pn) of conifers, largely pines, in relation to total O_3 dose, for several ranges of exposure concentration. (From Reich, 1987.)

Conifers have a wide range of photosynthetic and stomatal responses to O_3. Photosynthesis (*A*) usually declines in response to O_3. However, responses of *A* to O_3 depend on species, genotype, concentration, and duration of exposure (Reich, 1987). Decreases in *A* usually correlate with increasing external dose (concentration times duration of exposure), but much more rapidly at higher concentrations (Fig. 4). In general, conifers (except for *Larix*) have lower stomatal conductances and photosynthetic rates (Reich *et al.*, 1992), and thus have lower O_3 absorption rates (Reich, 1987) and are less affected by equal external doses of O_3 (Fig. 5) than are either hardwood trees or herbaceous vegetation. Thus, an observed lack of an O_3 response for conifers (e.g., Taylor *et al.*, 1986) may result from low O_3 uptake.

Slight increases in *A* have also been reported in response to O_3, although these are often small and statistically insignificant (Freer-Smith and Dobson, 1989; Eamus and Murray, 1991). Such an increase may be related to O_3-induced frost dehardening, causing an enhanced "metabolic potential" and thus raising *A* (Eamus and Murray, 1991). Alternatively, increased *A* of new needle flushes may compensate for O_3-caused photosynthetic decline or loss of other needle age cohorts, as found in conifers and other plants after defoliation (see Dickson and Isebrands, 1991; Reich *et al.*, 1993). In some cases, air is used to generate O_3, a procedure that can result in the inadvertent fumigation of plants with both O_3 and uncontrolled levels of nitrogen oxides (Brown and Roberts, 1988). Apparent O_3-caused stimulation of *A* may therefore actually be an artifact resulting from foliar N fertilization from NO_x (Wellburn, 1990).

Despite the numerous reports of O_3 impacts on photosynthetic rate, there are insufficient studies on responses of *A* to increasing CO_2 under

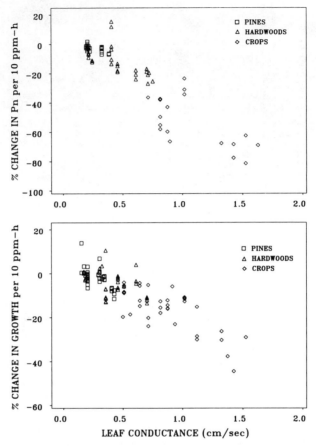

Figure 5 Percentage reduction in net photosynthesis and dry matter production due to equal ozone doses for conifers, hardwoods, and agricultural crops in relation to leaf diffusive conductance of each species. (From Reich, 1987.)

O_3 stress and of A to irradiance at high CO_2 concentration where there are no photosynthetic limitations from substrate or stomata. These types of gas exchange analyses are needed to understand the mechanistic basis for photosynthetic declines and can characterize the functional status of the photosynthetic apparatus in the mesophyll. Evidence from mechanistic gas exchange analysis shows O_3-caused reductions in A may be due to reduced photosynthetic capacity, reduced carboxylation efficiency (CE), and reduced quantum yield (QY). In addition, changes in stomatal conductance and/or carbon fixation processes in the mesophyll can alter C_i, which is the CO_2 concentration in the intercellular spaces of the leaf mesophyll.

Analysis of A/C_i curves shows that a declining A for *Pinus taeda*, *Picea* sp., and *Abies* sp. under O_3 stress can be related to lowered CE (Sasek and Richardson, 1989; Schweizer and Arndt, 1990). Reduction of CE is measured as a decline in the initial slope of the A/C_i curve (Fig. 6A) and may be taken as a measure of a depression in ribulose bisphosphate carboxylase/oxygenase (Rubisco) quantity and activity level. Such reductions have been observed for potato following O_3 exposure where both Rubisco activity and quantity declined in parallel with photosynthesis (Dann and Pell, 1989). Thus photosynthetic decline is due to O_3-caused changes in the leaf mesophyll and not due to changes in stomatal conductance caused by air pollution. Low activation of Rubisco due to O_3 stress may also be related to limited ribulose bisphosphate synthesis and result in a decreased ATP/ADP ratio (Hampp *et al.*, 1990). A shift in this ratio may result from reduced photophosphorylation, which in turn limits the resupply of ribulose bisphosphate and lowers A_{max} (the maximum rate of photosynthesis).

The inhibition of photochemical reactions and ribulose bisphosphate regeneration may explain one way that O_3 can lower the QY of *Pinus taeda* (Fig. 6B). However, the measurement in this study was conducted under the nonsaturating CO_2 concentration of ambient air, where the CE was found to differ for foliage in fumigated and control treatments. Thus, the decreased CE of the O_3-exposed plants may explain, in part, the lowered A under ambient CO_2 concentration in both low light and at saturating irradiances.

As A declines due to O_3 exposure, g_{H_2O} of conifers can remain stable (Sasek and Richardson, 1989; Schweizer and Arndt, 1990), which may indicate sluggishness in stomatal regulation (Keller and Häsler, 1987).

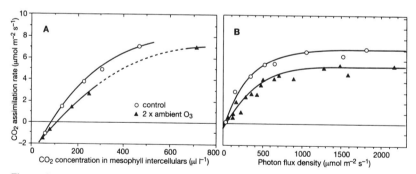

Figure 6 Net CO_2 uptake of the first needle flush of loblolly pine seedlings exposed to ozone (redrawn from Sasek and Richardson, 1989). (A) CO_2 dependence; (B) light dependence of the CO_2 assimilation rate. The control treatment was charcoal-filtered air with a daily 12-hr mean O_3 concentration of 29 nl/liter (about 50% of the O_3 level in nonfiltered ambient air). The O_3 concentration of the 2× ambient O_3 treatment was 92 nl/liter.

It seems that O_3 primarily affects the metabolism in the mesophyll and loosens the usually tight coupling between A and g_{H_2O}. Decreased water use efficiency (WUE = A/E, where E is transpiration rate) thus results when A declines proportionally more than g_{H_2O}. Decreased WUE has been measured following O_3 exposure (Reich and Lassoie, 1984; Matyssek et al., 1991) (Fig. 7). The adjustment between A and g_{H_2O} also influences the $\delta^{13}C$ isotope ratio in the plant (Farquhar et al., 1989). Ozone tends to increase $\delta^{13}C$ (toward less negative values) in leaves and wood (Martin et al., 1988; Greitner and Winner, 1988; Matyssek et al., 1992), either through stomatal limitation of photosynthesis (Greitner and Winner, 1988) or metabolic changes in the mesophyll (Luethy-Krause et al., 1990; Matyssek et al., 1991, 1992).

O_3 could possibly induce patchy stomatal behavior, which might affect patterns of ozone absorption and distribution within conifer needles. Although patchiness may not be a universal phenomenon or may simply be caused by experimental manipulation (Lauer and Boyer, 1992), heterogeneity in stomatal opening does occur across leaves, can affect A,

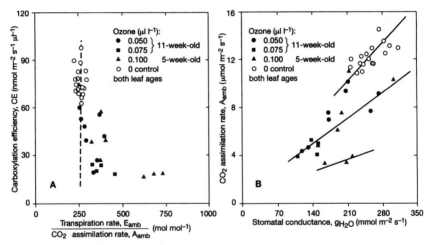

Figure 7 (A) Carboxylation efficiency, CE, of *Betula pendula* as related to the ratio, transpiration/photosynthesis (E_{amb}/A_{amb} = WUE^{-1}), for leaves under the following conditions: 340 µl/liter CO_2 concentration and nonlimiting light, temperature, and humidity. Each data point represents one individual leaf and plant from one of four O_3 treatments (the dashed line gives the mean ratio of the control). (B) Relationship between the CO_2 assimilation rate, A_{amb}, and stomatal conductance, g_{H_2O}, at A_{amb} (leaves, plants, and conditions as in A). The upper line describes leaves with E_{amb}/A_{amb} of 225–298 mol/mol (no visual symptoms of O_3 injury); values for the second line from the top are 300–455 mol/mol (with early visual O_3 symptoms up to established discoloration), and values for the lowermost line are 626–734 mol/mol (all with established O_3 discoloration). (From Matyssek et al., 1991.)

and may bias CE depending on leaf anatomy and the synchronization of stomatal movements. Conifers differ from angiosperms in the arrangement of stomata by rows. The compact mesophyll structure of needles may reduce intercellular gas diffusion, thereby enhancing any effects of stomatal heterogeneity (cf. Terashima *et al.*, 1988). Although evidence of stomatal patchiness does exist for *Picea* sp. and *Abies* sp. (Kresse, 1991), this phenomenon has not been investigated in the context of O_3 stress.

2. Biochemical Changes Due to O_3   Ambient levels of O_3, or levels below the United States Environmental Protection Agency (EPA) standard, are known to cause changes in CE and QY. However, O_3 concentrations much higher than ambient levels (as high as 450 nl/liter) are needed to induce fluorescence signals of altered photosynthetic electron transport and Calvin cycle in spruce (Urbach *et al.*, 1989). In fact, O_3 may hardly intrude into chloroplasts when leaves are exposed to low O_3 concentrations.

Most of the O_3 absorbed through stomata reacts in the cell wall and plasmalemma, where it is rapidly decomposed (Laisk *et al.*, 1989; Heath, 1980) by the oxidative impact by O_3 on these cell structures. Thus, photosynthesis may be affected indirectly by changed cell metabolism following disturbances in the plasmalemma rather than by direct O_3 impact on the chloroplasts. For example, O_3 can initiate potassium efflux and depolarize membrane potentials in algal cells, whereas photosynthesis does not change (Urbach *et al.*, 1989). Chloroplasts remained structurally intact in *Picea* sp. and *Abies* sp. exposed to O_3 (Schmitt and Ruetze, 1990), but displayed "crooked" shapes in the proximity of cell walls in O_3-fumigated *Pinus* sp. (Fink, 1989).

O_3 and its derivatives can oxidize unsaturated bonds in the fatty acids of membrane lipids (Heath, 1980), and thus can alter plasmalemma structure, membrane functions, and cell metabolism. Therefore, O_3 impact can be linked to frost hardiness, which requires large proportions of unsaturated bonds in membrane lipids (Wolfenden and Mansfield, 1991). Decreased frost hardiness of conifers may relate to membrane changes caused by O_3 together with other cellular disturbances such as delayed accumulation of the cryoprotectant raffinose in needles after O_3 exposure in summer (Alscher *et al.*, 1989a).

In conifer needles, impairment by O_3 can be delayed or prevented by the formation of antioxidants (e.g., ascorbate) (Bermadinger *et al.*, 1990) or the repair of membrane injury (see Wolfenden and Mansfield, 1991). The energy cost of this defense and repair is sometimes reflected in increased respiration (Skärby *et al.*, 1987; Adams *et al.*, 1990), and may depend on the developmental status of the tissue (cf. *Populus* sp.) (Reich,

1983). It is uncertain whether high nutrient supply would generally lower rather than increase O_3 injury (strengthening repair versus enhancing O_3 uptake via elevated g_{H_2O}).

O_3 increased pinitol but lowered inositol concentrations in *Pinus* sp. (Landolt *et al.*, 1989), but increased inositol in *Populus* sp., which lack pinitol (W. Landolt, personal communication). Although the mechanistic links between these substances and the whole-leaf impact of O_3 are unknown, the importance of inositol to metabolism is widespread and includes the synthesis of cell wall and plasmalemma (the main sites of O_3 attack). Pinitol may also be of importance because it can act in osmoregulation during drought. Thus plant responses to O_3 may be mediated by responses to drought stress. O_3 may cause cellular drought stress if water leakage from the cells initiates the cell collapse and decreases leaf water content. Such interactions between O_3 and leaf water relations are known to occur in *Betula* sp. (Matyssek *et al.*, 1991; Günthardt-Goerg *et al.*, 1993). Thus O_3 may induce drought stress in the leaf and contribute to lowered A and g_{H_2O}.

Although the compact mesophyll tissue in conifers may be resistant to structural breakdown by O_3 (Schmitt and Ruetze, 1990), cell collapse spreads from the substomatal cavity (Fink, 1989). Lowered frost hardiness of O_3-fumigated *Picea* sp. can lead to increased disruption of mesophyll cells during winter (Fincher *et al.*, 1989). Given the collapse of cells, A_{max}, CE, and QY may be found to decline in leaf gas exchange, but do not necessarily reflect metabolic changes in the remaining cells (Sasek and Richardson, 1989; Matyssek *et al.*, 1991). Apart from physiological causes, increasing g_{H_2O} can result from O_3-caused epidermal cell collapse (as seen in *Betula* sp.) (Günthardt-Goerg *et al.*, 1993), which is caused by O_3 impact from inside the leaf and marks the final stage of leaf decline. The epicuticular wax layer of conifer needles may render epidermis and stomata more resistant to O_3 than the mesophyll. O_3 does not affect the epicuticular wax (Günthardt-Goerg and Keller, 1987) or the permeability of the cuticle to water vapor (Kerstiens and Lendzian, 1989).

III. Organ Differentiation and Senescence in the Presence of O_3

A. Differentiation during Growth

O_3 has important effects on organ differentiation and ultimately alters plant architecture and productivity. For example, in conifers, O_3 stress (similar to other factors) may limit the length growth of needles (McLaughlin *et al.*, 1982; Schier *et al.*, 1990), cause foliar lesions, and accelerate leaf aging. In some cases, O_3 may cause differentiation that results in acclimation and compensation. In other cases, O_3 may simply cause incipient injury (cf. Mooney and Winner, 1988). O_3 can broadly affect leaf development, as shown for *Betula* sp. (Günthardt-Goerg *et al.*, 1993), causing reductions in leaf area, width of epidermal and mesophyll cells, and total number of cells as well as increases in stomatal density, leaf veins and hairs, leaf mass per area (LMA; i.e., leaf weight per leaf area), and leaf internal air space. The O_3-caused reductions in the total number of leaf cells may be either the result of a limited number of cells entering the division process during early primordial stages or a slow rate of mitosis during subsequent development. Even though O_3 reduces cell widths, other stresses that reduce leaf area may not affect cell sizes (Pieters, 1974). O_3 may alter leaf cell differentiation by disturbing early stages of membrane formation during leaf growth and enhancing water loss. Only slight changes in either cell water potentials or cell wall flexibility would be required to limit the extension growth of cells and tissues (Boyer, 1985). O_3 may also alter leaf cell differentiation by affecting the function of stem phloem (cf. Dickson and Isebrands, 1991; Günthardt-Goerg *et al.*, 1993). The effect of O_3 may not be constant because poplar leaves ozonated during growth had reduced stomatal density but unchanged size and LMA (Matyssek *et al.*, 1993b).

Changes in leaf differentiation due to O_3 may result in acclimation to O_3 stress. For example, birch leaves formed under O_3 delayed O_3 injury and premature leaf shedding due to continued O_3 exposure longer than did fumigated leaves formed in clean air (Günthardt-Goerg *et al.*, 1993). However, acclimation to O_3 developed in the presence of increased stomatal density, potentially increasing air pollution uptake by foliage. The example shows that acclimation can develop even though leaf structures seemingly predispose plants to O_3 injury.

B. O_3 during Senescence

Plant ontogeny is susceptible to O_3 impact during senescence. The process of senescence is driven by a declining capacity of the plant to cope with oxygen radicals released from metabolic processes. O_3 provides an additional strain by increasing the concentration of oxygen radicals (Pell and Dann, 1991). Because senescence is mediated at the biochemical level by processes similar to those mediating O_3 injury, the latter is often regarded to be "accelerated senescence" at the leaf level (Reich, 1983). However, equating the normal processes of senescence with those associated with O_3 injury may be questionable (Pell and Dann, 1991). For example, natural senescence is a genetically regulated process (Thomas and Stoddart, 1980) that requires metabolic and structural integrity to ensure controlled leaf degradation (as reflected, for example, in efficient nitrogen retranslocation) (Matyssek, 1986). However, plant responses to O_3 may not be under genetic control. In a specific contrast, WUE may decline while foliar nitrogen concentration either remains stable or increases in ozonated leaves. Such trends are opposed to those found in naturally aging leaves (Matyssek *et al.*, 1991, 1993a,b). Finally, O_3-caused breakdown of metabolism may limit the effectiveness of natural senescence.

IV. Scaling from the Leaf to the Whole-Plant Level

A. Carbon Gain

Although photosynthesis is the central physiological process in the plant, with the leaf as the primary productive unit, growth cannot be explained solely based on leaf photosynthetic capacity (Körner, 1991; Matyssek and Schulze, 1987; Reich *et al.*, 1992; Walters *et al.*, 1993). Thus, the ranking of growth sensitivity to O_3, i.e., evergreen conifers < deciduous hardwoods < annual crop plants (Fig. 8), may be partially explained by the similar ranking of photosynthetic sensitivity (Reich, 1987), but other factors likely play important roles as well. Such determinants of whole-plant production include total canopy net carbon gain (i.e., the multiplication of the proportion of achieved photosynthetic capacity in single leaves by the total mass of foliage), the structural and metabolic costs and life-span of all tissue types, the proportion of plant tissue allocated to each tissue type, and the crown and root architecture. In conifers, carbon gain must be viewed also in the context of evergreenness and needle longevity, and for purposes of this chapter, under the influence of O_3. We need to consider, therefore, how O_3 affects the carbon gain of conifer crowns, and how altered carbon gain may interact with whole-tree carbon allocation and growth rate.

In conifer needles, not only is the O_3 uptake rate low relative to other species, but also photosynthesis seems to be less O_3 sensitive than that of other leaf types when related to similar O_3 uptake (Reich, 1987). Conifers apparently display both avoidance and tolerance of O_3 stress. In fact, stress tolerance seems to be an inherent feature of evergreen leaves (Schulze, 1982), perhaps also predisposing conifers to delay or prevent leaf responses to O_3. However, such O_3 tolerance is mainly observed in small plants with young needles, over a relatively short and absolute time frame. When O_3 uptake is calculated for an estimated mean total needle life-span, conifer photosynthesis appears to be equally or more O_3 sensitive over this life-span-integrated period than other plant forms (Reich, 1987). Given this extrapolation, which has not been experimen-

exposed to both types of stresses (low N and high O_3) may be balanced in terms of carbon allocation pattern and thus be similar to plants experiencing no stress at all. However, patterns of carbon use will be achieved along with reductions in size, photosynthesis, stomatal conductance, and growth. Moreover, the impact of nutritional limitations will seldom be balanced exactly by the impact of the gaseous pollutants, and the response of plants will probably reflect this imbalance.

Greitner and Winner (1989) proposed a conceptual model for assessing both singular and combined effects on plants raised under various nutrient supplies and O_3 exposure regimes. They proceeded to test the model by studying the effects of three nutrient supply rates and O_3 doses on two types of broadleaf tree species, *Salix nigra* Marsh. and *Alnus serrulata* (Aiton.) Willdenow. The latter species has the capacity to link with the nitrogen-fixing actinomycete, *Frankia*, a rhizosphere symbiont that forms nodules on roots of the genus *Alnus*.

In *S. nigra* plants, the optimum nutrient supply for growth shifted from the medium to the high level as O_3 concentrations increased (Fig. 10). Carbon allocation clearly favored the shoot when plants were either exposed to high O_3 or amply supplied with nutrients. However, with increasing nutrient supply, the leaf-to-root ratio decreased more in high O_3, perhaps due to increased damage to the shoot.

In the *Alnus/Frankia* experiment, nodulated plants had about a fivefold greater photosynthetic rate and twice the stomatal diffusive conduc-

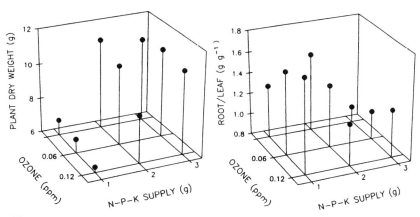

Figure 10 Ozone effect on dry weight and root/leaf ratio of *Salix nigra* cuttings as affected by the nutrient supply rate (slow-release fertilizer: 14-14-14 N-P-K). At high nutrient supply, ozone induced a shift in allocation toward root growth, without altering whole plant growth; at medium supply, high ozone reduced total growth and decreased allocation to the root; low nutrient supply caused a great reduction in growth and a high allocation to roots, whereas the effect of ozone was marginal. (From Greitner and Winner, 1989.)

tance of unnodulated plants while in clean air. Stomatal conductance generally increased with exposure to O_3, indicating a greater uptake of the pollutant. On return to clean air, however, only the stomata of the nodulated plants recovered whereas N deficiency of unnodulated plants may have reduced their capacity to recover to preexposure gas exchange behavior. Greitner and Winner (1989) concluded that for both species "the severity of one stress (low nutrients) can be so profound that the effects of a second, lesser stress (O_3 in this case) are overwhelmed." They speculated that species that do not respond to nutrient stress by altering leaf N concentration and gas exchange may reduce growth equally when exposed to O_3, regardless of the nutrient supply.

As with *S. nigra* (Fig. 10), a modeling simulation showed that the effect of simultaneous stress from low nutrition and high O_3 on *P. rubens* growth was less than the sum of each stress alone, in essence an avoidance of antagonistically interacting stresses (Weinstein *et al.*, 1991). However, the simulation resulted in a different pattern of response above- and belowground in the conifer compared with the broadleaf species. Nevertheless, the simulation predicted that *P. rubens* under combined stress will be smaller, but its shoot/root ratio will be balanced similarly to unstressed plants, as was shown for *S. nigra*. Weinstein *et al.* (1991) also concluded that plants under stress from nutrient deficiency are not likely to suffer severely from O_3 exposure. They caution, however, that this conclusion should not be generalized, because the interaction of multiple stresses depends on the action of each stress in relation to the pattern of resource use.

C. Soil Water

The response of plants in terms of growth, stomatal conductance, and photosynthesis under water stress (both flooding and drought) is in many ways similar to their response to nutrient stress. Stomatal conductance is low in plants in both water-saturated and dry soils. Therefore, given the role of stomatal conductance in O_3 response (e.g., Reich and Amundson, 1985), it should not be surprising if plants under water stress respond to gaseous pollutants similarly to plants under nutrient stress (Tingey and Hogsett, 1985).

Even flood-tolerant species reduce growth, stomatal conductance, and photosynthesis when inundated with moving water and, to a greater degree, with standing water (Shanklin and Kozlowski, 1985), due to impaired water uptake by the oxygen-starved roots and the closure of stomata with the accumulation of abscisic acid. Thus, SO_2 uptake by foliage of flood-tolerant broadleaf (*Betula nigra*) and coniferous (*Taxodium distichum*) species decreased under flooding, resulting in less injury and lower reduction in leaf growth compared with SO_2-exposed unflooded seedlings (Norby and Kozlowski, 1983; Shanklin and Kozlowski, 1985). Even differences in soil texture may influence soil aeration, helping to explain why O_3 uptake and injury in some crop plants were reduced due to the resulting oxygen deficiency in the root zone (Stolzy *et al.*, 1961). It is possible, therefore, that episodic pollutant exposure under flooded conditions may cause less damage to trees; the damage may even be less severe to flood-intolerant species, which are more likely than flood-tolerant species to close their stomata.

Growth, stomatal conductance, and foliar gas exchange also decrease under soil drought. The outcome may be a reduced uptake of gaseous pollutants (e.g., O_3) (Rich *et al.*, 1970; Rich and Turner, 1972) and potentially a greater availability of carbohydrates for repair of injured tissues. Soil moisture deficit causing high moisture stress can decrease the responses to chronic O_3 exposure (Amundson *et al.*, 1986). In *Populus* sp. and *Fagus* sp. O_3 injury was reduced in droughted plants with decreased leaf conductance (Harkov and Brennan, 1980; Taylor and Dobson, 1989). Similar results were found in several studies on conifers. Water stress lessened the intensity of foliage injury by O_3 in seedlings of three half-sib *P. taeda* families (Meier *et al.*, 1990). Drought reduced the photosynthetic rate and decreased stomatal conductance of *P. abies* and *Picea sitchensis* seedlings, consequently decreasing O_3 uptake (Dobson *et al.*, 1990). Thus, a mild water deficit may provide some protection from O_3 exposure via an exclusion mechanism (Tingey and Hogsett, 1985). However, as was stipulated regarding the effects on plants of both low nutrient supply and other gaseous pollutants, it is not clear that the potential avoidance of O_3 uptake and injury compensates well for the loss in carbon assimilation caused by drought (Amundson *et al.*, 1986; Dobson *et al.*, 1990).

Conditions in the field are always complex, involving interactions among more than two variables. Each variable may shift the balance in favor or against the plant health. For example, exposure of *P. sitchensis* seedlings to SO_2 reduced their root growth more when subjected to drought (Warrington and Whittaker, 1990). However, drought made the seedlings less desirable to the green spruce aphid *Elatobium abietinum* (Walker), a major pest of *Picea* sp. Thus, the combined effects of all three stresses resulted in a less than additive reduction of root growth, reminding us that the multiple interactions among stresses that occur in the field may result in surprising results compared to our more controlled experimental studies.

D. Atmospheric Humidity and Wind

To injure plants, gaseous pollutants must diffuse through both the boundary layer outside leaves and through the stomata. The boundary layer of conifer needles is relatively thin in comparison to that of broadleaf foliage, but clumps of needles may form a thicker, unstirred, boundary layer. The thickness of the boundary layer decreases with the rate of airflow near the leaf. Thus, with increasing wind speed the boundary layer should pose less resistance to pollutant flux. Ashenden and Mansfield (1977) found that the boundary layer provided a significant resistance for the entry of SO_2 into leaves of ryegrass.

Stomatal diffusive conductance generally decreases with the vapor pressure deficit caused by increasing air temperature and/or decreasing relative humidity. As humidity decreases, leaves of *P. virginiana* (Davis and Wood, 1973) and *B. papyrifera* (Norby and Kozlowski, 1982) appeared to become more resistant to ozone and SO_2, respectively. More specifically, *B. papyrifera* leaves absorbed less SO_2 and developed less necrosis than did plants fumigated at high humidity. In addition, seedlings fumigated at low humidity lost fewer leaves and grew faster than plants fumigated at high humidity (Norby and Kozlowski, 1982). Similar results were reported for three of four agricultural crop species (Mansfield and Majernick, 1970; Black and Unsworth, 1980).

The combined effect of both higher wind speed and lower atmospheric humidity may reduce the effect relative to that caused by each variable alone. Higher wind speed reduces the boundary layer resistance not only to the pollutant entry into the leaves but also to water loss out of leaves, often resulting in reduced stomatal opening. The effect is heightened when the atmospheric humidity is low, the leaves have a thick boundary layer at a low rate of air flow, and the species has stomata that are sensitive to atmospheric humidity. Thus, a careful assessment of several (not readily measurable variables) is required for predicting the relative change in the uptake rate of a pollutant in relation to different combinations of wind speed and atmospheric humidity.

E. Temperature

Apart from its role in affecting the atmospheric vapor pressure deficit, temperature may have a direct effect on plant sensitivity to air pollution. *Pinus virginiana* foliage exhibited increasing O_3 injury with decreasing temperature within a temperature range of 10 to 32° C (Davis and Wood, 1973). However, there is too little and inconsistent information on plant response to pollutants in relation to changing above-freezing temperature to make any general claims. For example, *P. strobus* (Costonis and Sinclair, 1969) and *F. americana* (cited in Davis and Wood, 1973) seedlings were injured more by ozone at high-exposure temperature.

Postfumigation temperature may mediate the response to pollutant exposure by affecting metabolism and, therefore, the recovery from exposure to pollutants. *Pinus rubra* formed new needles at 22 and 32°C after fumigation with SO_2 but not at 12°C. The new needles reduced the proportion of injured needles and allowed the seedlings at the higher temperature regime to grow faster (Norby and Kozlowski, 1981). In contrast, *P. strobus, P. virginia,* and *F. americana* were more severely injured when maintained at high compared to low postfumigation temperatures (Costonis and Sinclair, 1969; Davis and Wood, 1973). These studies, despite their conflicting results, demonstrate that wide variations in temperature following a pollution event may alter tree responses.

Several mechanisms may be involved in the increased sensitivity of conifers and other primarily evergreen species to gaseous pollutants under low, particularly subfreezing, temperature:

1. Changes in the saturation state of membranes lipids make the membranes more susceptible to oxidants.
2. The protective antioxidant enzyme systems in the needle chloroplast (Melhorn *et al.*, 1986; Hausladen *et al.*, 1990) may be less active at low temperature.
3. The photooxidation of chlorophyll increases and free radicals form independently under conditions of low temperature and high light (Öquist, 1986).

Thus, at low temperature, O_3 may cause a greater injury due to an enhanced sensitivity to oxidants, a greater amount of oxidants, and a reduced activity of the antioxidant system (Alscher *et al.*, 1989a). As was demonstrated for *P. rubens* (Alscher *et al.*, 1989b), the geographical distribution and evergreen nature of most conifers may annually expose many coniferous forests to this damaging combination of events.

F. Light

Few studies have examined the interaction of light and chronic O_3 stress with woody plants. One study compared shade-intolerant *P. taeda* and shade-tolerant *Fraxinus pennsylvanica* Marsh. seedlings (Oren and Schoeneberger, unpublished data), while a second compared shade intolerant *Populus* sp. to shade tolerant *Acer saccharum* seedlings (Tjoelker *et al.*, 1993; Volin *et al.*, 1993). In both studies, seedlings were grown in chambers at two contrasting light levels and in two contrasting O_3 regimes. When O_3 was elevated, the growth and leaf carbon assimilation of *P. taeda* was unaffected in the sun but was reduced in the shade, while the growth of *F. pennsylvanica* was reduced in both light conditions, but the reduction in the shade was more severe (Oren and Schoeneberger, unpublished data). In the other study (Tjoelker *et al.*, 1993; Volin *et al.*, 1993), CO_2 exchange and growth of the shade-intolerant *Populus* sp.

was more affected by O_3 in higher than in lower light (differing from *P. taeda*), but the shade-tolerant *Acer* reacted similarly to *Fraxinus* and was more adversely affected by O_3 in the shade. This tendency for greater O_3 sensitivity of shade-tolerant species in the shade was corroborated by a field experiment that examined gas exchange responses of sugar maple branches to O_3 at different positions in the canopy of a 15-m-tall *Acer saccharum* stand (Tjoelker *et al.*, 1994a, unpublished data). Again, leaves growing in more shaded positions showed a tendency to be more sensitive to O_3 than those in greater light microenvironments.

In all three studies, leaf conductance of shade foliage in the shade was lower than that of sun foliage in the sun, strongly suggesting lower O_3 uptake, and yet the shade leaves appeared to be more sensitive to O_3 in three of the four species. The reasons for this phenomena may include (1) lower availability of carbohydrates for repair of membranes under shaded conditions and (2) large differences in leaf morphology between shade and sun foliage, including fewer layers of mesophyll in shade foliage and shorter average pathways for oxidants to any cell. The combination of more vulnerable tissue and lower capacity for repair probably makes shade foliage more susceptible to injury. Thus, despite the relatively small contribution of shaded layers of a tree or a forest canopy to total net carbon balance (Matyssek, 1985; Reich *et al.*, 1990), the net carbon balance and persistence of the shaded branches could be significantly impacted by O_3, with important implications for nutrient storage capacity, especially in conifers.

G. Predisposition of Plants to Stress by Exposure to Pollutants

Air pollution exposure may reduce the capacity for plants to tolerate or acclimate to subsequent environmental stresses. There are potentially many situations in which plants chronically exposed to varying levels of gaseous pollutants are forced to suddenly face another stress, such as flooding, drought, or frost. For instance, air pollutants predispose plants to drought stress and may alter the behavior of plants under drought. *Pinus taeda* seedlings exposed to O_3 and then to two drought cycles showed a greater sensitivity to water stress during the second drought cycle (Lee *et al.*, 1990). As a result of O_3 exposure, net photosynthetic rate recovered more rapidly and reached a higher level at full hydration after one drought cycle relative to unexposed seedlings. However, net photosynthetic rate declined more rapidly in relation to needle water potential during the second cycle. In hybrid *Populus* sp. seedlings, O_3-exposed leaves showed no difference in leaf conductance on well-watered plants, but were markedly unable to control water loss once excised (Reich and Lassoie, 1984) and in general showed increasing uncoupling of stomatal aperture and net CO_2 uptake with increasing O_3

dosage, as did seedling and mature *A. saccharum* foliage (Volin *et al.*, 1993; Tjoelker *et al.*, 1993). SO_2-stressed *P. abies* seedlings exposed to drought experienced greater and earlier modification in the amounts of several proteins than did nonpolluted plants (Sieffert and Queiroz, 1989), and other impacts were found in the catalytic capacity of certain enzymes (Pierre and Queiroz, 1988). Under similar experimental conditions, 5-month-old *P. abies* seedlings not exposed to SO_2 recovered well after drought, whereas chronically exposed seedlings reached lower water potentials and content, and either died or showed impaired water relations (Macrez and Hubac, 1988).

Plants may be exposed to air pollutants during periods of low temperature. O_3 reduced the rate of raffinose accumulation, which may make trees more susceptible to winter injury (Alscher *et al.*, 1989a). Frost sensitivity of *P. abies* seedlings increased after exposure to O_3; the exposure affected their photosynthesis during the early part of the following season (Cape *et al.*, 1989). Other gaseous pollutants also appear to increase susceptibility of trees to low temperature. *Picea sitchensis* seedlings exposed during dormancy to relatively low concentrations of SO_2 showed poorer bud survival when brought to below-freezing temperatures in comparison to unexposed plants (Freer-Smith and Mansfield, 1987). A combination of SO_2 and NO_2 under similar experimental conditions increased frost injury. Impairment of frost hardening of *P. rubens* seedlings was also caused by sulfate and ammonium in mist (Cape *et al.*, 1991), but impairment of frost hardening in *P. abies* was mitigated by nitrate and ammonium in the mist (Cape *et al.*, 1989). Although some pollutants may increase the sensitivity of conifers to low temperature, predicting plant response to pollution exposure at low temperature may require information on the dose of all pollutants.

H. Spatial and Temporal Factors

Plant responses to pollutants vary in time and space, on a fine scale, in the context of other environmental factors (e.g., light, temperature, atmospheric humidity, and wind speed). Diurnal variation in pollutant uptake and antioxidant concentration have been shown in conifers (Schupp and Rennenberg, 1988). Based on such information, variation in sensitivity among trees within stands should also be expected among genotypes and, depending on stand structure and leaf area index, among age/canopy classes.

Using the information on the joint effect on *P. taeda* and other trees of light level and O_3, and the exponential decline in light level with height in closed-canopy forests, we may predict that suppressed and intermediate classes in forest stands will be damaged more severely than exposed, dominant trees. We may also predict that trees will be less affected by this interaction in open-canopy forests. An accurate prediction of the effects of air pollutants on trees and forests cannot rely solely on investigating processes in isolation; like most ecological questions, it requires a holistic approach.

VI. Scaling to Understand Mature Tree or Stand Level Responses

Much more is known about pollution responses of seedlings than about large trees. Seedling studies are more numerous because of the ease of plant propagation, treatment administration, response measurement, and harvest. Few air pollution studies address responses of mature trees or forests to experiments with air pollution. There are at least three different, but complementary, ways of attempting to assess mature tree and forest responses to air pollution:

1. Direct studies of mature tree responses to an array of air pollution treatments.
2. Modifying models of whole-stand or ecosystem processes to also simulate the responses to pollution.
3. Scaling up our understanding of seedling or branch responses to pollution and other factors to a whole-tree or forest level.

The first approach offers a direct method of addressing this issue, but such studies have been and will continue to be limited in number because of relatively high expenses and numerous logistical difficulties. The second and third methods offer contrasting approaches and involve the use of conceptual and quantitative models, or both. In the second method we begin by knowing a lot about other aspects of forests at a large scale (based on empirical data), with pollution responses at that scale being the largest unknown. Assumptions about the way in which forests at the stand level will respond to pollution must be built into the model, and these assumptions are a weak link in this approach. In the third approach, we begin by knowing a lot about how seedlings (or branches to a lesser extent) respond to pollution (again based on empirical data), and the assumptions built into the model involve how we scale up from leaf to canopy and/or plant to tree to stand, all of which entail uncertainties. At present, different researchers are using and combining aspects of all three approaches.

In beginning analysis of whole-forest responses to O_3, we return to the topic of O_3 uptake. If the water vapor surrogate method for calculating air pollution uptake by single leaves proves to be acceptable, then strategies must be developed for using it for plants in air pollution experiments, for large trees, for stands, and across the landscape. Application of the technique requires measurements of both stomatal conductance and ambient air pollution concentrations. Once stomatal conductance and ambient air pollution data are available to characterize landscapes, computer techniques, such as geographic information systems, can be used to create maps showing regional trends in air pollution absorption. Such maps could be used to identify those stands where air pollution absorption will likely be greatest and therefore stands at greatest risk from pollution. Regional scale estimates of air pollution uptake must include analysis of stomatal conductance measurements, air pollution adsorption, and ambient air pollution concentrations.

A. Measurements of Stomatal Conductance

Measurement of stomatal conductance can be made rapidly with porometers that readily clip on and off foliage. Even with a high potential rate of sampling, there are several issues of scale to resolve:

1. Stomatal conductance differs with leaf position in the canopy. To some extent this can be overcome by scaffolding and the use of lifts to bring the measurement to the foliage.
2. Stomatal conductance differs with leaf age, and such patterns may require intensive sampling for resolution. O_3 can accelerate leaf aging, leaf fall, and increase the rate of new leaf formation, making the analysis of leaf age effects on conductance more difficult to assess.
3. Stomatal conductance differs over the course of each day.
4. Stomatal conductance must be placed in the context of a growing season and leaf longevity. Because needles with low conductance can persist on trees for many years (commonly 3–8 years; in some species as long as 15–30 years), long-lived needles may absorb much more air pollution than do high-conductance deciduous leaves that commonly last only 5–8 months. Although long-term exposure of needles to O_3 results in accumulated impact, it does not result in the physical accumulation of O_3, or its transport to other plant tissues, because O_3 is too reactive and unstable. However, long-term exposure of needles to gaseous pollutants, such as SO_2, can result in high levels of S (usually as sulfate) in vacuoles, transport of S from needles to other tissues, and the foliar emission of reduced sulfur gases.
5. Stomatal conductance will differ between species. Assuming that stomatal conductance is a predictor of air pollution absorption capacity (Reich, 1987), evergreen conifer needles have generally lower stomatal conductance (Reich *et al.*, 1992; see also Chapter 8, this vol-

ume), and therefore lower O_3 absorption rates, than do broadleaf plants. Attempts to scale air pollution uptake rates from single leaves to stands must take into account differences in stomatal conductance for all forest species.

B. Air Pollution Adsorption

Adsorption of air pollution depends on total surface area of foliage as well as foliar surface characteristics, including morphology and chemistry. Total surface area is likely to be higher for coniferous than for deciduous forests, both because of generally greater projected leaf area (Reich *et al.*, 1992) and because needles of many coniferous species are not flat in cross section, and thus have greater total surface area than broad leaves of similar projected area. As with calculations of air pollution flux into leaves, the longevity of foliage plays an important role in defining total air pollution adsorption to leaf surfaces. The presence of needles year around, and their potential to live for more than 10 years for some species, enhance the capacity for conifer needles to adsorb air pollution.

C. Measurements of Air Pollution Concentrations

Measurements of air pollution concentrations in forest stands have not been sufficient for air pollution uptake analysis. There are problems associated with ambient air pollution measurements, including the fact that SO_2 and O_3 monitors are large, expensive, and require so much electricity that they cannot effectively be operated with batteries. Thus air pollution monitors are most commonly limited to well-protected sites where power lines have been installed. In addition, most air pollution monitoring has been done in urban areas, and these measurements may or may not be relevant to forests. Another problem is that air pollution concentrations may not be uniform through the canopy of a large tree. Foliage on the outside of the tree canopy has first contact with gaseous pollutants and resultant uptake may deplete air pollution concentrations as air moves through the canopy. Thus it may be necessary to monitor air pollution concentrations through canopies of conifers in order to analyze ambient pollution levels for all needles on a tree.

D. Relative Phytotoxicity of Absorbed Air Pollution

The analysis of air pollution uptake by vegetation is only the starting point for understanding the impacts of O_3, other gaseous pollutants, and wet and dry forms of acid deposition. There is a critical need for studies that quantify the metabolic changes due to specific amounts of absorbed air pollution. Thus the effects of absorbed air pollution may depend on needle age, season, plant water status, foliar nutrient levels, the onset of reproductive growth, herbivory, competition, and many other factors. Research on the effects of specific quantities of air pollutants in leaves has begun with analysis of physiological responses of single leaves in chambers and growth analysis of small plants. This is the beginning of what must be a much larger effort to learn about the ecological and physiological responses of large plants and stands to air pollution absorption.

E. Direct Studies of Mature Trees and Forests

To further address the issue of whole tree, stand, or ecosystem response to pollution, we will summarize results of field diagnostic surveys and/or experimental work with large and/or old trees (>10 m tall, >20 years old) or parts thereof. Few data of these kind exist for either conifers or hardwoods, and much of that is new and not yet in print (e.g., P. J. Hanson, unpublished data; R. O. Teskey *et al.*, unpublished data; P. B. Reich *et al.*, unpublished data). We will also discuss studies using large saplings or small trees (about 5 m tall), because these also provide information for larger plants than in the vast majority of seedling studies. Given the lack of data for large trees relative to pollution, we will compare "small versus tall tree" responses to O_3 stress, to see whether information obtained for seedlings will be useful in understanding whole-forest responses. Ideas about scaling from leaf to canopy or from seedling to tree or stand will then be explored, as will the subject of incorporating O_3 effects into ecosystem or stand dynamics models (which has been rarely attempted to date). To the best of our knowledge, no one has yet attempted a whole-ecosystem or stand-level experiment to assess directly forest response to O_3 or any other gaseous pollutant, such as CO_2 or SO_2, although attempts are being made in Europe and North America to find sufficient funding for such large, expensive studies. Direct studies of mature tree response to air pollutants have been made in two different ways, either in comparative studies that assess aspects of tree physiology, growth, and mortality for trees growing under contrasting pollutant stress (or in declining versus healthy trees), or, less frequently, by use of experimental branch studies in mature trees.

F. Investigations of Mature Trees under Ambient O_3 Stress

The longest history of comparative field studies of tree responses to O_3 can be found in urbanized regions of southern California, where investigators began studying coniferous tree health under chronic O_3 pollution over 20 years ago (Miller, 1984). These early studies focused on visible O_3 injury to coniferous tree foliage. In addition, a number of workers compared damaged versus undamaged trees in attempts to understand the processes involved (Coyne and Bingham, 1982). More recently, there have been a number of investigations of conifers in the southern Sierra Nevada mountains in California under regional influences of O_3 (e.g., Peterson et al., 1987; Peterson and Arbaugh, 1988). These forests are relatively distant from urban areas, but receive moderately high concentrations of O_3 during the summer, due to intermediate-range transport. At monitoring sites in the mountains, maximum daily hourly averages exceeded 0.10 ppm on over 30 days in 1985 and 50 days in 1987 (Peterson et al., 1987, 1991), and nighttime values remained between 0.04 to 0.07 ppm (Miller et al., 1988). Conifers such as *Pinus jeffreyi* and *P. ponderosa* display high levels of visible O_3 injury at many sites in the Sierra Nevada. Dendrochronological analyses suggest that radial growth in *P. jeffreyi* has been reduced since the late 1960s, probably as a result of ambient O_3, whereas *P. ponderosa* in the same region shows no such decline, despite substantial visible injury (Peterson et al., 1987; Peterson and Arbaugh, 1988). Explanations for these differences are not known at present, but may lie in differences in physiology or in site types where the trees are found.

Other comparative studies of large or mature trees have been made for *Picea* and *Pinus* species in Europe and North America under O_3 and/or acid deposition stress (e.g., Schulze et al., 1989; Amundson et al., 1992; McLaughlin et al., 1990, 1991). A large number of studies of declining and healthy *P. abies* trees were made in Europe during the 1980s. The conclusions drawn from many of these studies were that O_3 was not a likely causal agent in decline [as summarized in Schulze et al. (1989)], although in some regions where O_3 levels were relatively high it may play a greater role than elsewhere (e.g., Benner and Wild, 1987). The impacts of long-term acidic deposition were studied in comparative field studies of declining versus healthy trees (young or mature) by many groups of researchers (e.g., Oleksyn, 1988; McLaughlin et al., 1991; Schulze et al., 1989; Amundson et al., 1992). Discussion of the numerous acid deposition studies is beyond the scope of this chapter and readers are referred to the above references for further information.

G. Studies with Young, Small Trees

The principal method used to expose seedlings and young, small trees (trees less than 20 years old and less than 7 m tall) experimentally in the field has employed the open-top chamber, sometimes enlarged and modified to accommodate taller plant material. Among the most interesting of such studies (relevant to our discussion of long-term larger tree responses) were those carried out for several years at realistically low concentrations. In Sweden (Wallin et al., 1990) and New York (R. G. Amundson et al., unpublished data) young *P. abies* and *P. rubens* trees, respectively, were experimentally exposed to O_3 during the growing season for three consecutive years. Both studies used large open-top chambers. In Sweden trees were 5–7 years old during the study, and in New York they were 12–14 years old. Probably as a result of their low gas exchange rates and hence O_3 uptake (Taylor et al., 1986; Reich, 1987), neither species showed a rapid response in terms of photosynthesis, growth, or respiration during the first season. In the Swedish study, however, when shoots were exposed over a second and third growing season, O_3 induced progressively greater decreases in photosynthesis. So-called 2-year-old foliage (actually three growing-seasons old by the end of the study) exposed over three growing seasons to mean concentrations of 56 and 91 $\mu g/m^3$ (as contrasted with charcoal-filtered treatments with concentrations of 9 $\mu g/m^3$) had proportional declines of 35 and 59%, respectively. Wallin et al. (1990) pointed out that these decreases were larger than those seen in wheat exposed to greater O_3 concentrations of 100 nl/liter for one season (Lehnherr et al., 1987), thus supporting the idea of Reich (1987) that extended leaf life-span of many conifers offsets low O_3 uptake rates, such that over the leaf life-span conifers may be as adversely affected as agricultural crops or as fast-growing trees such as Populus sp. Wallin et al. (1990) added that this gradual response clearly points out the need for long-term exposures (more than one season) to O_3 of conifers such as Picea sp. to evaluate sufficiently the impact of ambient O_3 regimes. In the 3-year study of P. rubens the trees were relatively resistant to O_3 throughout the study (R. G. Amundson, personal communication).

H. Studies with Large, Old Trees

No chambers currently exist for exposing large, old trees (e.g., more than 10 m tall) to air pollution stress. A number of techniques have been developed to fumigate branches of mature trees artificially with controlled elevated O_3 doses, or with artificially reduced O_3 concentrations using charcoal-filtered air. For instance, Lovett and Hubbell (1991) used branch chambers to examine whether short-term (5-hour) O_3 fumigation and/or acid mist influenced foliar leaching in the upper canopy of 20-m-tall P. strobus or A. saccharum, and found little impact of O_3. Research at such heights is rare because of the difficulty of accessing the canopies of large trees, and the study of Lovett and Hubbell was apparently the first to examine O_3 or acid mist impacts on leaching in a mature tree crown. Moreover, long-term experimentation is even more difficult. For instance, because they had no microenvironmental control and were relatively fragile, Lovett and Hubbell's chambers were useful for short-term but not long-term studies. Chronic O_3 exposure is more representative of ambient stress than is short-term acute fumigation, but long-term fumigation in a mature canopy is technically difficult and there are numerous problems that must be dealt with. These include chamber artifacts (if a chamber is used), the physical difficulties of working up in a canopy, and the difficulty of constructing a fumigation and monitoring system that is durable and yet not destructive to the trees.

Vann et al. (1992) enclosed branches of mature 100-year-old high-elevation *P. rubens* trees on Whiteface Mountain, New York, in flow-through chambers designed to either include or exclude ambient O_3. They compared branches exposed in the chambers to ambient O_3 and mist with lesser O_3 and less acidic mist treatments to test for aspects of winter injury sensitivity. Chamber effects were significant (temperatures inside chambers were greater than ambient by 1–1.5°C and foliage inside control chambers had greater chlorophyll than that outside the chambers), indicating the need for often expensive control features. Treatment differences among chambers did suggest a trend for heightened sensitivity to winter injury of foliage exposed to ambient pollution; perhaps most significantly these data were relatively consistent with other experimental studies of seedlings and comparative survey studies of mature trees (DeHayes, 1992). Again demonstrating the difficulty of separating subtle effects in multiple-stress scenarios, despite hundreds of studies in total, it is still not clear to what extent climatic factors (low subfreezing temperatures following mild winter periods, winter desiccation), genetic factors (features of spruce), and pollution factors (acid deposition, O_3) each contribute to observed winter injury in red spruce (Hadley et al., 1991; Perkins et al., 1991).

We know of four studies in which foliage of large and/or old trees was experimentally exposed to O_3 for at least several weeks (Skärby et al., 1987; Tjoelker et al., 1994; R. O. Teskey et al., unpublished data; and P. J. Hanson et al., unpublished data). One of the critical questions to be answered from mature tree studies is whether their foliage behaves comparably to seedling foliage (with respect to O_3 responses), and results to date from these four studies are mixed: they show lesser, similar, and greater sensitivity of tree foliage compared with seedling foliage, suggesting that the issue will not have an easy resolution. Because studies of large trees are rare but of critical importance, we will briefly discuss each study.

Skärby et al. (1987) attached two gas exchange chambers to *P. sylvestris* shoots for 1 month at a far northern site in Sweden. The trees were 20 years old and slightly more than 3 m in height. One chamber served as a control, the other to fumigate a branch with O_3 levels ranging from 120 to 400 $\mu g/m^3$ for 16 days during a 21-day period. Although unreplicated, this was the first extended experimental fumigation in a relatively mature individual tree that we are aware of, and thus, worthy of mention. Results were relatively consistent with the observed responses of tree seedlings over this short time period.

Teskey *et al.* (1991) reported the development of a system of branch fumigation chambers to assess responses of *P. taeda* to long-term O_3 exposure. They used the chambers in a 2-year field study and found their performance comparable with that of open-top chambers. Air temperatures, relative humidities, and CO_2 concentrations were nearly identical inside and outside the chambers, although light levels were roughly 20% lower inside the chambers due to the light transmission properties of the plastic covering. They found relatively small impacts of long-term O_3 fumigation on CO_2 exchange rates of *P. taeda* (R. O. Teskey, personal communication).

Tjoelker *et al.* (1994a) developed a chamberless system to expose *A. saccharum* branches in a 35-year-old stand to season-long O_3 exposures. Their system produced zones of elevated O_3 concentration above delivery loops sited beneath experimental branches. They found relatively similar proportional dose responses of leaves from 15 m high in the canopy as for seedlings in laboratory experiments (Tjoelker *et al.*, 1993; Reich, 1987). Moreover, they also reported that in both the forest canopy and seedling studies, leaves growing in shaded microenvironments were somewhat more sensitive to O_3 than leaves in higher light microenvironments. Such results suggest that results from seedling studies may be both qualitatively and quantitatively useful in assessing responses of large trees or forests, assuming of course that other interactions occurring at these different levels of scale can be accounted for.

Hanson and colleagues (P. J. Hanson, personal communication) recently (1992) built a set of very large open-top chambers to expose *Quercus rubra* trees to long-term O_3 stress. After the first year of the study they report greater sensitivity of mature tree foliage than observed in seedling studies (Hanson *et al.*, 1993; P. J. Hanson, personal communication). Full reports of the results of the above studies are forthcoming (e.g., Teskey *et al.*, 1994; Tjoelker *et al.*, 1994b).

VII. Models That Incorporate O_3 Effects

A. Single-Tree Models

1. TREGRO A relatively complete model of O_3 impacts on individual whole plant processes is TREGRO (Weinstein *et al.*, 1991). This is a simulation model that addresses carbon assimilation, allocation, and use within the context of other resource availabilities (water and nutrients). The model employs three linked simulators, one each for the flow of water, carbon, and nutrients through the plant. The three simulators are linked via their influences on one another. In the model, the tree is divided into several compartments: a canopy (further subdivided into age classes), branches, stem, and coarse and fine roots. The model calculates the photosynthesis of the entire tree on hourly time steps as a function of ambient environmental conditions and the availability of light, water, nutrients, daily redistribution of carbon throughout the plant, and the loss of carbon by respiration and senescence (Weinstein *et al.*, 1991). The model was developed and used to simulate the response of individual open-grown *P. rubens* (3–4 m tall) to long-term O_3 stress, with or without accompanying nutrient stress. A strong point of this research was the close coordination and interaction of the modeling and direct physiological assessment efforts (D. A. Weinstein and R. G. Amundson, personal communication), probably improving both research endeavors. The model is equipped with the ability to examine the consequences of a variety of hypothesized O_3 effects. As an example, several simulations were run in which it was assumed that O_3 exerts its influence through instantaneous damage to the mesophyll cells (and hence impaired maximum mesophyll conductance and carbon assimilation), with no cost of repair (Weinstein *et al.*, 1991). There are currently ongoing efforts to test the model independently and improve its capacities both with respect to O_3 and in general (D. A. Weinstein, personal communication).

2. MAESTRO At a higher level of organization, the group working at the University of Georgia with branch chambers in loblolly pine also used a forest model, MAESTRO (Jarvis *et al.*, 1991), to simulate responses of large tree canopies to O_3. This modeling effort was also closely linked with active experimentation (R. O. Teskey, personal communication). MAESTRO focuses on the canopy of either a single tree or a group of neighboring trees of the same species. MAESTRO employs a radiation regime model for tree canopies in conjunction with crown structure, photosynthesis, and transpiration submodels. In the model used to estimate O_3 impacts on *P. taeda*, photosynthetically active radiation and vapor pressure deficits (VPD) were assumed to be the driving variables for photosynthesis, and the effect of O_3 was modeled to be a reduced maximum mesophyll conductance (similar to TREGRO). A simulation examined how variation in light might influence O_3 response of *P. taeda* canopies, using hourly time steps. Negative impacts of O_3 on 24-hour simulated carbon exchange were proportionally greater on a sunny than a cloudy day. The O_3-induced reduction of annual net carbon fixation by well-watered tall (more than 15 m) *P. taeda* trees in rural Georgia was estimated to fall between 2.4 and 9.3% (range from conservative to liberal estimates) for current ambient air, with the range increasing to 3.8–15.4% if O_3 levels were 1.5 times greater.

3. Canopy Models A model to simulate O_3 impacts on photosynthetic carbon gain at the stand canopy level was described by Reich et al. (1990). A forest canopy was divided into four horizontal strata. Canopy layer CO_2 exchange rates were predicted at 5-minute intervals using leaf nitrogen concentration, specific leaf area, O_3 exposure, PAR, and VPD as driving variables. O_3 effects on maximum photosynthetic rate were simulated both for dose-dependent and concentration-dependent relationships, and effects on dark respiration were a function of current and recent O_3 concentrations. Canopy stratification of leaf area, chemistry, and specific leaf area were determined by direct observations. Canopy CO_2 exchange was simulated for a 30-m-tall, mixed oak–maple forest in Wisconsin under contrasting O_3 pollution regimes for 1 day, following 90 prior days of O_3 exposure.

Simulations were run to evaluate the potential differences between canopy layers in terms of contributions to whole-canopy O_3 effects, and to assess the possible impact of O_3 extinction vertically through the canopy. There were proportionally similar impacts of O_3 on net carbon balance of different canopy layers in these simulations, despite orders of magnitude differences in light, largely because lesser reductions of photosynthesis in shaded canopy layers were offset by the proportionally more influential impact of heightened respiration. However, because of the importance of the upper layers of the canopy to total net carbon balance (estimated 86% of total net carbon gain in the top two layers), in absolute terms the major simulated effect of O_3 was to reduce carbon gain from these layers. Recent direct observations of O_3 impacts on foliar structure and function in a mature maple canopy (Tjoelker et al., 1994a), coupled with characterization of vertical zonation in canopy microenvironment, structure, and chemistry and further model development (Ellsworth and Reich, 1993), will be used to modify the previous model (Reich et al., 1990) and develop further simulations.

4. Model Comparisons All of the above models (Reich et al., 1990; Weinstein et al., 1991; Jarvis et al., 1991) assumed that O_3 responses of foliage in different canopy positions would be equal (and that differences in impact could ensue due to interactions with microenvironmental factors, but not due to different O_3 responses per se). However, among the issues that must be addressed in all models of closed-canopy forest response to O_3 are whether shade and sun foliage are equally affected by O_3. Evidence from Oren and Schoeneberger (unpublished) with *P. taeda* and Tjoelker et al. (1993) and Volin et al. (1993) with

A. saccharum suggests that shade foliage may be more sensitive to O_3 than sun foliage. To the extent that sun versus shade foliage of any species differs in sensitivity to O_3, such differences need to be incorporated into future models. Reich and colleagues estimate that the greater sensitivity of shade foliage in a mature maple forest canopy (Tjoelker *et al.*, 1993, 1994a) could increase the impact of O_3 [as compared with our previous estimates, assuming all foliage had similar dose response characteristics (Reich *et al.*, 1990)] on canopy net carbon gain by as much as 25%.

B. Stand and Landscape Models

At even larger landscape and regional scales, ecological risk assessments that incorporate O_3 pollution are needed. Ecosystem level models are being modified to incorporate ozone effects, and used in conjunction with GIS and spatial patterns of air quality and vegetation to estimate forest productivity responses to ozone at regional scales, such as for the northeastern United States (Ollinger *et al.*, 1994). This approach is similar to the risk assessment model of Hogsett *et al.* (1993) described at the start of this chapter, except the former approach uses an ecosystem model rather than the coupled single-tree and stand-simulator model of the latter. Also, we earlier mentioned the ongoing work in Europe to make a larger scale (continental) risk assessment in a related, but physiologically cruder, fashion.

An example of a different type of landscape model is provided by Graham *et al.* (1991). In this model, two levels of O_3 pollution and five different landscape features are considered. In this regional model (>300,000 ha), the impact of elevated O_3 was assumed to increase the probability that a given patch of coniferous forest would experience an initial bark beetle attack. This impact of O_3 is clearly different from that addressed in all of the prior discussed models, and suggests that we should be aware that assessments at different temporal and spatial scales may require and/or have available very different measures of O_3 impact. The report by Graham *et al.* (1991) was not actually intended to assess possible impacts of O_3 (because levels were intentionally chosen that were atypical), but instead as a first attempt at relativizing potential sensitivities of different regional features under the constraints of available regional data, and to demonstrate the potentials, hazards, and needs for such large-scale assessments.

In this section we have outlined a number of different modeling approaches to O_3 impacts at whole-plant to regional scales. As we cross that gradient, it is obvious that our ability to include important features and factors into any model decreases, and that direct data on O_3 impacts on conifers or other vegetation also decrease, leaving us with a dilemma.

Given the lack of understanding of O_3 at higher levels or organizations and larger temporal and spatial scales, we must use models to provide us with an organized framework for addressing such issues, but the identical problem plagues modeling in general across scales. Thus, in attempting to address O_3 as an ecological factor for coniferous forests at stand, landscape, and regional scales, we must join forces with others working to develop these relatively new, large-scale models that will be critically important to ecology and policy in the future.

VIII. Conclusions

1. Species at Risk Ample evidence exists that many conifer species grow in settings where ambient air pollution levels can affect their metabolism, growth, and ecology. Some common species of conifers known to respond to ambient levels of air pollution include *Pinus taeda*, *Pinus ponderosa*, *Pinus sylvestris*, *Pinus strobus*, *Picea rubens*, and *Picea abies*. Also other conifer species may be susceptible to air pollution stress.

2. Air Pollutants of Concern Conifers are known to respond to SO_2, O_3, NO_x, and acid deposition. Of these pollutants, O_3 is likely the most widespread and phytotoxic compound, and therefore of great interest to individuals concerned with forest resources.

3. Direct Biological Responses On entering cells, air pollutants can have direct toxicological effects on metabolism which can then scale to effects on tree growth and forest ecology, including processes of competition and succession. Air pollutants are known to have effects on all aspects of plant metabolism, as well as to disrupt interactions between metabolic processes. Air pollution can cause reductions in photosynthesis and stomatal conductance, which are the physiological parameters most rigorously studied for conifers.

4. Complex Biological Responses Some effects air pollutants can have on plants are influenced by the presence of co-occurring environmental stresses. For example, drought usually reduces vulnerability of plants to air pollution. In addition, air pollution sensitivity may differ among species and with plant/leaf age. Plants may make short-term physiological adjustments to compensate for air pollution or may evolve resistance to air pollution through the processes of selection.

5. Modeling Models are necessary to understand how physiological processes, growth processes, and ecological processes are affected by air pollutants. The most valuable models are based on equations that describe biological functions, simulate mechanisms and processes, can be verified, and can be linked to other models that function on different spatial and temporal scales.

6. Ecological Risk Assessment The process of defining the ecological risk that air pollutants pose for coniferous forests requires approaches that exploit existing databases, environmental monitoring of air pollutants and forest resources, experiments with well-defined air pollution treatments and environmental control/monitoring, modeling, predicting air pollution-caused changes in productivity and ecological processes over time and space, and integration of social values. The tools for all these activities exist or are in the process of development. Scientists and resource managers must continue working together to assemble the pieces necessary to predict effects of air pollutants on forests.

Acknowledgments

The critical comments by Drs. M. S. Günthardt-Goerg, T. Keller, and W. Landolt on parts of the manuscript are highly appreciated.

References

Adams, M. B., and O'Neill, E. G. (1991). Effects of ozone and acid deposition on carbon allocation and mycorrhizal colonization of *Pinus taeda* L. seedlings. *For. Sci.* 37:5–16.

Adams, M. B., Ewards, N. T., Taylor, G. E., and Skaggs, B. L. (1990). Whole-plant carbon-14 photosynthate allocation in *Pinus taeda*, seasonal patterns at ambient and elevated ozone levels. *Can. J. For. Res.* 20:152–158.

Alscher, R. G., Cumming, J. R., and Fincher, J. (1989a). Air pollutant–low temperature interaction in trees. *In* "Biological Markers of Air Pollution Stress and Damage in Forests," pp. 341–345. National Academy Press, Washington, DC.

Alscher, R. G., Amundson, R. G., Cumming, J. R., Fellows, S., Fincher, J., Rubin, G., van Leuken, P., and Weinstein, L. H. (1989b). Seasonal changes in pigments, carbohydrates and growth of red spruce as affected by zone. *New Phytol.* 113:211–223.

Amundson, R. G., Raba, R. M., Schoettle, A. W., and Reich, P. B. (1986). Response of soybean to low concentrations of ozone: II. Effects on growth, biomass allocation and flowering. *J. Environ. Qual.* 15:161–167.

Amundson, R. G., Hadley, J. L., Fincher, J. F., Fellows, S., and Alscher, R. G. (1992). Comparisons of seasonal changes in photosynthetic capacity pigments and carabohydrates of healthy sapling and mature red spruce and of declining and healthy red spruce. *Can. J. For. Res.* 22:1605–1606.

Andersen, C. P., and Rygiewicz, P. T. (1991). Stress interactions and mycorrhizal plant response: Understanding carbon allocation priorities. *Environ. Pollut.* 73:217–224.

Andersen, C. P., Hogsett, W. E., Wessling, R., and Plocher, M. (1991). Ozone decreases spring root growth and root carbohydrate content in ponderosa pine the year following exposure. *Can. J. For. Res.* 21:1288–1291.

Ashenden, T. W., and Mansfield, T. A. (1977). Influence of wind speed on the sensitivity of ryegrass to sulfur dioxide. *J. Exp. Bot.* 28:729–735.

Barnes, J. D., Eamus, D., and Brown, K. A. (1990). The influence of ozone, acid mist and soil nutrient status on Norway spruce (*Picea abies* (L.) Karst): I. Plant-water relations. *New Phytol.* 114:713–720.

Benner, P., and Wild, A. (1987). Measurement of photosynthesis and transpiration in spruce trees with various degrees of damage. *J. Plant Physiol.* 129:59–72.

Bermadinger, E., Guttenberger, H., and Grill, D. (1990). Physiology of young Norway spruce. *Environ. Pollut.* 68:319–330.

Black, V. J., and Unsworth, M. H. (1980). Stomatal response to sulfur dioxide and vapor pressure deficit. *J. Exp. Bot.* 31:667–677.

Blaschke, H., and Weiss, M. (1990). Impact of ozone, acid mist and soil characteristics on growth and development of fine roots and ectomycorrhizae of young clonal Norway spruce. *Environ. Pollut.* 64:225–263.

Boyer, J. S. (1985). Water transport in plants. *Annu. Rev. Plant Physiol.* 36:473–516.

Brown, K. A., and Roberts, T. M. (1988). Effects of ozone on foliar leaching in Norway spruce (*Picea abies* L. Karst): Confounding factors due to NO_x production during ozone generation. *Environ. Pollut.* 55:55–73.

Cape, J. N., Fowler, D., Eamus, D., Murray, M. B., Sheppard, L. J., and Leith, I. D. (1989). Effects of acid mist and ozone on frost hardiness of Norway Spruce seedlings. *CEC Air Pollut. Res. Rep.* 26:331–334.

Cape, N. J., Leith, I. D., Fowler, D., Murray, M. B., Sheppard, L. J., Eamus, D., and Wilson, R. H. F. (1991). Sulphate and ammonium in mist impair the frost hardening of red spruce seedlings. *New Phytol.* 118:119–126.

Chapin, F. S., III (1991). Effects of multiple environmental stresses on nutrient availability and use. *In* "Response of Plants to Multiple Stresses" (H. A. Mooney, W. E. Winner, and E. J. Pell, eds.), pp. 67–88. Academic Press, San Diego.

Chevone, B. I., Young, Y. S., and Reddick, G. S. (1984). Acidic precipitation and ozone effects on growth of loblolly and shortleaf pine seedlings. *Phytopathology* 74:756.

Costonis, A. C., and Sinclair, W. A. (1969). Relationship of atmospheric ozone to needle blight of eastern white pine. *Phytopathology* 59:1566–1574.

Coyne, P. I., and Bingham, G. E. (1982). Variation in photosynthesis and stomatal conductance in an ozone-stressed ponderosa pine stand: Light response. *For. Sci.* 28:257–273.

Dann, M. S., and Pell, E. J. (1989). Decline of activity and quantity of ribulose bisphosphate carboxylase/oxygenase and net photosynthesis in ozone-treated potato foliage. *Plant Physiol.* 91:427–432.

Davis, D. D., and Wood, F. A. (1973). The influence of environmental factors on the sensitivity of virginia pine to ozone. *Phytopathology* 63:371–376.

DeHayes, D. H. (1992). Winter injury and developmental cold tolerance of red spruce. *Ecol. Stud. (U.S.A.)* 96:295–337.

Dickson, R. E., and Isebrands, J. G. (1991). Leaves as regulators of stress response. *In* "Response of Plants to Multiple Stresses" (H. A. Mooney, W. E. Winner, and E. J. Pell, eds.), pp. 4–34. Academic Press, San Diego.

Dobson, M. C., Taylor, G., and Freer-Smith, P. H. (1990). The control of ozone uptake by *Picea abies* (L.) Karst. and *P. sitchensis* (Bong.) Carr. during drought and interacting effects on shoot water relations. *New Phytol.* 116:465–474.

Eamus, D., and Murray, M. (1991). Photosynthetic and stomatal conductance responses of Norway spruce and beech to ozone, acid mist and frost—A conceptual model. *Environ. Pollut.* 72:23–44.

Edwards, G. S., Edwards, N. T., Kelly, J. M., and Mays, P. A. (1991). Ozone, acidic precipitation, and soil Mg effects on growth and nutrition of loblolly pine seedlings. *Environ. Exp. Bot.* 31:67–78.

Edwards, N. T. (1991). Root and soil respiration responses to ozone in *Pinus taeda* L. seedlings. *New Phytol.* 118:315–322.

Ellsworth, D. S., and Reich, P. B. (1993). Canopy structure and vertical patterns of photosynthesis and related leaf traits in a deciduous forest. *Oecologia* 96:169–178.

Farquhar, G. D., Ehleringer, J. R., and Hubick, K. T. (1989). Carbon isotope discrimination and photosynthesis. *Annu. Rev. Plant Physiol. Plant Mol. Biol.* 40:503–537.

Fincher, J., Cumming, J. B., Alscher, R. G., Rubin, G., and Weinstein, L. (1989). Long-term ozone exposure affects winter hardiness of red spruce (*Picea rubens* Sarg.) seedlings. *New Phytol.* 113:85–96.

Fink, S. (1989). Pathological anatomy of conifer needles subjected to gaseous air pollutants or mineral deficiencies. *Aquilo, Ser. Bot.* 27:1–6.

Freer-Smith, P. H., and Dobson, M. C. (1989). Ozone flux to *Picea sitchensis* (Bong.) Carr. and *Picea abies* (L.) Karst. during short episodes and the effects of these on transpiration and photosynthesis. *Environ. Pollut.* 59:161–176.

Freer-Smith, P. H., and Mansfield, T. A. (1987). The combined effects of low temperature and SO_2 + NO_2 pollution on the new season's growth and water relations of *Picea sitchensis*. *New Phytol.* 106:237–250.

Gower, S. T., and Richards, J. H. (1990). Larches: Deciduous conifers in an evergreen world. *BioScience* 40:818–826.

Graham, R. L., Hunsaker, C. T., O'Neill, R. V., and Jackson, B. L. (1991). Ecological risk assessment at the regional scale. *Ecol. Appl.* 1:196–206.

Greitner, C. S., and Winner, W. E. (1988). Increases in $\delta^{13}C$ values of radish and soybean plants caused by ozone. *New Phytol.* 108:489–494.

Greitner, C. S., and Winner, W. E. (1989). Nutrient effects on responses of willow and alder to ozone. In "Transaction: Effects of Air Pollution on Western Forests" (R. K. Olson and A. S. Lefohn, eds.), pp. 493–511. Air & Waste Management Association, Anaheim, CA.

Guderian, R., ed. (1985). "Air Pollution by Photochemical Oxidants: Formation, Transport, Control, and Effects on Plants." Springer-Verlag, New York.

Günthardt-Goerg, M. S., and Keller, T. (1987). Some effects of long-term ozone fumigation on Norway spruce. II. Epicuticular wax and stomata. *Trees* 1:145–150.

Günthardt-Goerg, M. S., Matyssek, R., Scheidegger, C., and Keller, T. (1993). Differentiation and structural decline in the leaves and bark of birch (*Betula pendula*) under low ozone concentration. *Trees* 7:104–114.

Hadley, J. L., Friedland, A. J., Herrick, G. T., and Amundson, R. G. (1991). Winter desiccation and solar radiation in relation to red spruce decline in the northern Appalachians. *Can. J. For. Res.* 21:269–272.

Hampp, R., Einig, W., and Egger, B. (1990). Energy and redox status, and carbon allocation in one- to three-year-old spruce needles. *Environ. Pollut.* 68:305–318.

Hanson, P. J., Samuelson, L. J., Wullschleger, S. D., and Edwards, G. S. (1993). Impacts of ozone on photosynthesis and conductance of tree versus seedling *Quercus rubra* L. foliage. *Plant Physiol., Suppl.* 102:161.

Harkov, A., and Brennan, E. I. (1980). The influence of soil fertility and water stress on ozone response of hybrid poplar trees. *Phytopathology* 70:991–994.

Hausladen, A., Madamanchi, N. R., Fellows, S., Alscher, R. G., and Amundson, R. G. (1990). Seasonal changes in antioxidants in red spruce as affected by ozone. *New Phytol.* 115:447–458.

Heath, R. L. (1980). Initial events in injury to plants by air pollutants. *Annu. Rev. Plant Physiol.* 31:395–401.

Hogsett, W. E., Plocher, M., Wildman, V., Tingey, D. T., and Bennett, J. P. (1985). Growth response of two varieties of slash pine seedlings to chronic ozone exposures. *Can. J. Bot.* 63:2369–2376.

Hogsett, W. E., Herstrom, A. A., Laurence, J. A., Lee, E. H., Weber, J. E., and Tingey, D. T. (1993). Risk characterization of tropospheric ozone to forests. In "Comparative

Risk Analysis and Priority Setting for Air Pollution Issues," Proc. 4th U.S./Dutch Inter. Sym. Air Waste Manage. Assoc., Pittsburgh (in press).

Horton, S. J., Reinert, R. A., and Heck, W. W. (1990). Effects of ozone on three open-pollinated families of *Pinus taeda* L. grown in two substrates. *Environ. Pollut.* 65: 279–292.

Huber, S. C. (1983). Relation between photosynthetic starch formation and dry-weight partitioning between the shoot and root. *Can. J. Bot.* 61:2709–2716.

Jarvis, P. G., Barton, C. V. M., Dougherty, P. M., Teskey, R. O., and Masshedev, J. M. (1991). MAESTRO. In "Acidic Deposition: State of Science and Technology" (P. M. Irving, ed.), Vol. 3, pp. 167–178. U.S. National Acid Precipitation Assessment Program, Washington DC.

Keller, T., and Häsler, R. (1987). The influence of a fall fumigation with ozone on the stomatal behavior of spruce and fir. *Oecologia* 64:284–286.

Keller, T., and Matyssek, R. (1990). Limited compensation of ozone stress by potassium in Norway spruce. *Environ. Pollut.* 67:1–14.

Kerstiens, G., and Lendzian, K. J. (1989). Interactions between ozone and plant cuticles. II. Water permeability. *New Phytol.* 112:21–27.

Körner, C. (1991). Some often overlooked plant characteristics as determinants of plant growth: A reconsideration. *Funct. Ecol.* 5:162–173.

Kresse, F. (1991). Untersuchungen zur Heterogenität der stomatären Öffnungsweiten bei Koniferen am Beispiel von Fichte (*Picea abies* L. Karsten) und Tanne (*Abies alba* Mill.). Diploma Thesis, University of Würzburg, Germany.

Laisk, A., Kull, O., and Moldau, H. (1989). Ozone concentration in leaf intercellular air spaces is close to zero. *Plant Physiol.* 90:1163–1167.

Landolt, W., Pfenninger, I., and Lüthy-Krause, B. (1989). The effect of ozone and season on the pool sizes of cyclitols in Scots pine (*Pinus sylvestris*). *Trees* 3:85–88.

Lauer, M. J., and Boyer, J. S. (1992). Internal CO_2 measured directly in leaves, abscisic acid and low leaf water potential cause opposing effects. *Plant Physiol.* 98:1310–1316.

Lee, W. S., Chevone, B. I., and Seiler, J. R. (1990). Growth and gas exchange of loblolly pine seedlings as influenced by drought and air pollutants. *Water, Air, Soil Pollut.* 51: 105–116.

Lehnherr, B., Grandjean, A., Mächler, F., and Fuhrer, J. (1987). The effect of ozone in ambient air on ribulose bisphosphate carboxylase/oxygenase activity decreases photosynthesis and grain yield in wheat. *J. Plant Physiol.* 130:189–200.

Lovett, G. M., and Hubbell, J. G. (1991). Effects of ozone and acid mist on foliar leaching from eastern white pine and sugar maple. *Can. J. For. Res.* 21:794–802.

Luethy-Krause, B., and Landolt, W. (1990). Effects of ozone on starch accumulation in Norway spruce (*Picea abies*). *Trees* 4:107–110.

Luethy-Krause, B., Pfenninger, I., and Landolt, W. (1990). Effects of ozone on organic acids in needles of Norway spruce and Scots pine. *Trees* 4:198–204.

Luxmoore, R. J., Oren, R., Sheriff, D. W., and Thomas, R. B. (1994). Source–sink–storage relationships of conifers. In "Resource Physiology of Conifers: Acquisition, Allocation, and Utilization" (W. K. Smith and T. M. Hinckley, eds.), pp. 179–216. Academic Press, San Diego.

Macrez, U., and Hubac, C. (1988). Exposure to SO_2 and water stress in *Picea abies* (L.) Karsten. *Water, Air, Soil Pollut.* 40:251–259.

Mahoney, M. J., Chevone, B. I., Skelly, J. M., and Moore, L. D. (1985). Influence of mycorrhizae on the growth of loblolly pine *Pinus taeda* seedlings exposed to ozone and sulphur dioxide. *Phytopathology* 75:679–682.

Mansfield, T. A., and Majernick, O. (1970). Can stomata play a part in protecting plants against air pollutants? *Environ. Pollut.* 1:149–154.

Martin, B., Bytnerowicz, A., and Thorstenson, Y. R. (1988). Effects of air pollutants on the composition of stable carbon isotopes, $\delta^{13}C$, of leaves and wood, and on leaf injury. *Plant Physiol.* 88:218–223.

Matyssek, R. (1985). Der Kohlenstoff-, Wasser- und Nährstoffhaushalt der wechselgrünen Koniferen Lärche, Fichte, Kiefer. Doctoral Thesis, University of Bayreuth.

Matyssek, R. (1986). Carbon, water and nitrogen relations in evergreen and deciduous conifers. *Tree Physiol.* 2:177–187.

Matyssek, R., and Schulze, E.-D. (1987). Heterosis in hybrid larch (*Larix decidua* × *leptolepis*). II. Growth characteristics. *Trees* 1:225–231.

Matyssek, R., Günthardt-Goerg, M. S., Keller, T., and Scheidegger, C. (1991). Impairment of the gas exchange and structure in birch leaves (*Betula pendula*) under low ozone concentrations. *Trees* 5:5–13.

Matyssek, R., Günthardt-Goerg, M. S., Saurer, M., and Keller, T. (1992). Seasonal growth, $\delta^{13}C$ of leaves and stem, and phloem structure in birch (*Betula pendula*) under low ozone concentrations. *Trees* 6:69–76.

Matyssek, R., Keller, T., and Koike, T. (1993a). Branch growth and leaf gas exchange of *Populus tremula* exposed to low ozone concentrations throughout two growing seasons. *Environ. Pollut.* 79:1–7.

Matyssek, R., Günthardt-Goerg, M. S., Landolt, W., and Keller, T. (1993b). Whole-plant growth and leaf formation in ozonated hybrid poplar (*Populus* × *euramericana*). *Environ. Pollut.* 81:207–212.

McLaughlin, S. B., McConathy, R. K., Duvick, D., and Mann, L. K. (1982). Effects of chronic air pollution stress on photosynthesis, carbon allocation and growth of white pine trees. *For. Sci.* 28:60–70.

McLaughlin, S. B., Anderson, C. P., Edwards, N. T., Roy, W. K., and Layton, P. A. (1990). Seasonal patterns of photosynthesis and respiration of red spruce saplings from two elevations in declining southern Appalachian stands. *Can. J. For. Res.* 20:485–495.

McLaughlin, S. B., Anderson, C. P., Hanson, P. J., Tjoelker, M. G., and Roy, W. K. (1991). Increased dark respiration and calcium deficiency of red spruce in relation to acidic deposition at high-elevation southern Appalachian Mountain sites. *Can. J. For. Res.* 21:1234–1244.

Meier, S., Grand, L. F., Schoeneberger, M. M., Reinert, R. A., and Bruck, R. I. (1990). Growth, ectomycorrhizae and nonstructural carbohydrates of loblolly pine seedlings exposed to ozone and soil water deficit. *Environ. Pollut.* 64:11–27.

Melhorn, H., Seufert, G., Schmidt, A., and Kunert, K. J. (1986). Effect of SO_2 and O_3 on production of antioxidants in conifers. *Plant Physiol.* 82:336–338.

Michin, P. E. H., and Gould, R. (1986). Effect of SO_2 on phloem loading. *Plant Sci.* 43:179–183.

Miller, P. R. (1973). Oxidant-induced community change in a mixed conifer forest. *Adv. Chem. Ser.* 122:101–117.

Miller, P. R. (1984). Ozone effects in the San Bernardino National Forest. *In* "Symposium on Air Pollution and the Productivity of the Forest" (D. D. Davis, A. A. Millen, and L. Dochinger, eds.), pp. 161–197. Izaak Walton League of America, Arlington, VA.

Miller, P. R., Wilborn, R. D., Schilling, S. L., and Gormez, A. P. (1988). "Ozone Injury to Important Tree Species of Sequoia and Kings Canyon National Parks," Final Report. National Park Service Air Quality Division, Denver, CO.

Mooney, H. A., and Winner, W. E. (1988). Carbon gain, allocation, and growth as affected by atmospheric pollutants. *In* "Air Pollution and Plant Metabolism" (S. Schulte-Hostede, N. M. Darral, L. W. Blank, and A. R. Wellburn, eds.), pp. 272–287. Elsevier, London and New York.

Mooney, H. A., and Winner, W. E. (1991). Partitioning response of plants to stress. *In*

"Response of Plants to Multiple Stresses" (H. A. Mooney, W. E. Winner, and E. J. Pell, eds.), pp. 129–141. Academic Press, San Diego.

Mooney, H. A., Küppers, M., Koch, G., Gorham, J., Chu, C., and Winner, W. E. (1988). Compensating effects to growth of carbon partitioning changes in response to SO_2-induced photosynthetic reduction in radish. *Oecologia* 75:502–506.

Mortensen, L. M. (1990). Effects of ozone on growth and dry matter partitioning in different provenances of Norway spruce (*Picea abies* L. Karst.). *Norw. J. Agric. Sci.* 4:61–66.

Norby, R. J., and Kozlowski, T. T. (1981). Relative sensitivity of three species of woody plants to sulfur dioxide at high or low exposure temperature. *Oecologia* 51:33–36.

Norby, R. J., and Kozlowski, T. T. (1982). The role of stomata in sensitivity of *Betula papyrifera* seedlings to SO_2 at different humidities. *Oecologia* 53:34–39.

Norby, R. J., and Kozlowski, T. T. (1983). Flooding and sulfur dioxide stress interaction in *Betula papyrifera* and *B. nigra* seedlings. *For. Sci.* 29:739–750.

Oleksyn, J. (1988). Height growth of different European Scots pine *Pinus sylvestris* L. provenances in a heavily polluted and a control environment. *Environ. Pollut.* 55:289–299.

Ollinger, Reich, P. B., and Aber, (1994). In preparation.

Öquist, G. (1986). Effects of winter stress on chlorophyll organization and function in Scots pine. *J. Plant Physiol.* 122:169–179.

Oren, R., and Schulze, E.-D. (1989). Nutritional disharmony and forest decline: A conceptual model. *Ecol. Stud.* 77:425–443.

Payer, H. D., Pfirrmann, T., and Kloos, M. (1990). Clone and soil effects on the growth of young Norway spruce during 14 months exposure to ozone plus acid mist. *Environ. Pollut.* 64:209–227.

Pell, E. J., and Dann, M. S. (1991). Multiple stress-induced foliar senescence and implications for whole-plant longevity. *In* "Response of Plants to Multiple Stresses" (H. A. Mooney, W. E. Winner, and E. J. Pell, eds.), pp. 189–204. Academic Press, San Diego.

Perkins, T. D., Adams, G. T., and Klein, R. M. (1991). Desiccation or freezing? Mechanisms of winter injury to red spruce foliage. *Am. J. Bot.* 28:1207–1217.

Peterson, D. L., and Arbaugh, W. J. (1988). An evaluation of the effects of ozone injury on radial growth of ponderosa pine (*Pinus ponderosa*) in the southern Sierra Nevada (California USA). *J. Air Pollut. Control Assoc.* 38:921–927.

Peterson, D. L., Arbaugh, M. J., Wakefield, V. A., and Miller, P. R. (1987). Evidence of growth reduction in ozone-stressed Jeffrey pine (*Pinus jeffreyi* Grev. and Balf.) in Sequoia and Kings Canyon National Parks. *J. Air Pollut. Control Assoc.* 37:906–912.

Peterson, D. L., Arbaugh, M. J., and Robinson, L. J. (1991). Regional growth changes in ozone-stressed ponderosa pine (*Pinus ponderosa*) in the Sierra Nevada, California, USA. *Holocene* 1:50–61.

Pierre, M., and Queiroz, O. (1988). Air pollution by SO_2 amplifies the effects of water stress on enzymes and total protein content of spruce needles. *Physiol. Plant.* 73:412–417.

Pieters, G. A. (1974). The growth of sun and shade leaves of *Populus euramericana robusta* in relation to age, light intensity and temperature. *Meded. Landbouwhogesch. Wageningen*:74–111.

Pye, J. M. (1988). Impact of ozone on the growth and yield of trees: A review. *J. Environ. Qual.* 17:347–360.

Reich, P. B. (1983). Effects of low concentrations of O_3 on net photosynthesis, dark respiration, and chlorophyll contents in aging hybrid poplar leaves. *Plant Physiol.* 73:291–296.

Reich, P. B. (1987). Quantifying plant response to ozone: A unifying theory. *Tree Physiol.* 3:63–91.

Reich, P. B., and Amundson, R. G. (1985). Ambient levels of O_3 reduce net photosynthesis in tree and crop species. *Science* 230:566–570.

Reich, P. B., and Lassoie, J. P. (1984). Effect of low level O_3 exposure on leaf diffusive conductance and water-use efficiency in hybrid poplar. *Plant, Cell Environ.* 7:661–668.

Reich, P. B., Schoettle, A. W., Stroo, H. F., Troiano, J., and Amundson, R. G. (1987). Influence of O_3 and acid rain on white pine seedlings grown in five soils. I. Net photosynthesis and growth. *Can. J. Bot.* 65:977–987.

Reich, P. B., Schoettle, A. W., Stroo, H. F., and Amundson, R. G. (1988). Influence of O_3 and acid rain on white pine seedlings grown in five soils. III. Nutrient relations. *Can. J. Bot.* 66:1517–1531.

Reich, P. B., Ellsworth, D. S., Kloeppel, B. D., Fownes, J. H., and Gower, S. T. (1990). Vertical variation in canopy structure and CO_2 exchange of oak–maple forests: Influence of ozone, nitrogen, and other factors on simulated canopy carbon gain. *Tree Physiol.* 7:329–345.

Reich, P. B., Walters, M. B., and Ellsworth, D. S. (1992). Leaf lifespan in relation to leaf, plant and stand characteristics among diverse ecosystems. *Ecol. Monogr.* 62:365–392.

Reich, P. B., Walters, M. B., Krause, S. C., Vanderklein, D., and Raffa, K. F. (1993). Growth, nutrition and gas exchange of *Pinus resinosa* following artificial defoliation. *Trees* 7:67–77.

Rich, S., and Turner, N. C. (1972). Importance of moisture on stomatal behavior of plants subjected to ozone. *J. Air Pollut. Control Assoc.* 22:718–721.

Rich, S., Waggoner, P. E., and Tomlinson, H. (1970). Ozone uptake by bean leaves. *Science* 169:79–80.

Sasek, T. W., and Richardson, C. J. (1989). Effects of chronic doses of ozone on loblolly pine: Photosynthetic characteristics in the third growing season. *For. Sci.* 35:745–755.

Scherzer, A. J., and McClenahen, J. R. (1989). Effects of ozone or sulphur dioxide on pitch pine seedlings. *J. Environ. Qual.* 18:57–61.

Schier, G. A., McQuattie, C. J., and Jensen, K. F. (1990). Effects of ozone and aluminum on pitch pine (*Pinus rigida*) seedlings growth and nutrient relations. *Can. J. For. Res.* 20:1714–1719.

Schmitt, U., and Ruetze, M. (1990). Structural changes in spruce and fir needles. *Environ. Pollut.* 68:345–354.

Schulze, E.-D. (1982). Plant life forms and their carbon, water and nutrient relations. *Encycl. Plant Physiol., New Ser.* 12B:615–676.

Schulze, E.-D., Lange, O. L., and Oren, R., eds. (1989). "Forest Decline and Air Pollution: A Study of Spruce (*Picea abies*) on Acid Soils." Springer-Verlag, New York.

Schupp, R., and Rennenberg, H. (1988). Diurnal changes in the glutathione content of spruce needles (*Picea abies* (L.)). *Plant Sci.* 57:113–117.

Schweizer, B., and Arndt, U. (1990). CO_2/H_2O gas exchange parameters of one- and two-year-old needles of spruce and fir. *Environ. Pollut.* 68:275–292.

Shanklin, J., and Kozlowski, T. T. (1985). Effect of flooding of soil on growth and subsequent responses of *Taxodium distichum* seedlings to SO_2. *Environ. Pollut.* 38:199–212.

Sieffert, A., and Queiroz, O. (1989). Synergistic interaction of drought and SO_2 pollution on the protein pattern of *Picea abies* needles. *Plant Physiol. Biochem.* 27:269–274.

Skärby, L., Troeng, E., and Boström, C.-Ä. (1987). Ozone uptake and effects on transpiration, net photosynthesis and dark respiration in Scots pine. *For. Sci.* 33:801–808.

Smith, W. H. (1990). "Air Pollution and Forests: Interaction Between Air Contaminants and Forest Ecosystems." Springer-Verlag, New York.

Spence, R. D., Rykiel, E. J., Sharpe, P. J. H. (1990). Ozone alters carbon allocation in loblolly pine assessment with carbon-11 labeling. *Environ. Pollut.* 64:93–106.

Stitt, M. (1987). Fructose 2,6-bisphosphate and plant carbohydrate metabolism. *Plant Physiol.* 84:201–204.
Stolzy, L. H., Taylor, O. C., Letey, J., and Szuszkiewicz, T. E. (1961). Influence of oxygen diffusion rates on susceptibility of tomato plants to airborne oxidants. *Soil Sci.* 91: 151–155.
Stroo, H. F., Reich, P. B., Schoettle, A. W., and Amundson, R. G. (1988). Influence of O_3 and acid rain on white pine seedlings grown in five soils. II. Mycorrhizae. *Can. J. Bot.* 66:1510–1515.
Suter, G. W. (1990). Endpoints for regional ecological risk assessments. *Environ. Manage.* 14:923–935.
Taylor, G. E., and Dobson, M. C. (1989). Photosynthetic characteristics, stomatal response and water relations of *Fagus sylvatica*: Impact of air quality at a site in southern Britain. *New Phytol.* 113:265–273.
Taylor, G. E., Norby, R. J., McLaughlin, S. E., Johnson, A. H., and Turner, R. S. (1986). CO_2 assimilation and growth of red spruce seedlings in response to ozone, precipitation chemistry and soil type. *Oecologia* 70:163–171.
Taylor, G. E., Dobson, M. C., and Freer-Smith, P. H. (1989). Changes of partitioning and increased root lengths of spruce and beech exposed to ambient pollution concentrations in southern England. *Ann. Sci. For.* 46:573–576.
Taylor, G. E., Pitelka, L. F., and Clegg, M. T., eds. (1991). "Ecological Genetics and Air Pollution." Springer-Verlag, New York.
Terashima, I., Wong, S.-C., Osmond, C. B., and Farquhar, G. D. (1988). Characterisation of non-uniform photosynthesis induced by abscisic acid in leaves having different mesophyll anatomies. *Plant Cell Physiol.* 29:385–394.
Teskey, R. O., Dougherty, P. M., and Wiselogel, A. E. (1991). Design and performance of branch chambers suitable for long-term ozone fumigation of foliage in large trees. *J. Environ. Qual.* 20:591–595.
Teskey, R. O. *et al.* (1994). In preparation.
Thomas, H. T., and Stoddart, J. T. (1980). Leaf senescence. *Annu. Rev. Plant Physiol.* 31:83–111.
Tingey, D. T., and Hogsett, W. E. (1985). Water stress reduces ozone injury via stomatal mechanism. *Plant Physiol.* 77:944–947.
Tjoelker, M. G., Volin, J. C., Oleksyn, J., and Reich, P. B. (1993). Light environment alters response to ozone stress in seedlings of *Acer saccharum* Marsh. and hybrid *Populus* L. I. *In situ* net CO_2 exchange and growth. *New Phytol.* 124:627–636.
Tjoelker, M. G., Volin, J. C., Oleksyn, J., and Reich, P. B. (1994a). An open-air system for exposing a forest canopy to ozone pollution. *Plant, Cell Environ.* (in press).
Tjoelker, M. G. *et al.* (1994b). In preparation.
Urbach, W., Schmidt, W., Kolbowski, J., Rümmele, S., Reisberg, E., Steigner, W., and Schreiber, U. (1989). Wirkungen von Umweltschadstoffen auf Photosynthese und Zellmembranen von Pflanzen. *In* "Statusseminar der PBWU zum Forschungsschwerpunkt Waldschäden" (M. Reuther and M. Kirchner, eds.), Vol. 1, pp. 195–206. GSF München.
Vann, D. R., Strimback, G. R., and Johnson, A. H. (1992). Effects of air borne chemicals on freezing resistance of red spruce foliage. *For. Ecol. Manage.* 51:69–80.
Volin, J. C., Tjoelker, M. G., Oleksyn, J., and Reich, P. B. (1993). Light environment alters response to ozone stress in seedlings of *Acer saccharum* Marsh. and hybrid *Populus* L. II. Diagnostic gas exchange and leaf chemistry. *New Phytol.* 124:637–646.
Wallin, G., Skärby, L., and Selldén, G. (1990). Long-term exposure of Norway spruce, *Picea abies* (L.) Karst., to ozone in open-top chambers. I. Effects of the capacity of net photosynthesis, dark respiration and leaf conductance of shoots of different ages. *New Phytol.* 115:335–344.

Walmsley, L., Ashmore, M. R., and Bell, J. N. B. (1980). Adaptation of radish *Raphanus sativus* L. in response to continuous exposure to ozone. *Environ. Pollut.* 23:165–177.

Walters, M. B., Kruger, E. L., and Reich, P. B. (1993). Relative growth rate in relation to physiological and morphological traits for northern hardwood tree seedlings: Species, light environment and ontogenetic considerations. *Oecologia* 96:219–231.

Warrington, S., and Whittaker, J. B. (1990). Interaction between Sitka spruce, the green spruce aphid, sulphur dioxide pollution and drought. *Environ. Pollut.* 65:363–370.

Weinstein, D. A., Beloin, R. M., and Yanai, R. D. (1991). Modeling changes in red spruce carbon balance and allocation in response to interacting ozone and nutrient stresses. *Tree Physiol.* 9:127–146.

Wellburn, A. R. (1990). Why are atmospheric oxides of nitrogen usually phytotoxic and not alternative fertilizers? *New Phytol.* 115:395–429.

Winner, W. E., and Greitner, C. (1989). Field methods used for air pollution research with plants. *In* "Plant Physiological Ecology: Field Methods and Instrumentation" (R. W. Pearcy, J. Ehleringer, H. A. Mooney, and P. W. Rundel, eds.), pp. 399–425. Chapman & Hall, New York.

Winner, W. E., and Mooney, H. A. (1980). Ecology of SO_2 resistance: II. Photosynthetic changes of shrubs in relation to SO_2 absorption and stomatal behavior. *Oecologia* 44:296–302.

Winner, W. E., Mooney, H. A., and Goldstein, R. A., eds. (1985). "Sulfur Dioxide and Vegetation: Physiology, Ecology, and Policy Issues." Stanford Univ. Press, Stanford, CA.

Wolfenden, J., and Mansfield, T. A. (1991). Physiological disturbances in plants caused by air pollutants. *Proc. R. Soc. Edinburgh, Sect. B: Biol. Sci.* 97:117–138.

10

Potential Effects of Global Climate Change

Hermann Gucinski, Eric Vance, and William A. Reiners

I. Introduction

The detection of changes in global climate caused by worldwide human activities that result in the release of radiatively important trace gases remains elusive to date. Released gases include carbon dioxide (CO_2) from fossil fuel burning, chlorinated fluorocarbons (CFCs) from refrigeration machinery and other industrial uses, methane (CH_4) from indirect effects of increasing cattle production and other land-use alterations, and nitrous oxide (N_2O) and other oxides of nitrogen, sometimes collectively referred to as NO_x, from industrial and automotive sources. The difficulties of detecting climatic changes do not diminish the need to examine the consequences of a changing global radiative energy balance. In part, detecting global changes is difficult (even though many, though by no means all, theoretical climatic processes are well understood) because the potential effects of changes on the unmanaged ecosystems of the globe, especially forests, which may have great human significance, involve tightly woven ecosystems, inextricably linked to global habitat. Coniferous forests are of particular interest because they dominate high-latitude forest systems, and potential effects of global climate change are likely to be greatest at high latitudes [Intergovernmental Panel on Climate Change, (IPCC), 1990].

The degree of projected climate change is a function of many likely scenarios of fossil fuel consumption, and the ratios of manmade effects to natural sources and sinks of CO_2. Atmospheric CO_2 concentrations, taken from a pole-to-pole record, show that CO_2 levels rose from a globally averaged level of 315 ppmv (parts per million by volume) in 1958 to 354 ppmv by 1988 (Keeling *et al.*, 1989). Because CO_2, like water vapor, CH_4, CFCs, and other gases, absorbs infrared energy, it will alter the radiation balance of the global atmosphere (IPCC, 1990). The consequences of this alteration to the radiation balance cannot simply be translated into changing climate because (1) the existence of large energy reservoirs (the oceans) can introduce a lag in responses, (2) feedback loops between atmosphere, oceans, and biosphere can change the net rate of buildup of greenhouse gases in the atmosphere, (3) complex interactions in the atmospheric water balance can change the rate of cloud formation with their persistence, in turn, changing the global albedo and the energy balance, and (4) there is intrusion of other global effects, such as periodic volcanic gas injections to the stratosphere.

Intensive efforts are ongoing to understand, quantify, and model these interactions. General circulation models (GCMs) use conservation of momentum and energy principles in conjunction with mass data to determine energy balance and the energy distribution within the global atmosphere and oceans (GFDL, Manabe and Wetherald, 1987; NCAR, Dickinson *et al.*, 1993; OSU, Schlesinger and Zhao, 1989; GISS, Hansen *et al.*, 1983; UKMO, Wilson and Mitchell, 1987). Work is ongoing that will address the effects of the terrestrial biosphere on energy balance. At the time of writing, GCMs can solve the energy balance equations only on coarse scales, in terms of large "boxes" into which the globe has been divided, presently several hundred kilometers on edge in the horizontal dimension. Moreover, as mentioned, GCMs may fail to include all the processes in the energy balance because of limitations in either fundamental understanding or in the data required to represent such processes adequately, such as cloud feedbacks (Ramanathan and Collins, 1991; Heymsfield and Miloshevich, 1991; Cess, 1989; reply by Mitchell, 1991). Ongoing efforts to include terrestrial biospheric effects suffer from input limitations due to lack of data (i.e., measurements) regarding the global distribution of vegetation and its capacity to take up CO_2 from the atmosphere (Post *et al.*, 1990). They may also suffer from lack of process-level understanding of the role of feedbacks that may either stabilize the atmosphere against changes resulting from increased greenhouse concentrations or speed their manifestations, including the direct effects of increased atmospheric CO_2 on plant productivity. GCMs may also behave with deterministic chaos (Tsonis, 1991), wherein a slight change in initial conditions used to commence a GCM run can lead to considerably different outcomes. This could add uncertainty in specifying regional conditions.

Increases in global mean surface temperature, changes in temperature extremes, and alterations in precipitation, potential evapotranspiration, cloudiness, and winds, both in their mean as well as in their variability, have been inferred from GCM outputs. It should be noted that an increase in mean global temperature of $3°C$ will greatly exceed interglacial high temperatures seen in recent geologic history (IPCC, 1990).

II. Effects on and Responses of Conifers and Coniferous Forests

The effects on, and responses of, coniferous forests to climate change may be expressed by processes that act at the tissue, organism, stand, and ecosystem levels. Because the carbon storage and exchange potential of coniferous forest systems is large, responses of these systems can provide feedback to biospheric processes (Neilson and King, 1992). To understand and predict resulting effects requires linking ecophysiological principles across multiple scales. Hence, global change poses a challenge to ecophysiologists and ecologists alike to provide the necessary insight to meet such demanding objectives.

The response of coniferous forests to potential global climate change is likely to occur as combinations of direct effects and secondary effects, which are complex and difficult to predict. The rise in atmospheric CO_2 concentration is more certain than other global change factors and deserves special attention. However, many of the pathways discussed for CO_2 effects are also directly relevant to those associated with changes in air temperature, precipitation, or other environmental conditions.

In this chapter, we intend to examine selected primary, or first-order, effects of global climate change as well as some of the secondary effects and interactions. Direct effects include the effect of increased atmospheric CO_2 and changes in ambient solar visible and ultraviolet-B radiation, and effects of changes in climate variables (e.g., temperature, precipitation, humidity, and winds, both in magnitude and variability and in seasonal distribution) that lead to physiological and phenological responses by trees. These variables can alter forest productivity by changes in key physiological and developmental processes. For example, forest productivity may be modified by changes in carbon acquisition and allocation as well as water and resource use efficiency (Zasada *et al.*, 1992). Altered pollen–ovule phenological compatibility in some conifers will affect both the ability to respond and the rate of response to changing environmental conditions. Climatic extremes affecting the phenology of the tree (e.g., late and early frosts), cumulative climatic events (e.g., prolonged droughts, extended warm or cold periods), and linked climatic extremes (e.g., a 50-year drought followed by the second coldest winter on record) will place multiple stresses on trees and result in reduced productivity and increased mortality (Hinckley *et al.*, 1979; Cannell *et al.*, 1989; Haenninen, 1991). Milder winters would reduce the number of chilling hours and thus impact the release of vegetative (and reproductive) buds from dormancy in species such as *Pseudotsuga menziesii* (Leverenz and Lev, 1987). Developmental processes, particularly those associated with cone and flower development, fertilization, seed maturation, ripening, and germination, appear particularly sensitive to changes in cumulative heat exposure, moisture stress, and extremes in temperature (Zasada *et al.*, 1992). As yet, knowledge is insufficiently advanced to allow determination of the ecosystem-level responses of these processes.

Secondary effects include plant–plant, plant–animal, and plant–microbe interactions (Strain, 1985), as well as the potential feedback effects between changing coniferous ecosystems and changes to the atmosphere and climate system. It is difficult to even posit an exhaustive list of secondary interactions. Other events, such as feedbacks of nutrient availability on plant physiological processes, the role of reproductive success, seed dispersal, and corridors for conifer migration should be viewed in terms of the scale at which the response becomes important. These include changes in competition, herbivory, disease resistance, and fire regimes as a result of climate-induced stresses, including those related to changing UV-B doses (Stenberg *et al.*, 1994). Moreover, the range of secondary effects is highly scale dependent, with many effects being expressed at the tissue or even subtissue level, and others, such as changes in disturbance regime (i.e., wildfire, insect pest, and disease outbreaks) occurring at the stand or landscape level.

III. Direct Effects of Elevated CO_2

Potential effects at the organ and tissue levels include CO_2-induced changes in photosynthesis, productivity, water use efficiency, resource use efficiency, carbon allocation, and resistance to stress (i.e., water/nutrient/disease stress). Effects at the whole-tree level and higher levels may be factors in the potential redistribution of coniferous forests, with accompanying mortality in incompatible life zones, and in the invasion of conifers into currently nonconifer landscapes. The extent of total response will be constrained by seasonality and light availability at high latitudes, and competitive interaction elsewhere. These processes may mediate the unique abilities of conifers with respect to N translocation, organism–soil interactions in relation to changes in mineralization, carbon–nitrogen ratios in leaf, litter, and organic matter decomposition, and linkages of C (carbon) allocation and litter quality (see Chapter 4, this volume).

Elevated CO_2 appears to act as a "fertilizer" and thus enhances photosynthesis and productivity. A large number of growth chamber studies have demonstrated such effects (see Strain and Cure, 1985; Drake, 1992). However, other tissue level effects may occur concomitantly and may have significant consequences for ecological interactions. Alteration in plant tissue chemistry resulting from elevated CO_2 is a primary mechanism through which ecological interactions are effected. Elevated CO_2 can result in carbohydrate accumulation leading to higher tissue C/N ratios of plants (O'Neill *et al.*, 1987; Lindroth *et al.*, 1993; Lincoln, 1993; Grodzinski, 1992). The reduction in plant tissue N concentration appears to result from the CO_2-induced increase in plant growth, rather than from changes in N use efficiency (Coleman *et al.*, 1993). There is evidence that CO_2-induced nutrient limitation in plants can also alter concentrations of secondary metabolites (e.g., tannins and other phenolic compounds) in plant tissues (Lambers, 1993; Lindroth *et al.*, 1993). Concentrations of secondary compounds in tissues of both coniferous and deciduous trees have also been found to be inversely related to the availability of nutrients and other resources (Coley *et al.*, 1985; Bryant, 1981; Tiarks *et al.*, 1989). Consequently, any global change factor that alters resources or affects growth of trees has the potential to alter the secondary chemistry of their tissues. Tropospheric ozone concentrations and UV-B radiation have also been found to affect plant tissue chemistry (Lechowicz, 1987; Caldwell *et al.*, 1989).

The integrative response of forests to changing plant tissue chemistry is uncertain, due to multiplicity of effects and the suspected and demonstrated presence of multiple feedback mechanisms. Tissue chemistry affects resistance of living and dead plant tissues to microbial decomposition and herbivory by mammals and insects (Bryant, 1981; Choudbury, 1988; Lawrey, 1989). Tannins, for example, reduce plant palatability to mammalian and insectivorous herbivores by inactivating enzymes and precipitating proteins (Tiarks *et al.*, 1989). Lower N and water concentrations in leaves exposed to elevated CO_2 have also been found to have negative effects on herbivorous larvae (Fajer *et al.*, 1989). Herbivory may also be affected by changes in insect populations that rapidly adapt or migrate in response to changing temperature, precipitation, or growing season length (Hedden, 1989; Elias, 1991). For example, local environmental and biotic factors are believed to affect the transition between a stable, endemic phase and a transient, epidemic phase of the southern pine beetle (Mawby *et al.*, 1989). Likewise, environmental conditions can affect the susceptibility of conifers to disease (Froelich and Snow, 1986).

Nutrient cycling is another process altered by elevated CO_2, through a variety of mechanisms. Altered plant tissue chemistry results in changing nutrient mineralization rates, which, in turn, feed back to affect tissue chemistry (Coley *et al.*, 1985; Tiarks *et al.*, 1989; Aber and Melillo, 1991). Changes in herbivory can also affect nutrient cycling through changed mycorrhizal mutualism caused by the removal of photosynthetic tissue (Gehring and Witham, 1991). Elevated CO_2 can directly alter tree nutrient uptake (O'Neill et al., 1987) and internal nutrient withdrawal from leaves prior to abscission (Graham et al., 1990). Changes in litter production resulting from elevated CO_2 could affect nutrient cycling directly by altering the substrate available for forest floor decomposition and nutrient release, or indirectly by altering litter accumulation and soil moisture regimes. Elevated CO_2 also affects nutrient cycling by increasing plant C allocation belowground. Studies have shown that elevated CO_2 can increase root and mycorrhizal densities in shortleaf pine (*Pinus echinata*) and other species (Norby et al., 1987; O'Neill et al., 1987). Increased root exudation could also stimulate rhizosphere bacteria, altering nutrient availability to the host plant (O'Neill et al., 1987).

Plant competition is affected by elevated CO_2 primarily through changes in relative growth rates (Long and Hutchin, 1991). The degree to which photosynthesis is stimulated depends to a large extent on the pathway used by the plant. Conifers use the C_3 photosynthetic pathway whereas many herbaceous competitors use the C_4 pathway. Because the C_4 photosynthesis system is saturated by present atmospheric CO_2 concentrations, C_4 species should theoretically respond less to elevated CO_2 concentrations than C_3 species (Kramer and Sionit, 1987), a contention supported by empirical evidence (Rogers et al., 1983; Johnson et al., 1993). Changes in water use efficiency should also increase the drought tolerance of conifers and other C_3 species relative to C_4 species (Long and Hutchin, 1991). Changes in temperature and precipitation could complicate these interactions. Species using the C_4 pathway could be favored as the result of increased temperatures or precipitation. Although photosynthesis and productivity may be enhanced in C_3 plants by elevated CO_2, internal and external feedbacks may serve to limit the length and extent of response. Potential feedbacks through changes in factors such as litter quality may actually maintain or reduce ecosystem productivity. Competitive interactions between conifers and leguminous and actinorhizal tree species could also be altered by enhanced N fixation due to increased belowground allocation of photosynthate (Norby, 1987). Changes in UV-B radiation could also impact competitive interactions due to varying plant sensitivities (Gold and Caldwell, 1983).

IV. Large-Scale Responses of Coniferous Forests to Climate Change

Two general aspects need to be considered when attempting to understand how changing climate might affect coniferous forests and how these forests may contribute to changing climate via biospheric feedbacks. The first relates to the effect potential climate change has on changing the distribution of coniferous forests, and the other relates to the role such forests play in affecting the carbon cycle that largely drives the rate and magnitude of potential change. There is also the problem of scale. Stand management, species migration, and species distribution are all issues posed at the stand and landscape scales of biological information. In contrast, physiologists study processes that operate at much lower levels of biological organization. The mechanisms by which stresses, whose primary effects occur at the level of molecular or cellular physiology, are propagated up through progressively more integrative levels of organ physiology (e.g., leaf, branch, root), whole-plant physiology, and stand dynamics to the landscape level are complex and are only partially understood (Rykiel, 1985). This situation accounts for at least some of the intense debate about the causes of, for example, "forest decline" in various parts of the world (Pitelka and Raynal, 1989; Stout, 1989) and the uncertainty regarding the impact of climate change on physiological processes in forest trees and stands (Amthor, 1991; Bazzaz and Fajer, 1992).

One of the problems in integrating the large-scale effects of low-level stresses is that only a small fraction of the stress at molecular and cellular levels is propagated up to become a disturbance at the tree, stand, or landscape level. For example, a small-scale insect defoliation may severely reduce the growth of one or several branches, whereas tree growth apparently remains unaffected (see references in Sprugel *et al.*, 1991). Hence, past assessments of how changing climate will affect terrestrial biota, including conifers, have been extremely generalized and can perhaps be illustrated with the biome-scale correlational modeling approach. For example, Holdridge (1967) mapped vegetation types against temperature, precipitation, and potential evapotranspiration (but see also Martin, 1993; Smith *et al.*, 1992). He then argued that a change in climate will change the dominant vegetation consistent with the climate–vegetation correlation. Smith *et al.* (1992) used this approach to simulate changes in global vegetation, King and Tingey (1992) summarize their results for the Pacific Northwest, and Neilson (1987) and Neilson and King (1992) have summarized results for the United States. This is an equilibrium approach that predicts the distribution of vegetation after sufficient time has elapsed to permit its full equilibration. Transition dynamics, including seedling dispersion, propagation, and survival, and the influence of soil properties and nutrient availability, as well as dieback of the previous vegetation, are obviously not addressed by this process, and the direct effects of CO_2 are often ignored or handled in a very empirical and rudimentary fashion. Research is in progress to address these complex problems (D'Arrigio and Jacoby, 1993).

Efforts to improve the biome-scale correlational modeling approach are ongoing (see, for example, Woodward, 1987; Prentice et al., 1992) [for a somewhat different approach, see Rastetter et al. (1991)]. One example of such efforts is the Mapped Atmosphere Plant Soil System (MAPSS) model, chosen here as generally indicative of these modeling approaches. MAPSS predicts changes in leaf area index (LAI), site water balance, runoff, and changes in biome boundaries in response to climate change (Neilson and King, 1992; Neilson et al., 1994), although it builds in constraints in addition to water balance. It treats the direct effects of increased CO_2 in a simplistic manner, represented by an arbitrary change in water use efficiency. Relationships between seasonal water balance and thermal limits are used as the primary constraints that determine much of the world's vegetation. The distribution of biomes such as forest, savanna, grassland, and desert (i.e., major physiognomic types, but not floristic associations or species) can then be modeled using leaf area as the physiologic response to the available water and thermal regime in the absence of other constraints. Three soil layers are used in the model. The top layer provides for water uptake by grasses, shrubs, and trees, the middle supplies water only for deeper tree roots, and the deepest provides for underground water transport. Thus, competitive interactions between grasses, shrubs, and trees, resulting from water limitations, can be inferred, although competitions for light, not explicitly treated, must enter into these relations also. In MAPSS, the rate of canopy conductance is coupled to transpiration as a function of leaf surface area and stomatal conductance. The maximum LAI possible before soil water depletion occurs is determined iteratively. LAI thus integrates effects from snow formation and melt, canopy interception, evaporation, transpiration, surface runoff, and deep soil drainage. MAPSS simulations are based on GCM outputs, and may also accept weather records and climatological data that are spatially distributed. These include changes in precipitation and potential evapotranspiration (PET), in MAPSS determined via a turbulent transfer model that uses wind speed, surface roughness, temperature, precipitation, and runoff data as input (see Marks, 1990).

Outputs describe the relocation, if any, of biome boundaries. Some scenarios imply large-scale changes such as potential relocation of temperate forests poleward, with possible dieback in the forests' low-latitude range, which may be envisioned to include catastrophic disturbance such as wildfire. Invasion and ingrowth at the poleward boundary may result, as may general forest expansion, particularly in tropical zones when increased water use efficiency from direct effects of CO_2 are included (Neilson and King, 1992). Temperate coniferous forests are seen to either expand or shrink, depending on the scenario used. Outputs also allow quantification of changes in LAI that may arise from increased water stress, which could be accompanied by significant changes in the system's carbon balance. MAPSS handles changes in water use efficiency in a somewhat simplistic manner, by assigning a 35% reduction in stomatal conductance for all vegetation types. As with other global change factors, ongoing research on the processes suggest that data from chamber and limited field studies do not as yet allow confident prediction of responses to increased atmospheric CO_2 concentrations, in part because of complex nutrient feedbacks (Telewski and Strain, 1994; Tissue et al., 1993). Plants may be able to increase their nutrient uptake ability in step with rising atmospheric CO_2 and realize significant gains in ecosystem production (Drake, 1992).

Clearly, uncertainties in GCMs will create uncertainties in the MAPSS outputs. One example may be winds. MAPSS uses wind speeds from GCM outputs to arrive at potential evapotranspiration. Thus, uncertainties in GCM wind speeds will propagate as uncertainties of water availability terms in MAPSS. Order-of-magnitude estimates of the discrepancies may be made when GCMs for current conditions are compared to monthly wind averages from climatological data. Similarly, MAPSS outputs can be compared to current climate vegetation distribution. Correspondence is good except in zones where overlap of vegetation is found in the real world. This includes the zone where short-grass and tall-grass prairies overlap, and in boreal and subboreal zones where hardwoods mingle with conifers. Moreover, where soil characteristics and nutrient availability are the determinants of climax vegetation in place of water availability, MAPSS cannot resolve the dynamics at present. The latter problems are typically scale dependent—at global levels the small-scale heterogeneity introduced by the dominance of locally important processes will be obscured. MAPSS also makes no attempt to arrive at vegetation distribution in a species-specific way. Resolution of species-specific responses to climate change must await the development of process models coupled to responses at scales larger than the stand level. Such work is in progress, but results are not available for inclusion here.

Coniferous forests cover approximately 3.6×10^9 ha, or 24% of the terrestrial surface of the globe. A significant shift in the distribution and extent of coniferous forests because of climate change may release large amounts of carbon to the atmosphere, which in turn may constitute a significant positive feedback to the greenhouse gas concentration. On the other hand, MAPSS modeling runs show that such effects could be offset by expansion of forests in low latitudes (Neilson and King, 1992). The model, like others, provides a step in assessing potential effects. Table I summarizes modeled change in coniferous biome translocation for global circulation models based on MAPSS runs using two GCM scenarios as inputs. Two of five commonly used GCMs were chosen because

Table I Modeled Change in Areal Extent of Global Coniferous Forests for Two GCM Scenarios

Category	Control area (Mha)	Control (% cover)	UKMO area (Mha)	UKMO difference (Mha)	UKMO (% cover)	OSU area (Mha)	OSU difference (Mha)	OSU (% cover)
Direct effects of CO_2 not included								
Taiga tundra	1025	10.2	365	−660	3.6	636	−390	6.3
Boreal conifer	1073	10.7	1069	−3	10.6	1246	173	12.4
Temperate conifer	498	4.9	488	−10	4.8	439	−59	4.4
Mixed conifer/broadleaf	601	6.0	377	−224	3.7	358	−244	3.5
Coniferous savanna	384	3.8	489	105	4.9	451	67	4.5
Nonconiferous	6488	64.4	7283	794	72.3	6942	453	68.9
Total	10,069	100.0	10,071	—	100.0	10,071	—	100.0
Water use efficiency due to CO_2 fertilization increased by 35%								
Taiga tundra	1031	10.2	374	−657	3.7	647	−384	6.4
Boreal conifer	1110	11.0	1125	15	11.1	1275	165	12.6
Temperate conifer	757	7.5	848	90	8.4	785	28	7.8
Mixed conifer/broadleaf	733	7.3	608	−125	6.0	575	−158	5.7
Coniferous savanna	334	3.3	444	110	4.4	421	88	4.2
Nonconiferous	6141	60.8	6707	566	66.4	6403	261	63.4
Total	10,106	100.0	10,106	—	100.0	10,106	—	100.0

they represent the greatest range between little and considerable climate change; these illustrate the potential magnitude as well as present uncertainties regarding such investigations. As can be seen, the UKMO GCM leads to MAPSS outputs that suggest a significant decline in mixed conifer/broadleaf forests, 224 Mha, whereas the OSU GCM scenario yields a large increase in the boreal conifer biome (173 Mha), although showing a decline in the mixed conifer/broadleaf biome similar to that shown by inputs from the UKMO GCM (244 Mha). The contrast in results is but one indication of the uncertainties still resident in GCMs, and propagated into the vegetation models. Moreover, the scientific foundations of models, and the consequent domain of their utility, have recently been called into question (Oreskes *et al.*, 1994).

The category of mixed conifer was included in the data of Table I without further segregation because the model parameters that determine northern deciduous forms matching those supporting conifers in these climatic zones are virtually indistinguishable. In addition to using an index derived from the occurrence of $-40°C$ over significant portions of a given month, which limits the survival of woody plants, the assumption in MAPSS is that closed boreal forests and open taiga/tundra and taiga regions are energy rather than water limited. MAPSS does not include consideration of effects of nutrient limitations and other soil-related constraints.

The MAPSS approach provides a means of assessing the direction and magnitude of presumptive changes in WUE, and may serve to allow a first cut of the magnitude of possible responses. Table I illustrates the modeled magnitude of effects occasioned by a 35% increase in water use efficiency. In this case, there is a gain in the boreal conifer biome for both UKMO and OSU models, whereas the decline in mixed conifer/broadleaf forests is much reduced from earlier outputs. For temperate coniferous forests, simulations range from a decline of 59 Mha (OSU GCM, no change in WUE) to a gain of 90 Mha (UKMO model, increased WUE).

The geographic extent of such redistribution and potential change of coniferous forest is shown in Fig. 1 (for current climate, with and without WUE "benefits"), in Fig. 2 for equilibrium distributions after a doubling of CO_2 (no WUE effects), and in Fig. 3 with WUE "benefits." As may be expected, the model simulations show the potential expansion of boreal forest into higher latitudes. For the UKMO GCM, this comes at the apparent cost of significant decline in the equatorward boundary of these systems and is reflected by a general expansion for the OSU GCM with and without WUE considerations. There exist potential changes in forest LAI even within zones that do not shift geographically. This is further illustrated by the "change" map, Fig. 4. Here the geographic

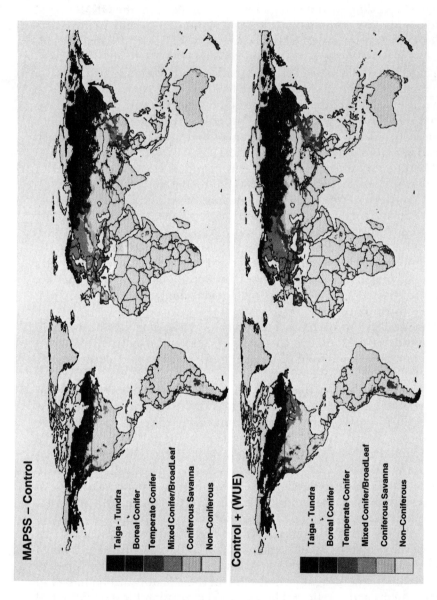

Figure 1 MAPSS output of coniferous forest distribution for current climate with and without a water use efficiency increase of 35%.

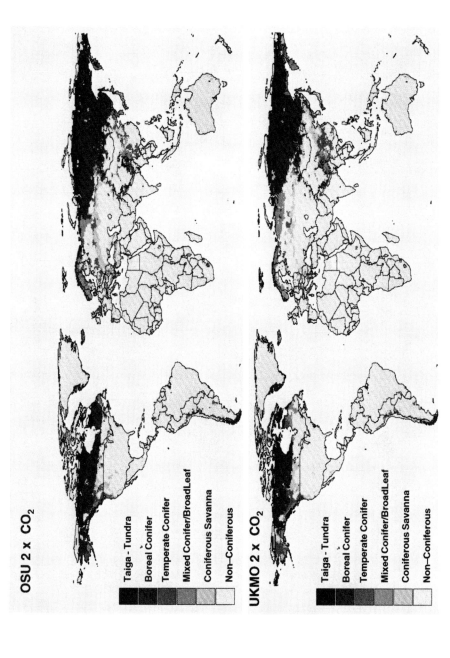

Figure 2 MAPSS output showing equilibrium distribution of coniferous forests after a doubling of CO_2 on the basis of OSU and UKMO GCM model runs.

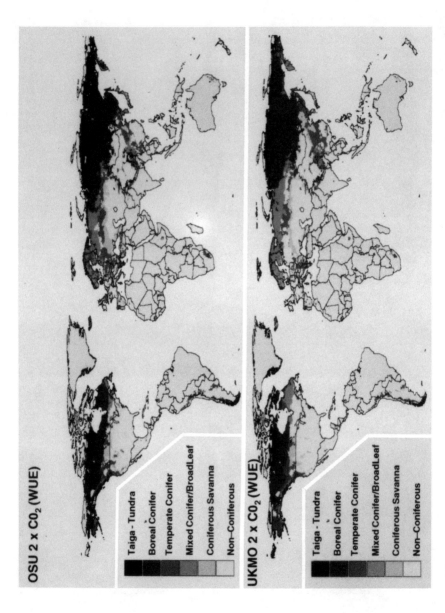

Figure 3 MAPSS output showing equilibrium distribution of coniferous forests after a doubling of CO_2 on the basis of OSU and UKMO GCM model runs with a 35% increase of water use efficiency.

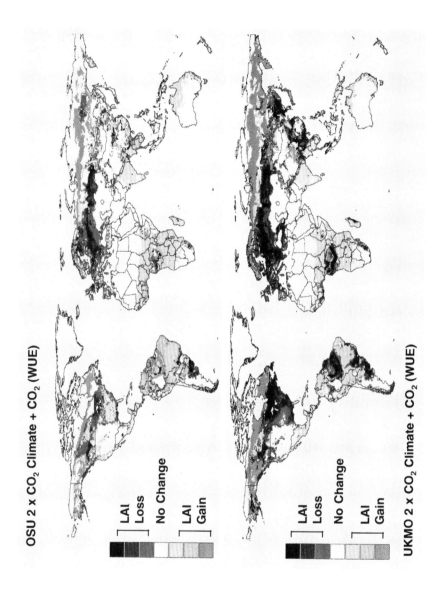

Figure 4 A difference map produced by subtracting the MAPSS output of Fig. 3 from that of Fig. 1 for scenarios with increased water use efficiency. Gray areas show ingrowth of coniferous forests into nonconifer biomes, black areas show loss of conifer biomes due to changed water availability and other limitations.

Table II Modeled Change in Leaf Area Index of Global Coniferous Forests with and without the Effect of CO_2 Fertilization on Water Use Efficiency

Category	Leaf area index		UKMO (% change)		OSU (% change)	
	No WUE	With WUE	No WUE	With WUE	No WUE	With WUE
Taiga tundra	0	0	95	96	77	74
Boreal conifer	12.4	13.4	−50	−23	−28	−8
Temperate conifer	9.5	9.7	−56	31	−26	1
Mixed conifer/broadleaf	14.5	14.7	−54	34	−34	−16
Coniferous savanna	2.7	2.7	−87	59	−79	−25
Nonconiferous	2.2	2.7	−6	15	−2	21

zones expected to show conifer forest decline and those showing conifer ingrowth can be delineated. Because the MAPSS model "generates" LAI from the inferred climate drivers, a revised distribution of LAI values will also be generated and may serve as an indicator of additional stresses or gains from changing climate. Table II illustrates the changes in LAI for the basic coniferous forest types categorized in this discussion, with and without gains in WUE. The areal extent of coniferous forests in the Northern Hemisphere, and the preponderance of conifers in high-latitude regions, immediately suggest the potential importance of these effects. Considerable uncertainty associated with present models limits application of their simulations to primarily heuristic uses. In particular, great care must be exercised in attempting to use the outputs for policy analysis and even more so in management applications.

V. Pools and Flux of Carbon in Coniferous Biomes

An immediate consideration is the role coniferous forests may play in climate feedback. The magnitude and direction of such potential feedback effects on the rate of potential climate change are not presently known. This can be illustrated in terms of the global carbon cycle. Major reservoirs of carbon are as follows: oceans, 37,000 Pg (1 Pg = 1 billion metric tons) inorganic, plus 1000 Pg organic carbon; belowground fossil reserves, 4000 Pg; terrestrial soil and litter, 1200–1600 Pg; the atmosphere (in 1988), 750 Pg; terrestrial vegetation, 420–830 Pg (Post et al., 1990). Flux estimates are not nearly as well known; for example, the net flux from terrestrial biota has been derived using a difference method, its value assigned by the amount needed to achieve a global balance. The magnitude of this flux, a 100- to 120-Pg uptake through photosynthesis, and a comparable release through respiration and decay of residues, are roughly comparable to the ocean–atmosphere exchange, estimated at 100 to 120 Pg/yr, with a net ocean sink of 1.6 to 2.4 Pg/yr (Box, 1988; Esser, 1989). This leaves us with small differences between large numbers, a notorious problem; the net biospheric flux is within an order of magnitude of the fossil fuel flux to the atmosphere, about 5.3 Pg/yr, and of the flux from land use-related releases, estimated to be 0.6 to 2.6 Pg/yr. We do know that anthropogenic releases of CO_2 from the burning of fossil fuels are not matched by an equivalent rise in atmospheric CO_2. This suggests that small alterations in any system that regulates carbon pools could have important consequences for the atmospheric balance (Houghton, 1993).

Carbon pools of coniferous forest are difficult to estimate, in part because forest inventories have typically focused on timber yield, which omits carbon storage in understory vegetation (minor), coarse woody debris (major) (see Harmon *et al.*, 1990), and belowground pools (vast) (Botkin and Simpson, 1990). Recent work designed to arrive at carbon pools and flux estimates from global forests is not species specific (Dixon *et al.*, 1994). One may get a flavor for the relative role of coniferous forests in the global carbon cycle by extrapolating from more limited data. For the conterminous United States (i.e., excluding Alaska), data for aboveground softwood carbon content are extant (Cost *et al.*, 1990) and can be used to infer total carbon in coniferous forests by assuming that soil carbon pools are similar for softwoods and hardwoods. Coniferous forests of the conterminous United States may thus be estimated to contain 5.2 Pg of carbon, or 45% of the total forest pool (Turner *et al.*, 1993; Gucinski *et al.*, 1992). The corresponding net flux of carbon is approximately 52 Tg/yr (1 Tg = 1 million metric tons) from atmosphere to forest (Birdsey, 1992).

Estimates from Canadian work allow inferences for coniferous forests by assuming that the Boreal, Cordilleran, and Pacific Cordilleran bioclimatic zones are conifer dominated. With that assumption, the conifer forest carbon pool of Canada can be estimated at 74.3 Pg, with a net flux into the forest of 56 Tg/yr (Kurz *et al.*, 1992). Continuing with this somewhat simplistic approach, Russian conifer forests can be estimated to have carbon pools totaling 82 Pg, with a net flux of 439 Tg/yr into the forests (Kolchugina and Vinson, 1994). The large differences in flux per unit pool size are partially accounted for by differences in age structure of Canadian and Russian forests, the inclusion of coarse woody debris in the Russian analysis (Kolchugina and Vinson, 1993), and differing harvest practices in these regions. Estimating net ecosystem productivity is notoriously difficult, and the error range of that factor in the Russian forest flux estimates is broad (T. P. Kolchugina, personal communication). Harvest removes carbon from forests and returns a fraction to the atmosphere because of short-term conversion during processing, decay of waste, and slower decay from products having long-term use (Schlesinger, 1993).

Conifer-dominated forests of the United States, Canada, and the former Soviet Union account for approximately 42% of total forests in area in those regions, 27% of the carbon pools, and a large fraction of the total flux. The flux estimates contain considerable uncertainty because of the lack of reliable data for Alaska, among other estimation errors. Significantly, *in toto* these forests represent a net sink of carbon of about 550 Tg/yr (excluding Alaska). Global forests as a whole have been estimated to be a net source of carbon to the atmosphere at 900 Tg/yr because of deforestation (Dixon et al., 1994). This quantity is approximately 17% of the carbon from global fossil fuel emissions, and suggests that the Northern Hemisphere conifer carbon sink (~10% of fossil fuel emission) may be a significant factor in slowing the increase of atmospheric CO_2. Climate change-induced decline could lead to a significant positive feedback to the rate of change, whereas expansion, possibly resulting from CO_2 fertilization, could provide a negative feedback, slowing the rate of carbon accumulation in the atmosphere. A full understanding of the dynamics of these trends in light of direct responses to CO_2 enrichment, management practices, and responses to changing climate is necessary to attain reliable forecasts of climatic variation on human time scales.

VI. Final Comments

The complexity of ecological interactions affecting forest response to potential global change illustrates the need for research that links processes across a range of scales. Although simulation models should be useful tools for integrating whole-system responses, their predictions are only as strong as the scientific data on which they are based. The potential of using management intervention to conserve or sequester carbon or to minimize negative impacts may be kept in mind as a tool to preserve the natural resources residing in global coniferous forests. This potential remains dubious, however, as many of the interactions are not yet sufficiently well understood to make predictions with the degree of reliability one would want. The significant uncertainty associated with such predictions should be emphasized in developing any models for resource management.

Acknowledgments

We gratefully acknowledge the thoughtful contributions of Bob Teskey, Evan DeLucia, Richard Thomas, Dave Hollinger, and Merrill Kaufmann during the conceptual development of the chapter. We would like to thank Ron Neilson for making conifer-specific simulations from MAPSS available to us. In particular, we thank Tom Hinckley and Bill Smith for their editorial assistance, cogent criticism, and productive suggestions.

References

Aber, J. D. and Melillo, J. M. (1991). "Terrestrial Ecosystems." Saunders College Publishing, Philadelphia.
Amthor, J. S. (1991). Respiration in a future, higher-CO_2 world. *Plant, Cell Environ.* 14:13–20.
Bazzaz, F. A., and Fajer, E. D. (1992). Plant life in a CO_2-rich world. *Sci. Am.* 264:68–74.
Birdsey, R. A. (1992). Carbon storage and accumulation in United States ecosystems. *USDA For. Serv. Gen. Tech. WO Rep.* WO-59.
Botkin, D. B., and Simpson, L. G. (1990). Biomass of the North American boreal forest. A step toward accurate global measures. *Biogeochemistry* 9:161–174.
Box, E. O. (1988). Estimating the seasonal carbon source–sink geography of a natural, steady-state terrestrial biosphere. *J. Appl. Meteorol.* 27:1109–1124.
Bryant, J. P. (1981). Phytochemical deterrence of snowshoe hare browsing by adventitious shoots of four Alaskan trees. *Science* 213:889–890.
Caldwell, M. M., Teramura, A. H., and Tevini, M. (1989). The changing solar ultraviolet climate and the ecological consequences for higher plants. *Trends Ecol. Evol.* 4:363–367.
Cannell, M. G. R., Grace, J., and Booth, A. (1989). Possible impacts of climatic warming on trees and forests in the United Kingdom: A review. *Forestry* 62:337–364.
Cess, R. D. (1989). Gauging water-vapour feedback. *Nature (London)* 342:736–737.
Choudbury, D. (1988). Herbivore induced changes in leaf-litter resource quality: A neglected aspect of herbivory in ecosystem nutrient dynamics. *Oikos* 51:389–393.
Coleman, J. S., McConnaughay, K. D. M., and Bazzaz, F. A. (1993). Elevated CO_2 and plant nitrogen use: Is reduced tissue nitrogen concentration size-dependent? *Oecologia* 93:195–200.
Coley, P. D., Bryant, J. P., and Chapin, F. S. (1985). Resource availability and plant antiherbivore defense. *Science* 230:895–899.
Cost, N. D., Howard, J. O., Mead, B., McWilliams, W. H., Smith, W. B., Van Hooser, D. D., and Warton, E. H. (1990). The biomass resource of the United States. *USDA For. Serv. Gen. Tech. Rep. WO* WO-57.
D'Arrigio, R., and Jacoby, G. C. (1993). Tree growth–climate relationships at the northern boreal forest tree line of North America: Evaluation of potential response to increasing carbon dioxide. *Global Biogeochem. Cycles* 7:525–535.
Dickinson, R. E., Kennedy, P., and Henderson-Sellers, A. (1993). "Biosphere-Atmosphere Transfer Scheme (BATS) Version 1e as coupled to the NCAR Community Climate Model," NCAR Tech. Note, NCAR/TN-xxx. National Center for Atmospheric Research, Boulder, CO (in press).
Dixon, R. K., Brown, S., Houghton, R. A., Solomon, A. M., Trexler, M. C., and Wisniewski, J. (1994). Carbon pools and flux of global forest ecosystems. *Science* 263:185–190.

Drake, B. G. (1992). The impact of rising CO_2 on ecosystem production. *Water, Air, Soil Pollu.* 64:25–44.

Elias, S. A. (1991). Insects and climate change: Fossil evidence from the rocky mountains. *BioScience* 41:552–559.

Esser, G. (1989). Zum Kohlenstoff-Haushalt der Terrestrischen Biosphäre. *Poster Trans. Soc. Ecol.* 17.

Fajer, E. D., Bowers, M. D., and Bazzaz, F. A. (1989). The effects of enriched carbon dioxide atmospheres on plant–insect herbivore interactions. *Science* 243:1198–1200.

Froelich, R. C., and Snow, G. A. (1986). Predicting site hazard to fusiform rust. *For. Sci.* 32:21–35.

Gehring, C. A., and Whitman, T. G. (1991). Herbivore-driven mycorrhizal mutualism in insect-susceptible pinyon pine. *Nature (London)* 353:556–557.

Gold, W. G., and Caldwell, M. M. (1983). The effects of ultraviolet-B radiation on plant competition in terrestrial ecosystems. *Physiol. Plant.* 58:435–444.

Graham, R. L., Turner, M. G., and Dale, V. H. (1990). How increasing CO_2 and climate change affect forests. *BioScience* 40:575–587.

Grodzinski, B. (1992). Plant nutrition and growth regulation by CO_2 enrichment. *BioScience* 42:517–525.

Gucinski, H., Turner, D., Peterson, C., and Koerper, G. (1992). Carbon pools and flux on forested lands of the United States. *Proc. Workshop Carbon Cycl. Boreal For. Sub-arctic Ecosys.*, Oregon State University, Corvallis: EPA/600R-93/084.

Haenninen, H. (1991). Does climatic warming increase the risk of frost damage in northern trees? *Plant, Cell Environ.* 14:449–454.

Hansen, J., Russell, G., Rind, D., Stone, P., Lacis, A., Lebedeff, S., Ruedy, R., and Travis, L. (1983). Efficient three-dimensional global models for climate studies: Models I and II. *Mon. Weather Rev.* 3(4):609–662.

Harmon, M. E., Ferrell, W. K., and Franklin, J. F. (1990). Effects of carbon storage on conversion of old-growth forests to young forests. *Science* 247:699–702.

Hedden, R. L. (1989). Global climate change: Implications for silviculture and pest management. In "Proceedings of the Fifth Biennial Southern Silvicultural Research" (J. H. Miller, ed.), pp. 555–562.

Heymsfield, A. J., and Miloshevich, L. M. (1991). Limit to greenhouse warming? *Nature (London)* 351:14–15.

Hinckley, T. M., Dougherty, P. M., Lassoie, J. P., Roberts, J. E., and Teskey, R. O. (1979). A severe drought: Impact on tree growth, phenology, net photosynthetic rate and water relations. *Am. Midl. Nat.* 102:307–316.

Holdridge, L. R. (1967). "Life Zone Ecology," rev. ed. Trop. Sci. Cent., San Jose, Costa Rica.

Houghton, R. A. (1993). Is carbon accumulating in the northern temperate zone? *Global Biogeochem. Cycles* 7:611–617.

Intergovernmental Panel on Climate Change (IPCC) (1990). Climate Change: The IPCC scientific assessment. In "The IPCC Scientific Assessment: Working Group I Report" (J. T. Houghton, G. J. Jenkins, and J. J. Ephraums, eds.). Cambridge University Press, Cambridge, MA.

Johnson, H. B., Pollex, H. W., and Mayeux, H. S. (1993). Increasing CO_2 and plant–plant interactions: Effects on natural vegetation. *Vegetatio* 104/105:157–170.

Keeling, C. D., Bacastow, R. B., Carter, A. F., Piper, S. C., Whorf, T. P., Heimann, M., Mook, W. G., and Roeloffzen, H. (1989). A three-dimensional model of atmospheric CO_2 transport based on observed winds: 1. Analysis of observational data. *Geophys. Monogr., Am. Geophys. Union* 55:165–236.

King, G. A., and Tingey, D. T. (1992). Potential impacts of climate change on Pacific Northwest Forests. *U.S. Environ. Prot. Agency Rep.* EPA/600/R-92/095.

Kolchugina, T. P., and Vinson, T. S. (1993). Carbon sources and sinks in forest biomes of the former Soviet Union. *Global Biogeochem. Cycles* 7:291–304.

Kolchugina, T. P., and Vinson, T. S. (1993). Comparison of two methods to assess the carbon budget of forest biomes of the former Soviet Union. *J. Water, Air, Soil Pollution* 70(1–4):207–221.

Kramer, P. J., and Sionit, N. (1987). Effects of increasing carbon dioxide concentration on the physiology and growth of forest trees. *In* "Greenhouse Effect, Climate Change and U.S. Forests" (W. E. Shands and J. S. Hoffman, eds.), pp. 219–246. Conservation Foundation, Washington, DC.

Kurz, W. A., Apps, M. J., Webb, T. M., and McNamee, P. J. (1992). "The Carbon Budget of the Canadian Forest Sector: Phase I," Inf. Rep. NOR-X-326. Forestry Canada Norwest Region, Northern Forestry Center, Edmonton, Alberta.

Lambers, H. (1993). Rising CO_2, secondary plant metabolism plant–herbivore interactions and litter decomposition: Theoretical considerations. *Vegetatio* 104/105:263 Z71.

Lawrey, J. D. (1989). Lichen secondary compounds: Evidence for a correspondence between antiherbivore and antimicrobial function. *Bryologist* 92:326–328.

Lechowicz, M. J. (1987). Resource allocation by plants under air pollution stress: Implications for plant–pest–pathogen interactions. *Bot. Rev.* 53:281–300.

Leverenz, J. W., and Lev, D. J. (1987). Effects of carbon dioxide-induced climate changes on the natural ranges of six major commercial tree species in the western United States. *In* "The Greenhouse Effect, Climate Change, and U.S. Forests" (W. E. Shands and J. S. Hoffman, eds.), pp. 123–155. Conservation Foundation, Washington, DC.

Lincoln, D. E. (1993). The influence of plant carbon dioxide and nutrient supply on susceptibility to insect herbivores. *Vegetatio* 104/105:273–280.

Lindroth, R. L., Kinney, K. K., and Platz, C. L. (1993). Responses of deciduous trees to elevated atmospheric CO_2: Productivity phytochemistry and insect performance. *Ecology* 74:763–777.

Long, S. P., and Hutchin, P. R. (1991). Primary production in grasslands and coniferous forests and climate change: An overview. *Ecol. Appl.* 1:139–156.

Manabe, S., and Wetherald, R. T. (1987). Large scale changes in soil wetness induced by an increase in carbon dioxide. *J. Atmos. Sci.* 44:1211–1235.

Marks, D. (1990). A continental-scale simulation of potential evapotranspiration for historical and projected double-CO_2 climate conditions. *In* "Biospheric Feedbacks to Climate Change: The Sensitivity of Regional Trace Gas Emissions, Evapotranspiration, and Energy Balance to Vegetation Redistribution. Status of Ongoing Research" (H. Gucinski, D. Marks, and D. P. Turner, eds.), EPA/600/3-90/078. U.S. Environ. Prot. Agency, Washington, DC.

Martin, P. (1993). Vegetation responses and feedbacks to climate: A review of models and processes. *Clim. Dyn.* 8:201–210.

Mawby, W. D., Hain, F. P., and Doggett, C. A. (1989). Endemic and epidemic populations of southern pine beetle: Implications of the two-phase model for forest managers. *For. Sci.* 35:1075–1087.

Mitchell, J. F. B. (1991). No limit to global warming? *Nature (London)* 353:219–220.

Neilson, R. P. (1987). Biotic regionalization and climate controls in western North America. *Vegetatio* 70:135–147.

Neilson, R. P., and King, G. A. (1992). Continental scale biome responses to climate change. *In* "Ecological Indicators" (D. H. McKenzie, D. E. Hyatt, and V. J. McDonald, eds.), pp. 1015–1040, vol. 2. Elsevier, Amsterdam.

Neilson, R. P., King, G. A., and Lenihan, J. (1994). Modeling forest response to climatic change: The potential for large emissions of carbon from dying forests. *In* "Proceedings of the IPCC Workshop, Carbon Balance of the World's Ecosystems: Toward a Global Assessment" (M. Kaaninen, ed.), pp. 150–162. Publication of the Academy of Finland, Helsinki.

Norby, R. J. (1987). Nodulation and nitrogenase activity in nitrogen-fixing woody plants stimulated by CO_2 enrichment of the atmosphere. *Physiol. Plant.* 71:77–82.

Norby, R. J., O'Neil, E. G., Hood, W. G., and Luxmoore, R. J. (1987). Carbon allocation, root exudation and mycorrhizal colonization of *Pinus echinata* seedlings grown under CO_2 enrichment. *Tree Physiol.* 3:203–210.

O'Neill, E. G., Luxmoore, R. J., and Norby, R. T. (1987). Elevated atmospheric CO_2 effects on seedling growth, nutrient uptake, and rhizosphere bacterial populations of *Liriodendron tulipifera* L. *Plant Soil* 104:3–11.

Oreskes, N., Shrader-Frechette, K., and Belitz, K. (1994). Verification, validation, and confirmation of numerical models in the earth sciences. *Science* 263:641–646.

Pitelka, L. F., and Raynal, D. J. (1989). Forest decline and acid deposition. *Ecology* 70:2–10.

Post, W. M., Peng, T.-H., Emanuel, W. R., King, A. W., Dale, V. H., and DeAngelis, D. L. (1990). The global carbon cycle. *Am. Sci.* 78:310–326.

Prentice, I. C., Harrison, S. P., Leemans, R., Monserud, R. A., and Solomon, A. M. (1992). A global biome model based on plant physiology and dominance, soil properties and climate. *J. Biogeg.* 19:117–134.

Ramanathan, V., and Collins, W. (1991). Thermodynamic regulation of ocean warming by cirrus clouds deduced from observations of the 1987 El Niño. *Nature (London)* 351:27–32.

Rastetter, E. B., Ryan, M. G., Shaver, G. R., Melillo, J. M., Nadelhoffer, K. J., Hobbie, J. E., and Aber, J. D. (1991). A general biogeochemical model describing the responses of the C and N cycles in terrestrial ecosystems to changes in CO_2, climate, and N deposition. *Tree Physiol.* 9:101–126.

Rogers, H. H., Bingham, G. E., Cure, J. D., Smith, J. M., and Surano, K. A. (1983). Responses of selected plant species to elevated carbon dioxide in the field. *J. Environ. Qual.* 12:569–574.

Rykiel, E. J., Jr. (1985). Towards a definition of ecological disturbance. *Aust. J. Ecol.* 10:361–365.

Schlesinger, M. E., and Zhao, Z. C. (1989). Seasonal climate change introduced by doubled CO_2 as simulated by the OSU atmospheric GCM/mixed layer model. *J. Clim.* 2:429–495.

Schlesinger, W. H. (1993). Response of the terrestrial biosphere to global climate change and human perturbation. *Vegetatio* 104/105:295–305.

Smith, T. M., Shugart, H. H., Bonan, G. B., and Smith, J. B. (1992). Modeling the potential response of vegetation to global climate change. *In* "Global Climate Change: The Ecological Consequences" (F. I. Woodward, ed.), pp. 93–116. Academic Press, London.

Sprugel, D. G., Hinckley, T. M., and Schaap, W. (1991). The theory and practice of branch autonomy. *Annu. Rev. Ecol. Syst.* 22:309–334.

Stenberg, P., DeLucia, E. H., Schoettle, A. W., and Smolander, H. (1994). Photosynthetic light capture and processing from the cell to canopy. *In* "Resource Physiology of Conifers: Acquisition, Allocation, and Utilization" (W. K. Smith and T. M. Hinckley, eds.), pp. 1–38. Academic Press, San Diego.

Stout, B. B. (1989). Forest decline and acid deposition—A commentary. *Ecology* 70:11–14.

Strain, B. R. (1985). Physiological and ecological controls on carbon sequestering in terrestrial ecosystems. *Biogeochemistry* 1:219–232.

Strain, B. R., and Cure, J. D. (1985). "Direct Effects of Increasing Carbon Dioxide on Vegetation," DOE/ER-0238. U.S. Dept. of Energy, Carbon Dioxide Res. Div., Washington, DC.

Telewski, F. W., and Strain, B. R. (1994). The response of trees to global change. In preparation.

Tiarks, A. E., Bridges, J. R., Hemingway, R. W., and Shoulders, E. (1989). Condensed tannins in southern pines and their interactions with the ecosystem. In "Chemistry and Significance of Condensed Tannins" (R. W. Hemingway and J. J. Karchesy, eds.), pp. 369–390. Plenum, New York.

Tissue, D. T., Thomas, R. B., and Strain, B. R. (1993). Long-term effects of elevated CO_2 and nutrients on photosynthesis and rubisco in loblolly pine seedlings. *Plant, Cell Environ.* 16:859–865.

Tsonis, A. A. (1991). Sensitivity of the global climate system to initial conditions. *Eos* 72:313, 328.

Turner, D. P., Lee, J. J., Koerper, G. J., and Barker, J. T. (1993). The forest sector carbon budget of the United States: Carbon pools and flux under alternative policy options. *U.S. Environ. Prot. Agency Rep.* EPA/600/3-93/083.

Wilson, C. A., and Mitchell, J. F. B. (1987). A doubled CO_2 climate sensitivity experiment with a GCM including a simple ocean. *J. Geophys. Res.* 92:13315–13343.

Woodward, F. I. (1987). "Climate and Plant Distribution." Cambridge Univ. Press, London.

Zasada, J. C., Sharik, T. L., and Nygren, M. (1992). The reproductive process in boreal forest trees. In "Systems Analysis of Global Boreal Forest" (H. H. Shugart, R. Leemans, and G. B. Bonan, eds.), pp. 85–125. Cambridge Univ. Press, Cambridge, UK.

Index

Acclimation to cold, 100–101
Acid deposition effects on ozone response, 279–280
Air pollution, response of conifers, 255
 allozyme effects, 286–287
 assessing risk, 255–257
 conclusions, 299
  biological responses, direct, 299
  complex biological responses, 299
  ecological risk assessment, 300–301
  modeling approach, 299
  pollutants of risk, 299
  risk species, 299
 genetic resistance, 18
 herbivory effects, 129
 multiple stress effects on ozone responses, 278–279
  humidity and wind, 283–284
  light, 285–286
  nutrients, 279
  partitioning of carbon, 275–278
  predisposition from pollutant exposure, 286–287
  spatial and temporal factors, 287–288
  temperature, 284–285
 objectives and focus, 257–261
 ozone response models, 295
  single tree, 295
   canopy, 297
   comparisons, 295–298
   MAESTRO, 296–297
   TREGRO, 295–296
 ozone uptake and impact, 261–262
  calculation of, using stomatal conductance, 262–263
  leaf level, 262
  organ differentiation and senescence, 270–271
  scaling from leaf to cell level, 264
   biochemistry, 269–270
   gas exchange, 264–269
   gaseous pollutants, 280–282
   ozone and acid deposition, 279–280
   soil water, 282–283
   past research, 261–262
  scaling from leaf to whole plant, 272
   carbon gain, 272–275
  scaling from tree to stand level, 288–289
   air pollution
    absorption, 280
    measurement, 290
   case studies of mature trees and forests, 291
    large, old trees, 293–295
    mature trees under ozone stress, 291–292
    young, small trees, 292–293
   phototoxicity of absorbed air pollutants, 290–291
   stand and landscape models, 298–299
   stand conductance, 289–290
Alaskan polar forests, paleoecology, 51–53
Allometry of leaf area dynamics, 182–183
Allozymes, variation in conifers
 geographic, 8–9
 interspecific, 4
Autumn decline in stomatal conductance and winter dormancy, 105–106

Biochemical changes for frost resistance, 100–101
Biochemistry of ozone uptake, 269–270
Biological response complexity in air pollution studies, 299
Black pine, effects of rapid climate change, 55–56
Boreal zone conifers
 paleoecology of, 53–54
 species outside of, 87–88

Canopy level
  genetic variation, effect on crown architecture and leaf area, 200–204
  leaf-to-plant scaling, air pollution effects, 272
  ozone response model, 297
Carbon
  balance during winter, 112–114
  partitioning in response to ozone pollution, 275–276
Carbon balance during winter dormancy, 109–114
Carbon gain or assimilation, *see also* Photosynthesis
  effects of herbivory, 162–166
  response to ozone pollution, 278–279
Carbon metabolism during winter, 109
Carbon partitioning in response to ozone, 275–276
Chemical signals of woody plants, 66–68
Climate
  and the distribution of conifers, 96–99
  as a limitation to the geographic distribution, 80–83
  correlation with vegetation changes, 41–51
  rapid changes, implications to ecosystems, 54–55
  short-term changes, implications to conifer distribution, 56–57
  variation in, 37–38
    glacial/interglacial, 38–41
    last millennium, 45–51
Climate change , *see* Climate
Coniferous forest zones
  distribution versus climate, 96–99
  implications for rapid climate change
    black pine, 55–56
    scotch pine, 55
  temperature relations, 99–100
  western forests of North America, winter dormancy, 44–45
Conifers
  Alaskan boreal zone, 51–53
  forest zone, 96
  outside boreal zone, 87–88
  western North America, 44–45
Crown architecture, genetic effects, 200–204
Cuticular transpiration
  effects on winter water relations, 108

Desiccation
  effects on winter water relations, 105–109
  needles in winter, 108
Distribution, conifers, 79–80
  boreal zone, species outside of, 87–88
  climate limits, 80–83
  conifer forest zone, 96
  growth effects, 84–87
  xylem structure, correlations with, 83–84
DNA (deoxyribonucleic acid)
  cpDNA and mtDNA variation, 25–26
  variation in conifers, 4–5
Dormancy, winter
  carbon relations, 109–114
  organs, 101–102
  respiration, 113–114
  water relations, 105–109
  winter, 95–96
Drought, *see* Desiccation

Ecological and evolutionary significance of genetic variation, 12–14
Ecological risk assessment, air pollution, 300–301
Ecophysiological implications from paleoecology, 51
Ecophysiology and hormones, 66
Ecosystem level
  effects of rapid climate and vegetation change, 54–55
  leaf area dynamics, 238
  leaf life-span effects on ecosystems, 238
  stand and landscape models of air pollution response, 298–299
Endogenous control of winter dormancy, 102–103
Environmental control of winter dormancy, 102–103
Evolutionary and ecological significance of genetic variation, 12–14

Foliage level, *see* Leaf level
Forest level
  case studies of air pollution effects, 291
  conifer forest distribution, 96, 311–312
  scaling air pollution effects from tree to stand levels, 288–289
  stand and landscape models of ozone response, 298–299

Frost
  acclimation to, 100–101
  damage, occurrence in field, 103–105
  resistance to, 100
Frost resistance and winter dormancy, *see* Winter dormancy, ecophysiology

Gas exchange, *see also* Photosynthesis, and Transpiration
  and ozone uptake, 264–269
Gaseous pollutants, effects on ozone response, 280–282
Genetics
  ecological and evolutionary significances, 12–14
  air pollution resistance, 18
  DNA variation, cpDNA and mtDNA, 25–26
  growth rates, 16–18
  herbivore resistance, 18
  mating system analysis, 24–26
  microgeographic variation, 14–16
  natural selection, 14
  oleoresin pressure, 18–21
  respiration, 21–23
  viability, 23–24
  geographic variation, 5
  allozymes 8–9
  discordant patterns, 9–12
  provinces, 5–8
  variation, 2–3
  allozymes, 4
  DNA, 4–5
  provinces, 3–4
Geographic variation in conifer allozymes, 8–9
Global climate change effects on conifer distribution, 41–51, 80–83
  carbon flux and pools in coniferous biomes, 324–325
  elevated CO_2 effects, 314–315
  large scale responses, conifer forests, 314–324
Growth, *see also* Leaf area dynamics
  hormone regulators, 64–66, 68–71
  effects on
    growth, 166–167
    nutrient acquisition, 125–126, 148–151
    photosynthesis and carbon gain, 162–166
    plant water relations, 154–158

Herbivory, *see also* Insect herbivory
  effects on
    nutrient acquisition and allocation, 148–151
    plant water relations, 158–159
  nutrient effects on, 137–140
  resistance to, in conifers, 18
Hormones
  ABA (abscisic acid) importance to stomatal response and climate variables, 71–73
  chemical signaling, woody plants, 66–68
  growth and development regulation, 73–74
  model of chemical regulation of stomata, water relations and development, 68–71
  plant growth regulators, 64–66
  implications to ecophysiology, 66
Humidity effects on ozone response, 283–284
Hydraulic model and leaf area dynamics, 188
  empirical evidence, 189
  sapwood permeability, 189–191
  theory, 188–189

Insect herbivory, 125
  carbon acquisition and allocation, 159
  effects of herbivores, 162
    on growth, 166–167
    on photosynthesis and carbon gain, 162–166
  effects on insect herbivores, 162
  factors influencing effects of conifer-herbivore interactions, 159–162
  herbivory effects on nutrient acquisition and allocation, 148–151
  herbivory effects on plant water relations, 158–159
  influence of plant water relations on, 154–158
  nutrient acquisition and allocation, 125–126
  influencing factors
    air pollution, 129
    nutrients, 127–129
    plant age, 129–132
    plant genotype, 134
    stand age, density, and structure, 137

tissue age, 132–134
  water, 126–127
nutrient effects on insect herbivores,
    137–140
  as influenced by
    environmental stress, 140–144
    plant and tissue age, 144–146
    plant genotype, 146–147
    stand age, density, and structure,
      147–148
  water relations
    effects of
      environment, 151
      plant genotype, 153–154
      plant tissue age, 152–153
      stand age and structure, 154
Insects, see Insect herbivory
Interglacial vegetation change, 42–44
Interspecific variation in allozymes, 4

Landscape level
  model of ozone response, 298–299
  stand and landscape model of air pollution responses, 298–299
Leaf area dynamics, 181
  effects on leaf, plant, ecosystem, 238
    nutrient use efficiency, 248–249
    species variation and plant traits,
      238–244
    stand and ecosystem traits, 244–248
  foliage and stem area in relation to stand perturbation, 195–197
  genetic variation, effects on crown architecture and leaf area, 200–204
  leaf area in natural and managed stands,
    204–208
  senescence of leaf area, causes and consequences, 208–210
  spatial scales of leaf area, 210–211
  structural and functional relationships,
    182
    allometry, simple, 182–183
    foliage and roots, functional relationships, 191–193
      empirical evidence, 189
      sapwood permeability, influence of site and age, 189–191
      theory, 188–189
      wood anatomy effects on sapwood permeability, 191–193
    pipe model theory, 183–188

Leaf area, genetic effects, 200–204
Leaf level, see also Leaf area dynamics, and Leaf life-span
  ozone uptake, 262
Leaf life-span, 225–227
  variation due to
    insects and disease, 237–238
    internal and external factors,
      227–228
    light, 232–233
    temperature, 228–232
    water and nutrients, 233–236
Light
  effects on ozone response, 285–286
  phototoxicity of air pollutants, 290–291

MAESTRO, model of ozone response,
    296–297
Mating system of conifers, 24–25
Microgeographic variation, genetic, 14–16
Models
  as an approach to air pollution assessment, 299
  hydraulic, 188
  of ozone response
    canopy, 297
    comparisons of, 297–298
    MAESTRO, 296–297
    single tree, 295
    stand and landscape, 298–299
    TREGRO, 295–296
  pipe model theory, 183–188
Multiple stress effects in evaluating our pollution responses, 278–279
  carbon partitioning, 275–278
  humidity and wind, 283–284
  light, 285–286
  nutrients, 279
  spatial and temporal factors, 287–288
  temperature, 284–285

Natural selection and evolution, genetic factors, 14
Nutrient acquisition and allocation,
    125–126
  effects of
    air pollution, 129
    plant age, 129–132
    plant genotype, 134
    nutrients, 127–129
    stand age, density, and structure, 137

tissue age, 132–134
  water, 126–127
Nutrients
  effects on
    insect herbivory, 127–129, 137–140
    ozone response, 279

Oleoresin pressure, 18–21
Organ differentiation and senescence, effects on ozone impact, 270–271
Osmotic potentials, adjustment during winter, 105
Ozone
  and acid deposition, 279–280
  landscape model of conifer response to, 298–299
  response models, 295
    canopy, 297
    comparisons of, 297–298
    MAESTRO, 296–299
    TREGRO, 295
Ozone response
  effects of
    humidity, 283–284
    light, 285–286
    nutrients, 279
    soil water, 282–283
    temperature, 284–285
    wind, 283–284

Paleoecology
  climate variation, 37–38
    glacial/interglacial period, 38–41
    last millennium, 41
  ecosystem implications for rapid climate and vegetation change, 54–55
    black pine, 55–56
    scotch pine, 55
  implications for ecophysiological studies, 51
  short-term events, implications of, 56–57
  vegetation change, 41
    Alaskan polar forests, late glacial, 51–53
    boreal and thermophilous species, eastern North America, 53–54
    glacial/interglacial period, 42–44
      boreal forest, 42–44
      western conifer forests, 44–45
    last millennium, 54–56

Permeability of sapwood, 189–191
Photosynthesis, *see also* Carbon gain
  air pollution effects, 264–269, 272–275
  herbivore effects, 159, 162–166
  ozone effects on, 272–275
  spring recovery, 112–113
  winter depression, 109–112
Phototoxicity, of our pollutants, 290–291
Pipe model theory of leaf area dynamics, 185–188
Plant genotype effects on herbivores, 146–147
Plant growth regulators, 64–66
Polar forests of Alaska, paleoecology, 51–53
Pollutants, risk assessment, 299
Pollution, *see also* Air pollution, response of conifers, 255
Providences, genetic, 3–4

Resistance to
  frost, 100–102
  ozone uptake, 18
Respiration during winter dormancy, 113–114
Risk pollutants and species, 299–300

Sapwood permeability, 189–191
Scaling of
  air pollution responses
    leaf-to-cell, 266
    leaf-to-plant, 272
    leaf-to-stand, 288–289
  leaf area dynamics
    leaf, plant, ecosystem, 238
    plant and species, 238–244
    spatial scales, 210–211
    stand and ecosystem, 244–248
Scotch pine, effects of rapid climate change, 55–56
Senescence and winter dormancy
  autumn decrease in stomatal conductance, 105–106
  depression of carbon metabolism, 109–112
  endogenous and environmental control, 103–105
  osmotic potentials, 105
  spring recovery, 112–113
  temperature effects, 99–100
  water uptake and movement, 106–108

Short-term climate change events, implications, 56–57
Soil water effects on ozone response, 282–283
Spatial scales in leaf area dynamics, 210–211
Species with high risk to pollutants, 299
Spring recovery of photosynthesis, 112–113
Stand level
  age, density, structure, 147–148
Stomatal conductance
  ABA (abscisic acid), 71–73
  autumn in 105–106
  calculations of ozone uptake, 261–262
  hormone effects, 68–71
Stress effects on
  ozone responses, 278–279

Temperature
  effects on
    ozone response, 284–285
    winter dormancy, 99–100
Thermophilous species, paleoecology of, 53–54
Transpiration
  cuticular, effects on winter water relations, 108
Tree level
  case studies, ozone stress, 291–295
  models of ozone response, 295, 298–299
TREGRO model of ozone response, 295–296

Water relations
  during winter, 105–109
  effects of herbivory, 158–159
Water uptake and movement in winter, 106–108
Western conifer forests of North America, winter dormancy, 44–45
Winter dormancy, ecophysiology, 95–96
  carbon metabolism, 109
    carbon balance, respiration, 113–114
    spring recovery, 112–113
    winter depression of, 109–112
  coniferous forest zone, 96
    distribution versus climate, 96–99
    temperature relations, 99–100
  frost resistance, 100
    cold acclimation, biochemical and structural changes, 100–101
    endogenous and environmental control of hardness, 102–103
    frost damage, occurrence in field, 103–105
    of organs and limitations to distribution, 101–102
  winter water relations, 105
    cuticular transpiration, 108
    osmotic potentials, adjustment of, 105
    stomatal conductance, autumn decreases in, 105–106
    water uptake and movement, 106–108
    winter desiccation, 108
Woody plants, chemical signals of, 66–68

Xylem structure and conifer distribution patterns, 83–84

Physiological Ecology
A Series of Monographs, Texts, and Treatises

Continued from page ii

F. S. CHAPIN III, R. L. JEFFERIES, J. F. REYNOLDS, G. R. SHAVER, and J. SVOBODA (Eds.). Arctic Ecosystems in a Changing Climate: An Ecophysiological Perspective, 1991

T. D. SHARKEY, E. A. HOLLAND, and H. A. MOONEY (Eds.), Trace Gas Emissions by Plants, 1991

U. SEELIGER (Ed.). Coastal Plant Communities of Latin America, 1992

JAMES R. EHLERINGER and CHRISTOPHER B. FIELD (Eds.). Scaling Physiological Processes: Leaf to Globe, 1993

JAMES R. EHLERINGER, ANTHONY E. HALL, and GRAHAM D. FARQUHAR (Eds.). Stable Isotopes and Plant Carbon–Water Relations, 1993

E.-D. SCHULZE (Ed.). Flux Control in Biological Systems, 1993

MARTYN M. CALDWELL and ROBERT W. PEARCY (Eds.). Exploitation of Environmental Heterogeneity by Plants: Ecophysiological Processes Above- and Belowground, 1994

WILLIAM K. SMITH and THOMAS M. HINCKLEY (Eds.). Resource Physiology of Conifers: Acquisition, Allocation, and Utilization, 1994

WILLIAM K. SMITH and THOMAS M. HINCKLEY (Eds.). Ecophysiology of Coniferous Forests, 1994